Thermodynamics and Statistical Mechanics

An introduction for physicists and engineers

Thermodynamics and Statistical Mechanics

An introduction for physicists and engineers

Samya Zain
Department of Physics, Susquehanna University, Selinsgrove, PA, USA

IOP Publishing, Bristol, UK

Contents

Preface

This text is intended to be an introduction to the subject of thermodynamics and statistical mechanics. It has sufficient material for one semester long course. During my years as an undergraduate and a graduate student, thermodynamics and statistical mechanics were the subjects I found most difficult to grasp. So when I started writing this book I pictured myself sitting one-on-one with a student and discussing concepts that they also may find difficult.

In this book I have tried to establish wherever possible context and background before introducing new physics concepts and/or quantities. I feel that this manner of learning makes it easier for the reader to understand and grasp the validity of the particular physics concept/topic. A reader with some math background should be able to pick up this book and learn on their own just as easily as they would in a classroom environment.

At first glance this book may appear simplistic compared to most thermodynamics and statistical mechanics textbooks. This is by design; I have actively tried to minimize the use of more ostentatious physics vocabulary.

For instructors

I realize that you may see some concepts repeated in multiple chapters. This repetition is intentional. I feel that reviews and re-reviews of these important concepts can never hurt.

Some of the questions at the end of the sections may go just beyond the material discussed in the chapter. Such content is intended for students to apply the concepts that they have learnt during the course of the book.

For students

Before we begin, I have a few additional comments before you begin reading, please understand physics must be understood. Additionally, know that understanding does not come quickly or easily, it takes patience and resilience. Recall how you learnt to play an instrument? Did you become an expert overnight? Or did it take hours and hours of practice? Physics is just the same, you need to actively learn it, cramming overnight before the exam is not the ticket to understanding and success in physics. Remember, someone very wise once said:

'*None of the tricks of success will work unless you do*'.

Samya Zain.
January 31, 2021

Acknowledgments

2020 has been a very interesting year to say the least and I am indebted to many people that have been along for the ride this past year.

Special thanks to my ever patient editor, Professor Ken Brakke, one of the smartest people I know. Dr Brakke performed the Herculean task of editing the entire project from start to finish. I know I can go to him for help whenever I get stuck on a problem, whether it is a physics problem or a math problem. I have always appreciated your every suggestion and correction, as I do so now.

I also thank Dr Jamila Butt, a brilliant mind and a selfless human being. Thank you for letting me bounce my ideas off of you for statistical mechanics. It has been my honor to be able to call you my friend for more than half of my life. Thank you also to Mr Robert Everly, for making me chuckle, smile and laugh out loud with your comments throughout my time at SU. To Drs Fed Grosse, Koz, M L Klotz and V Martin, thank you for all your support and encouragement in the past twelve years. Last but not least thank you to all the students in my thermodynamics and statistical mechanics classes these past few rounds, there are too many to name but you all know who you are!

Special thanks to my husband Zain and my children Hareem, Sami and Eman, for putting up with me these past few months, both when I could not see you in the COVID-19 ward at the ICU and after I came home. You guys have been taking care of me and I have no words to express my appreciation. I also want to thank my parents Dr Iqbal and Safia Bano, for raising me with patience and understanding, my brother, Dr Aamir Iqbal, Sofia, Zoha, Lena for always supporting me and my sister-in-law Shazia and M Shuaib, Warisha, Khabbab and Huzaifa for always supporting me and keeping up the vigil. To all my friends from central PA that have become more than family, I want to say you all are awesome and words cannot express how much I appreciate all that you have done for us in the last couple of months. You made sure my family were supported and well fed in the past couple of months and that has made such a big difference in our lives.

My heartfelt gratitude also gets extended to the doctors and nurses at the ICU at the Geisinger medical hospital, Danville, PA for hosting me for about a month. I may not know all of your names or even recognize your faces if I see you walking down the street, but please know that I appreciate all you have done for me during my long stay at the ICU. I am also grateful to the doctors that have been taking care of me post-hospitalization. I might not be here without your efforts and I appreciate every single one of you.

Author biography

Samya Zain

Samya Zain grew up in Lahore, Pakistan in the 1980s. Her interest in science began in seventh grade when her science teacher, who was actually from the United States, explained how the atom worked. It fascinated her because we are made of atoms. If every atom only has electrons, neutrons and protons, as their basic constituents how is it that there are so many different things around us? How deep can we dig into for answers?

Like most of the girls in Pakistan she was married after high school, but she continued to pursue her interest to find the secrets of the universe. She completed her undergraduate work at the University of the Punjab, Lahore, Pakistan and was awarded the gold medal for the highest score in the annual final exam. Then with her husband and four-year old son she went on to graduate work in physics at the University at Albany, State University of New York in 2001. She received her Master's in physics in 2002 and her PhD in experimental particle physics in 2006. She continued as a postdoctoral fellow at the Albany High Energy Physics group working on the BaBar experiment at Stanford linear accelerator center (SLAC) in California and at the ATLAS detector at CERN, Geneva, Switzerland. In the fall of 2008, she joined Susquehanna University as an assistant professor, where she currently serves as an associate professor and chair of the Department of Physics.

When not teaching or working on some physics problem, she can often be found with her nose in a book of some kind. She makes time to read every day, embroiders and paints when the mood strikes her. She loves collecting books and absolutely refuses to let any of her collection go!

Section I

Classical thermodynamics

Chapters 1–10.

When you study thermodynamics you inevitably have to start at the beginning and as always we will start with classical thermodynamics. In this section we lay down the foundation of the basic concepts of thermodynamics and discuss the historical aspects of its development.

For this purpose, we will subdivide the study of classical thermodynamics into the three main subdivisions for each concept:

1. Historical development of concepts of thermodynamics.
2. The fundamental laws of thermodynamics.
3. Theoretical and practical applications including chemical thermodynamics.

Chapter 1

Introduction and background

1.1 Thermodynamics: what is it all about?

Thermodynamics is a branch of science and an engineering tool that engineers and physicists use to describe processes involving transformations of energy, changes in temperature, and the relationship between heat and work. The word 'thermodynamics' was first used formally by Lord Kelvin in 1849 CE and it is a combination of two Greek words, '*therme*' (heat) and '*dynamis*' (power).

Before we delve into the study of thermodynamics, it is essential to note that classical thermodynamics is an extremely general subject, in that physicists and engineers alike do not hypothesize about the structure and types of matter under consideration. Rather, thermodynamics describes '*what is possible*' and '*what is impossible*' during an energy conversion process. Thermodynamics describes the '*direction*' of a process and is used to describe the performance of propulsion systems, power generation systems, refrigerators, fluid flow, combustion, and many other phenomena.

1.2 Thermodynamics: a very brief introduction

Even though the principles of thermodynamics have existed since the beginning of the Universe, the field we know now as thermodynamics really began with the industrial revolution in the 17th Century with the study of steam engines. Let's begin with a bit of a historical perspective on the steam engine.

1.2.1 The steam engine

In 1698 CE, Thomas Savery, a British military engineer, invented a machine that used steam to pump water out of coal mines. This was the beginning of a very simple steam engine which converted heat into mechanical power. However, the principles behind its operation were not understood well during Savery's time.

Two decades later, in 1712 CE, Thomas Newcomen, also a British engineer, came up with an improved steam engine design. The Newcomen engine was exceptionally expensive to build and not very reliable. The next major change in the design of the

doi:10.1088/978-0-7503-3083-1ch1

steam engine came in 1765 CE when James Watt invented the first inexpensive and rather reliable steam engine. Watt's steam engine was more than six times more efficient than Newcomen's steam engine.

But still the first steam engines were very, very inefficient, converting perhaps at best 2% of their fuel into useful work.

1.2.2 Development of laws of thermodynamics

In 1824 CE, more than fifty years after the Watt's steam engine, Sadi Carnot, a French physicist, introduced the analysis of an ideal gas cycle in his book 'Reflections on the Motive Power of Fire'. The Carnot cycle is not an actual thermodynamical cycle, but is rather a theoretical ideal thermodynamic cycle which provides to this date the upper limit on the efficiency of all classical thermodynamical engines during the conversion of heat into work or vice versa; it gives the upper limit on the efficiency of a refrigeration system, in which by the application of work to a system a temperature difference is created.

In the 1840s Mayer, Joule, and Helmholtz developed the idea that energy could not be created or destroyed. This principle is now known as the First law of thermodynamics and we shall learn more about it in chapters 2 and 3. Then in the 1850s Rudolf Clausius came up with the term 'entropy' in a paper on the efficiency of steam engines. Clausius stated that no cycle can transfer heat from a low temperature region to a high temperature region without interacting with its surroundings. This principle is known as the Second law of thermodynamics, and we will learn a great deal more about it in chapters 5–9. Clausius also defined enthalpy, which is discussed in chapter 3.

With help from Clausius, James Maxwell formulated the statistical mechanical branch of thermodynamics in 1871 CE. Ludwig Boltzmann connected entropy and molecular motion in 1875 CE. Boltzmann's formulation of entropy is perhaps the easiest formulation to understand. The contribution of these scientists and many others led to the development of the broad field of thermodynamics that we know today.

1.2.3 Branches of thermodynamics

In the years since the industrial revolution, the study of thermodynamical systems has developed into several branches, each using a different fundamental model as a basis. Most broadly, we can divide the study of thermodynamics into the following sub-fields:

1. **Classical thermodynamics:** Classical thermodynamics is based on experimental observations and is the description of thermodynamical systems at near-equilibrium states, using properties that are directly measurable in the laboratory. Classical thermodynamics is discussed in chapters 2–9.
2. **Chemical thermodynamics:** Chemical thermodynamics is the study of the interrelation of energy with chemical reactions and the physical change of state. Chemical thermodynamic concepts are discussed in chapter 10.

3. **Statistical thermodynamics:** Statistical thermodynamics relates the microscopic properties of individual atoms to the macroscopic properties of materials. Statistical thermodynamics is based on the average behavior of large groups of particles. Statistical thermodynamics is discussed in detail in chapters 11–13.

1.2.4 Classical thermodynamics

Classical thermodynamics takes a macroscopic approach to the behavior of matter. In classical thermodynamics, we characterize the behavior of a large group of molecules based on properties of the entire group of molecules, such as temperature and pressure. These help us predict other properties and the behavior of matter as it undergoes a process.

The keys to classical thermodynamics are the First and Second laws of thermodynamics. I will describe them very briefly here,

1. **First law of thermodynamics:** First law of thermodynamics is also known as the conservation of energy principle. Very simply, it states that energy cannot be created or destroyed; energy can only be changed from one form to the other.
2. **Second law of thermodynamics:** There are many equivalent statements of the Second law, but the simplest one says, energy in the form of heat flows spontaneously only from higher temperature regions to lower temperature regions. For example, a hot cup of tea left on the countertop will always cool to room-temperature. On the other hand, room-temperature tea will never heat up by itself sitting on the counter. Another way of thinking about the Second law of thermodynamics is to say that the Second law asserts that the high-temperature energy of the tea is degraded to the lower-temperature by transferring the energy to the surrounding air.

These two laws taken together form the basis of classical thermodynamics.

1.2.5 Statistical thermodynamics

Statistical thermodynamics is a more elaborate approach than the classical thermodynamics approach. In statistical thermodynamics the aim is to characterize the behavior of a collection of mass particles based on the following two inherent properties of matter and the laws of probability:

1. The properties of each individual molecule that makes up the system, and
2. The way by which the molecules of the system interact with each other.

1.3 Dimensions and units

All physical quantities are characterized by dimensions, which specify the types of quantities. Units are standard measurement magnitudes assigned to dimensions. Basic dimensions include mass (M), length (L), and time (T) with SI units kilogram (kg), meter (m) and second (s), respectively. Secondary or derived dimensions include velocity ($v = \text{m s}^{-1}$), area ($A = \text{m}^2$) and volume ($V = \text{m}^3$), etc.

1.3.1 The SI system

In 1960, the General Conference of Weights and Measures (CGPM) produced the International system of units, from Système International in French, abbreviated as the SI system, which was first based on the first six fundamental quantities, given in table 1.1. Then in 1971 CE, CGPM added a seventh fundamental quantity and unit: mole (mol) for the amount of matter.

1.3.2 SI unit prefixes

The SI is based on a decimal relationship between units. The prefixes used to express the multiples of the various units are listed in table 1.2.

1.3.3 SI and English units

As seen above, the units of mass, length, and time are the kilogram (kg), meter (m), and second (s), respectively. The units in the English system are the pound-mass (lbm) or the pound (lbs)[1], foot (ft), and second (s). In SI, the force unit is the newton (N), and it is defined as the force required to accelerate a mass of 1 kg at a rate of 1 m s^{-2}.

Table 1.1. SI system fundamental units.

Dimension	Units
Length	meter (m)
Mass	kilogram (kg)
Time	second (s)
Temperature	kelvin (K)
Electric current	ampere (A)
Amount of light	candela (cd)
Amount of matter	mole (mol)

Table 1.2. SI unit prefixes.

10^{12} = tera (T)	10^{-1} = deci (d)
10^{9} = giga (G)	10^{-2} = centi (c)
10^{6} = mega (M)	10^{-3} =milli (m)
10^{3} = kilo (k)	10^{-6} = micro (μ)
10^{2} = hecto (h)	10^{-9} = nano (n)
10^{1} = deca (da)	10^{-12} = pico (p)

[1] The pound symbol *lb* is actually the abbreviation of the ancient Roman unit of weight, *libra.* The English retained this symbol even after the end of the Roman occupation of Britain in 410 AD.

In the English system, the force unit is the pound-force (lbf) and is defined as the force required to accelerate a mass of 32.174 lbm (1 slug) at a rate of 1 ft s^{-2}.

The mass (m) of a body remains the same regardless of its location. Its weight ($W = mg$), which is a force, however, changes with a change in gravitational acceleration. A body weighs less on the surface of the Moon than on the surface of Earth, since $6g_{\text{Moon}} = g_{\text{Earth}}$.

Work is a form of energy and is defined as force times distance. The SI unit of work is the joule (J) defined as newton-meter (N m). In the English system, the energy unit is the Btu (British thermal unit), defined as the energy required to raise the temperature of 1 lbm of water at 68 °F by 1 °F. The magnitudes of the kilo-joule and Btu are almost identical (1 Btu = 1.0551 kJ).

Pressure is defined as a normal force[2] exerted by a fluid per unit area; it has the SI unit of newtons per square meter (N m^{-2}), which is called a pascal (Pa). We speak of pressure only when we deal with fluids, a gas or a liquid. The counterpart of pressure in solids is called '*normal stress*'. In English units, the unit of pressure is pound per square inch (psi or lbf in^{-2}). Two other pressure units commonly used in practice, especially in Europe, are *bar*, and standard atmospheric pressure (atm). 1 atm = 14.696 psi = 1.013 25 bar = 101 325 Pa ($\approx$ 101 kPa).

The units in the two systems are related to each other as shown in table 1.3. For a more detailed explanation of various units please refer to appendix A.

Worked example 1.1. Find the weight of a unit mass at sea level in SI and English units.

Solution: The weight of an object is the gravitational force applied to a body, and its magnitude is determined from Newton's second law,

$$W = mg \tag{1.1}$$

Table 1.3. Relation between SI and English units.

	SI Units	English units
Mass	0.453 59 kg	1 lbm
	1 kg	2.2 lbm
Length	0.304 8m	1 ft
	1 m	3.28 ft
Force	newton: 1N = 1 kg m s^{-2}	pound-force: lbf = 32.174 lbm ft s^{-2}
Acceleration due to gravity	g = 9.807 m s^{-2} = 1 kgf	31.174 ft s^{-2} = 1 lbf
Work	joule (J) = N m	British thermal unit (Btu)
	1.055 kJ	= 1 Btu

[2] The word 'normal' means perpendicular to the surface. Normal force is the perpendicular force that the surface exerts on an object because of surface contact. Hence, normal force is a contact force between two objects, say a book and a table it rests on. When the two objects are separated, even by a very small distance, the normal force disappears.

where m is the mass of the body, and g is the local gravitational acceleration, approximately 9.8 m s^{-2} or 32.2 ft s^{-2} at sea level and 45° latitude.

1. **SI units:** The unit of mass in SI units is 1 kg and the acceleration due to gravity is about 9.8 m s^{-2}. Using equation (1.1), the weight will be calculated to be 9.8 N.

$$W = mg = (1\text{kg})(9.8 \text{ m s}^{-2}) = 9.8 \text{ N}.$$

2. **English units:** The unit of mass in English units is 1 lbm and the acceleration due to gravity is about 32.2 ft s^{-2}. Using equation (1.1), the weight will be calculated to be 32.2 lbm ft s^{-2} = 1 lbf (= 4.5 N).

$$W = mg = (1 \text{ lbm})(32.2 \text{ ft s}^{-2})\left(\frac{1 \text{ lbm}}{32.2 \text{ lbf ft s}^{-2}}\right) = 1 \text{ lbf}.$$

At sea level a mass of 1 kg weighs 9.8 N, whereas a mass of 1 lbm weighs 1 lbf, which misleads many people and students to believe that pound-mass and pound-force can be used interchangeably as pound (lb), a unit of mass. This is a major source of error in the English system.

Rule for notations:

The two basic rules for the notation of the units:

1. The degree symbol (°) is not used with the absolute temperature (K).
2. All unit names are written without capitalization. For example, we use newton as unit of force (not Newton); pascal (not Pascal) for pressure, and kelvin (not Kelvin) for absolute temperature. However, if the abbreviation of the unit derived from a proper name is used, the first letter of the abbreviation must be capitalized.
   - 10 newton (not 10 Newton) but 10 N (not 10 n).
   - 5 ampere (not 5 Ampere) but 5 A (not 5 a).
   - 1 pascal (not 1 Pascal) but 1 Pa (not 1 pa).
   - 298 kelvin (not 298 Kelvin) but 298 K (not 298 k).

1.4 Thermodynamic systems

As we have discussed above, thermodynamics is the science that relates the concepts of heat and work and the respective changes that result in the properties of the working substances.

Before we begin, we should understand some commonly used terms in thermodynamics.

1. **System:** A system is a collection of matter within an identifiable boundary. The boundary may enclose a solid, a liquid, a gas, or a batch of photons in a vacuum, etc. The boundary may also be either **real**, like the inner surface of a tank containing a compressed gas or **imaginary**, like a surface that bounds

a certain mass of fluid flowing along a pipe. This imaginary boundary is called imaginary because it is followed in imagination only as the mass progresses down the pipe. The boundary is also not limited to be necessarily fixed in either shape or volume, and thus the fluid expands or contracts the volume enclosed by the boundary increases or decreases. For example a balloon expands when heat is added. There are three basic types of systems in thermodynamics:

(a) **Open system:** When mass is allowed to transfer across a boundary of the system, the system is called an open system or control volume.

(b) **Closed system:** When mass is not allowed to be transferred across a system boundary it is called a closed system or control mass system.

(c) **Isolated system:** When both mass and energy cannot be interchanged between a system and its surroundings the system is called an isolated system.

Three types of systems are shown in figure 1.1.

There are two main types of boundaries:

(a) **Diathermic boundary:** A diathermic (also called diathermal) boundary is one through which heat can pass even when matter cannot. A diathermic boundary is made from a thermal conductor and is thermally conductive. When two objects at different temperatures having diathermic boundaries are brought in thermal contact with each other we observe a change of state in both objects. An example of a diathermic boundary is a metal wall.

(b) **Adiabatic boundary:** An adiabatic boundary between two systems is one that does not allow heat or matter to pass through. An adiabatic boundary is a thermal insulator. When two objects at different temperatures having adiabatic boundaries are brought in thermal contact with each other, no change in state of either object will be observed. A vacuum flask is a good example of an adiabatic boundary.

2. **Surroundings:** Particles of matter external to the system that may or may not be affected by the changes within the system are called the surroundings of the system.

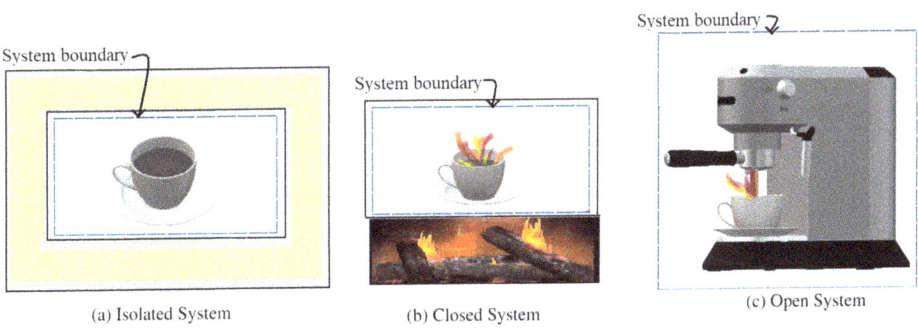

Figure 1.1. Three types of systems.

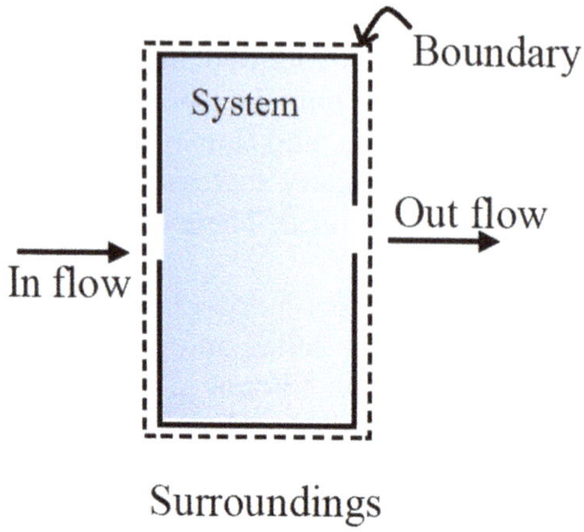

Figure 1.2. System and its surroundings.

A generalized system and its surroundings are shown in figure 1.2. The choice of the proper system, along with the proper form of the thermodynamic laws, are decisions that you, as the analyst, must make whenever solving a thermodynamics problem. As a first step, drawing out the system with a clearly defined system boundary goes a long way toward making these decisions. When doing this, keep in mind that the sketch of a system in thermodynamics is equivalent to the free body diagram in classical mechanics.

1.5 Temperature

From a very young age we all know if an object is hot or cold when we touch it. It is surprising that this common sensation took so long to be scientifically defined. This sensation is called 'temperature'. Let's quickly look at some history behind the definition of temperature.

Galileo (1564–1642 CE): In 1592 CE, Galileo built a fluid based 'thermo-scope', a device made up of a glass tube in which a liquid rises and falls as the temperature changes, as shown in figure 1.3. The thermo-scope was meant to show changes in temperature. However, his thermo-scope had no fixed points because readings from different thermo-scopes could not be compared.

Daniel Fahrenheit (1686–1736 CE): A century later, Fahrenheit used alcohol thermometers and was the first scientist to develop and use mercury thermometers. He also proposed a temperature scale that has his name. Fahrenheit was the first to use the following three fixed points for temperature:

1. 0°→ for a mixture of ice, water, and salt.
2. 32°→ for a mixture of ice and water.
3. 97°→ body temperature of a healthy person.

Anders Celsius (1701–1744 CE): The Celsius scale was introduced by Anders Celsius, a Swedish scientist. In 1742 CE Celsius described his thermometer to the Swedish Academy of Sciences, using 0 °C as the freezing point of water and 100 °C as

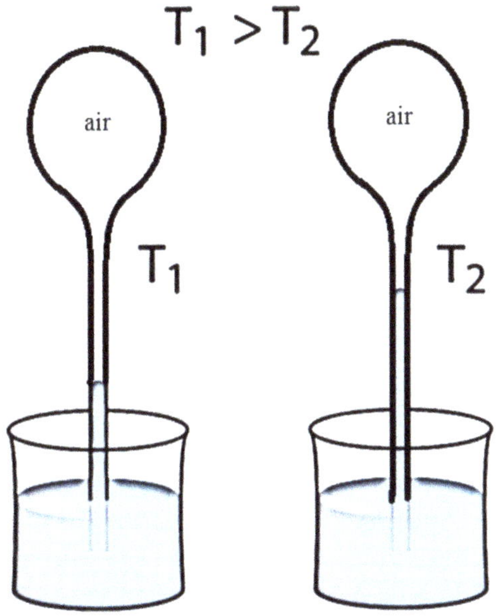

Figure 1.3. Principle of the thermo-scope. Higher temperature expands the air in the bulb and lowers the level in the tube.

the boiling point with 100 divisions between the freezing and boiling points of water. The unit for temperature in the SI system of units is the Celsius and is commonly used around the world to mark temperature as well as in conducting scientific experiments.

1.5.1 So what is temperature?

The simplest definition of '*temperature*' might really be that temperature is something you measure with a thermometer. Measuring temperature is simple; stick a mercury thermometer into the object whose temperature needs to be measured, wait a bit until the mercury in the thermometer stabilizes and stops changing. Look at the thermometer scale and read the temperature of the object. This is called an '*operational definition of temperature*', since it tells you how to simply '*measure*' the temperature.

The reason why this works is because the mercury in thermometer expands or contracts depending on the 'temperature' of the object it is in thermal contact with. Boiling water causes the mercury to expand from room temperature until it equals the temperature of the water. The volume occupied at this point by mercury tells us the temperature of object. Similarly, if the mercury thermometer is inserted into an ice bucket, the mercury in the thermometer contracts until it equals the temperature of ice. The volume occupied at this point by the mercury tells us the temperature of the object. Please note, both of these processes take some time to reach thermal equilibrium. This time delay is defined as '*relaxation time*', discussed in section 1.7.2.

Please keep in mind that temperature is a scalar quantity[3]; it is a measure of the tendency of any object to spontaneously give up energy to its surroundings or another object. Remember, when two objects are in thermal contact the one that tends to spontaneously 'lose' energy is the one at the higher 'temperature'.

1.6 Temperature scales

In other words, we can say that temperature is an inherent property of matter and various temperature scales developed over the years are a way to measure temperature quantitatively. In practical terms, a temperature scale is based on a particular single, measurable and repeatable physical property of the thermodynamic system, for example, the volume expansion of a liquid (usually mercury) or a change in pressure of a gas.

Having said that, there is the philosophical problem of how to set the temperature scale. The square or the logarithm of the thermometer reading would theoretically work, from a purely descriptive point of view. However, in practice linear properties are used for practical applications, as we see in section 1.6.2. Temperature scales that are purely based on physical measurements are called '*experimental temperature scales*'.

1.6.1 Thermometer

A device used to measure the 'temperature' of an object or system is known as a thermometer. In any fluid thermometer, two important temperature factors must be specified,

1. A temperature sensor which contains a bulb filled with a fluid, such as mercury, that is sensitive to temperature changes and will expand when the temperature rises and contract equally when the temperature decreases.
2. A visible scale with predefined numerical values that an observer can use to note the thermal expansion and contraction of the liquid.

It is worth mentioning here that gases can be also used in thermometers. However, precise measurement of temperature with a gas thermometer is a difficult task that requires many months of painstaking laboratory work. Once completed it results in an international event once every 20 years. In practice, therefore, the gas thermometer is used to establish certain fixed points that can then be used to calibrate other, more convenient secondary thermometers.

1.6.2 Measuring temperatures

We know that in order to measure the temperature of any object we must begin with a temperature scale and hence a thermometer of some sort. In principle, any property that varies with temperature can be used as a basis for a thermometer. Most commonly used properties are:

1. The volume of a liquid.
2. The pressure of a gas at constant volume.

[3] A scalar is a physical quantity that is completely described by only specifying its magnitude or a numerical value.

3. The electrical resistance of a wire.
4. The length of a strip of metal.
5. The color of a lamp filament.

However, each of these choices above may give us thermometers that may or may not agree with thermometers based on other properties. Scientists generally call thermometers scales made from various properties listed above as 'private' temperature scales and are defined only for that property.

Even though all thermometers can be calibrated around the triple point (T_{TP})[4], they will not agree as we increase or decrease temperature. However, 'private' thermometers, when properly calibrated, are useful as a secondary standard for measuring temperature.

Now let's try to make our own thermometer. We start by assuming that our thermometer is based on a single measurable property of our material, 'X'. This means that the temperature is a function of the property X, such that $T = T(X)$. The simplest relationship between temperature and the respective property would be a linear function that can be mathematically expressed as ($T^{*}=AX$), where A is a constant that can be calculated by measuring 'X' at the triple point of water to give us an X_{TP}. Hence we can write:

$$T_{TP} = AX_{TP} \tag{1.2}$$

$$\Longrightarrow A = \frac{T_{TP}}{X_{TP}} \tag{1.3}$$

Since our derived value of X will be device sensitive for any value of temperature (T), we introduce T^{*} because it is not a true Kelvin temperature.

$$\begin{aligned} T^{*}(X) &= AX = \frac{T_{TP}}{X_{TP}}(X) \\ \Longrightarrow T^{*}(X) &= \frac{X}{X_{TP}}(T_{TP}) = \frac{X}{X_{TP}}(273.15\ \text{K}) \end{aligned} \tag{1.4}$$

Worked example 1.2. Suppose we make a thermometer from a coil of wire whose resistance increases by factor of 1.75 between the triple point of water and the normal boiling point of water at atmospheric pressure. What temperature will this thermometer show for the normal boiling point of water?

---

[4] The term '*triple point*' was first used in 1873 CE by James Thomson, the brother of Lord Kelvin. The triple point of a substance is the temperature and pressure at which the three phases (gas, liquid, and solid) of that substance coexist in thermodynamic equilibrium. All thermometers will agree at the triple point of water at T_{TP} = 273.16 K = 0.01 °C and 0.006 027 atm = 0.611 7 kPa. The triple point of mercury is −38.83°C at 0.2 mPa. Pure water is placed in a U-shaped manometer, which is lowered into a water-ice bath. The system is at the triple point when ice, water vapor are present at equilibrium inside the cell.

Solution: In this thermometer, we are utilizing the resistance (R) of the wire as the generalized thermal property X and are given $R = 1.75R_{TP}$. We can use equation (1.5) as,

$$T^*(R) = \frac{1.75R_{TP}}{R_{TP}}(T_{TP}) = \frac{1.75}{1}(273.15\text{ K}) = 478.012\text{ K}. \quad (1.5)$$

Please keep in mind that the actual boiling point of water in the Kelvin scale is 373 K[5].

Worked example 1.3. The Fahrenheit temperature scale defines the freezing point of water as 32 °F and the boiling point as 212 °F, whereas Celsius temperature scale defines the freezing point of water as 0 °F and the boiling point as 100 °F. Derive a formula that connects °F to °C.

Solution:

	°F	°C
Water freezes	32 °F	0 °C
Water boils	212 °F	100 °C
Divisions	180	100

We know that in order to find a formula that connects the Fahrenheit scale to the Celsius scale we want to find a linear function that will take 32 °F to 0 °C and 212 °F to 100 °C while still maintaining the proper divisions in each of the temperature scales. The linear relationship between the two temperature scales implies that we can use the equation of a straight line ($y = mx + b$); in this case the y-axis is T(°F) and the x-axis is T(°C). Also recognize here that the intercept (b) will be 32°. Keeping these things in mind we can write the slope (m) as,

$$m = \frac{y}{x} = \frac{T(°\text{F})}{T(°\text{C})} = \frac{(212 - 32)\ °\text{F}}{(100 - 0)\ °\text{C}} = \frac{180}{100} = \frac{9}{5}$$

Hence the function will be:

$$y = mx + b$$
$$T(°\text{F}) = \frac{9}{5}T(°\text{C}) + 32$$

1.6.3 Accepted temperature scales

All temperature scales are calibrated according to thermal properties of substances. The International Temperature Scale, which consists of a selection of procedures

[5] Also other thermometers may give different values, for example, normal boiling point of water is given as 412.5 K when measured with a thermometer based on electrical voltage generated by two joined dissimilar wires.

providing the best possible approximation to the Kelvin scale has been adopted for the calibration of thermometers for scientific and industrial use. The International Committee of Weights and Measures reviews and redefines the scales about every 20 years. An acceptable scale consists of the following three qualities:

1. A set of fixed points.
2. Specific devices that can be used for interpolating between these fixed points.
3. The potential for reasonably accurate extrapolation beyond the highest fixed point.

The three main temperature scales used in the present era are given below and shown in figure 1.4. Please note that the benchmarks for these scales could have been placed at any other numerical value, since the scales are of our own creation.

1. **Fahrenheit (°F):** Fahrenheit was the first scientist to use mercury in a liquid-in-glass thermometer. On the Fahrenheit scale, the freezing point of water is 32 degrees Fahrenheit (°F) and the boiling point is 212 °F, at standard atmospheric pressure conditions. Fahrenheit is the official temperature scale in the United States.
2. **Celsius (°C):** Also called the centigrade scale; it was developed in around the 1740s by Anders Celsius. The Celsius scale is used worldwide and on the Celsius scale, the freezing point of water is 0 degrees Celsius (°C) and the boiling point is 100 °C, at standard atmospheric pressure conditions. The Celsius system is used today all over the world because it uses a decimal-based counting system and is easier to use.
3. **Kelvin (K):** In the 19th century, Lord Kelvin developed the Kelvin temperature scale, which is founded on the theoretical existence of absolute zero. Absolute zero temperature is said to be the point when a system has no thermal energy

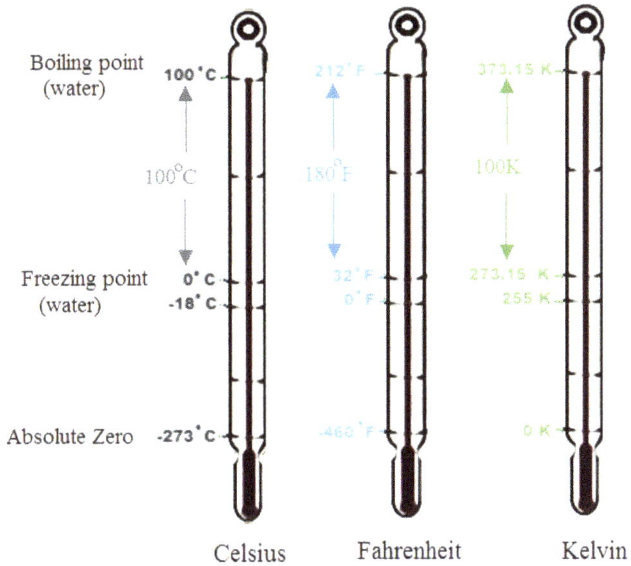

Figure 1.4. Temperature scales, Fahrenheit (°F), Celsius (°C), and Kelvin (K).

and all molecular motion in the system stops. It is not possible for any temperature to be colder than the absolute zero temperature. To establish the size of a degree on the Kelvin scale, the *triple point* of water is used as a calibration. The Kelvin temperature for the triple point of water has been set by international agreement since 1954 CE, placing the triple point at $T_{TP} = 273.16$ K. The Kelvin scale is the only one that is founded on the principles of thermodynamics and not on material behaviors, such as the freezing and boiling of water. Hence Kelvin is used as the SI unit for temperature.

Conversion between Fahrenheit (°F) and Celsius (°C) scales is given as:

$$T_F = \frac{9}{5}T_C + 32, \tag{1.6}$$

whereas the conversion between Kelvin (K) and Celsius (°C) scales is given as:

$$T_C = T_K - 273.15. \tag{1.7}$$

The conversion formulas for the various temperature scales in use today are given in table 1.4.

Worked example 1.4. Rankine (°R) is a thermodynamic temperature scale named after engineer and physicist William Rankine who proposed it in 1859 CE. Rankine temperature (°R) uses the same size degrees as °F, but is measured up from absolute zero like the Kelvin scale.

1. Find the conversion between °R and °F, and vice versa.

   Solution: Rankine temperature (°R) is measured from absolute zero on the Kelvin scale hence absolute zero (°R) = −273 °C = −459.7 °F.

   This implies that to convert from Fahrenheit to Rankine scale we need to simply add 459.7 or 460° to temperature. For example room temperature is generally taken to be 72 °F, which will be 532 °R, as shown in figure 1.4.
2. Find the conversion between °R to Kelvin scale.

   Solution: Both of the Rankin and the Kelvin scales are measured from the same absolute zero point, however the Kelvin degrees are larger, that is $1(°R) = \frac{5}{9}1(°F)$, hence,

$$T(°R) = \frac{9}{5}T(°F)$$

Table 1.4. Conversion formulas for temperature scales.

	From °C	From Kelvin	From °F	From Rankine
To °C	–	$T_C = T_K - 273.15$	$T_C = \frac{5}{9}(T_F - 32)$	$T_C = \frac{5}{9}T_R - 273.15$
To K	$T_K = T_C + 273$	–	$T_K = \frac{5}{9}(T_F + 459.67)$	$T_K = \frac{5}{9}T_R$
To °F	$T_F = \frac{9}{5}T_C + 32$	$T_F = \frac{9}{5}T_K - 459.67$	–	$T_F = T_R - 459.67$
To R	$T_R = \frac{9}{5}T_C + 491.67$	$T_R = \frac{9}{5}T_K$	$T_R = T_F + 459.67$	–

This implies that room temperature (300 K) will be,

$$T(^{\circ}\text{R}) = \frac{9}{5}(300\,K) = 540\ ^{\circ}\text{R}$$

Concept questions 1.1

1. What is the main difference between classical and statistical approaches to thermodynamics?
2. What is the difference between pound-mass and pound-force? kg-mass and kg-force?
3. At what temperatures do the Fahrenheit and Celsius scales coincide?
4. At what temperatures do the Fahrenheit and Kelvin scales coincide?
5. Can you define temperature in terms of density of water? Is that a good idea?
6. Can one object be hotter than another if a thermometer reads the same temperature for both?

Exercises 1.1

1. Determine Kelvin temperature for each:
   (a) Human body temperature.
   (b) Boiling point of water (at standard atm).
   (c) The coldest day you can remember.
   (d) Boiling point of liquid nitrogen (−196 °C)
   (e) Melting point of lead (327 °C)
2. Does it make sense to say one object is 'twice as hot' as another?
   Does it matter whether one is referring to a Celsius or a Kelvin scale? Explain in your own words.
3. At what temperature is the Fahrenheit scale reading equal to,
   (a) Twice that of the Celsius reading.
   (b) Half that of the Celsius reading.

1.7 Zeroth law of thermodynamics

We shall discuss the state of a system in detail in chapter 2, section 2.3. Here I will start the discussion by just saying that the state of any system is defined by specifying values of certain experimentally measurable quantities that are called 'state variables of the system'. The State variables of the system are generally taken to be uniform throughout the system.

1.7.1 Equilibrium

A particular state of a system is said to be in *equilibrium* when it no longer experiences a change in any of its properties and the state remains unchanged. Equilibrium can be of several different types as defined below:

1. **Thermal equilibrium:** When two objects are brought into thermal contact with each other, heat energy spontaneously travels from the object at the higher temperature to the object at the lower temperature. This spontaneous flow stops when both objects are at the same temperature and at this point the system is said to be in 'thermal equilibrium'.
2. **Mechanical equilibrium:** Systems can exchange volumes between them. Any two systems are said to be in mechanical equilibrium, when the exchange of volume between the system becomes zero, i.e., $\Delta V = 0$. Examples of mechanical equilibrium include an expanded balloon.
3. **Diffusive equilibrium:** Systems can also exchange particles when they are brought in contact with each other. Any two systems are said to be in diffusive equilibrium, when the exchange of particles between the systems becomes zero, i.e., $\Delta N = 0$. An example of diffuse equilibrium includes mixing of milk in coffee. The molecules of milk and coffee eventually have the same kinetic energies and each substance has a constant concentration throughout the other. When this occurs the systems are said to be in diffusive equilibrium.

1.7.2 Thermal equilibrium

After two objects have been in thermal contact long enough, they are said to be in thermal equilibrium with each other. The time for objects to reach thermal equilibrium is called the '*relaxation time*'. Cold creamer and hot coffee when mixed together while stirring will have a short relaxation time, but warm creamer and hot coffee without any stirring will have a long relaxation time.

What we define as '*thermal contact*' is a bit tricky to answer. It could be either a physical contact between the bodies or it could be long-distance contact, where energy is transferred between bodies via electromagnetic waves, such as those between the Sun and the Earth. In general, thermal contact between two systems is defined when the two systems are allowed to spontaneously exchange heat between them. Systems in physical contact have a small relaxation time and systems with long-distance contact have a longer relaxation time. It is important to note, relaxation times may vary from a few seconds to a few eons, however all systems will eventually come to thermal equilibrium.

Worked example 1.5. Your friend has a fever and you take her temperature with a mercury thermometer. What is the approximate relaxation time?

Solution: In general, doctors tell you to put the thermometer in your mouth under your tongue for about one minute. This allows them to find if you have a fever or not because this is the time that it takes your body and the thermometer to reach thermal equilibrium. This means that the relaxation time of the mercury thermometer is about 60 seconds.

1.7.3 Zeroth law of thermodynamics

The definition of thermal equilibrium helps us define the Zeroth law of thermodynamics. Suppose systems A and B are each in thermal equilibrium with a third system C, then we can conclude systems A and B are also in thermal equilibrium with each other; this is called the Zeroth law of thermodynamics. In figure 1.5 we see glass (at temperature T_G), fluid (at temperature T_S) and mercury (at temperature T_M), Zeroth law states that they will be in thermal equilibrium when,

$$T_G = T_S = T_M$$

The Zeroth law of thermodynamics seems trivial on its surface, however it has far-reaching consequences.

The Zeroth law of thermodynamics, in terms of temperature, can be stated as: *There exists a scalar quantity called 'temperature', which is a property of all thermodynamic systems in equilibrium. Two systems are considered to be in thermal equilibrium iff*[6] *their temperatures are equal.* Hence, the Zeroth Law defines the concept of temperature and allows us to experimentally determine temperature by the use of thermometers.

1.8 Thermal expansion

In designing houses, buildings, roads, bridges, and other structures engineers and architects must always take into account that most things expand when they are heated and contract when they are cooled. This is known as thermal expansion and is the tendency of matter to change its shape when there is a change in temperature.

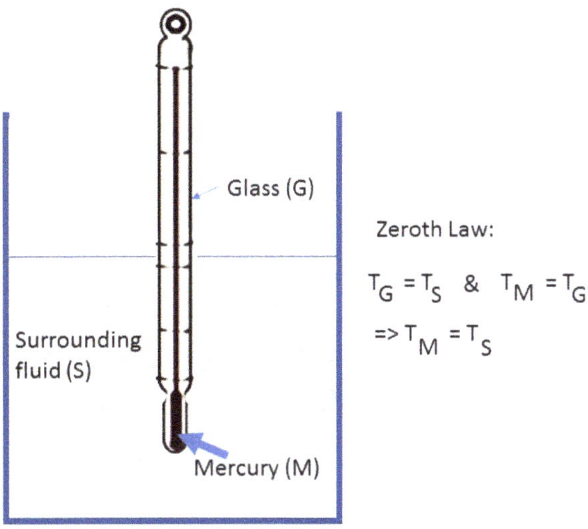

Figure 1.5. The Zeroth law of thermodynamics applied to mercury in a glass thermometer.

[6] In the fields of mathematics, logic and philosophy, **if and only if** is shortened as *iff*.

If you put a jam jar with a metal top and a glass body in the refrigerator, it is hard to open the lid; but if you hold it upside down in hot water it becomes easier to open. This happens because cooling makes the metal lid contract more than the glass and tightens on the glass body, which makes it hard to open it. But if you hold the jam jar in hot water the metal lid will expand more than the glass and make it easier to open. This is an example of thermal expansion. Other examples include;

1. Power lines contain loops to account for the difference between summer and winter temperatures.
2. Bridges have cement slabs that are connected with sliding metal connectors. The metal connectors allow the cement slabs in the bridges to expand and contract during the different seasons without damaging the bridge.
3. Tooth fillings have expansion properties just like tooth enamel to account for the difference between hot and cold foods we eat every day.

Normally, as the temperature decreases the sizes of objects also decrease. Very few substances expand when they are cooled and contract when they are heated. The most common example of this phenomenon is water. For most temperature ranges water behaves normally; that is, it expands when it is heated and contracts when it is cooled. However, when cooled to below 4 °C above its freezing point water begins to expand. This means one gram of ice takes up more 'space' then one gram of liquid water. This is why ice floats on top of the water surface in a lake and this also explains the fact that a filled plastic water bottle breaks in a freezer, however a half-filled water bottle does not.

1.8.1 Coefficient of thermal expansion

A metal wire expands when it is heated. The amount by which a particular wire expands depends on the length of wire, the amount of temperature change and thermal properties of the particular metal the wire is made of. Some metals of the same size under similar conditions will expand more than others. Wires made of aluminum, iron and tungsten will all expand to various lengths when they are heated even when we keep all other conditions the same.

The degree of expansion divided by the change in temperature is called the '*coefficient of thermal expansion*' of the material and it generally varies with temperature. The degree of expansion is the fractional increase, for example going from 5 m to 5.10 m is a degree of expansion of $(5.10 - 5)/5 = 0.02$. The coefficient of thermal expansion describes how the size of an object changes with a change in temperature.

Many types of coefficients of thermal expansion have been developed over the years depending on the application under consideration. These include linear coefficients of expansion, area coefficients of expansion and volumetric coefficients of expansion. For example, linear coefficients of expansion are usually used for solids, where we are mostly only concerned about the change in the length of the solid. Whereas for fluids, the volumetric thermal expansion coefficient is used as the basic thermal expansion coefficient.

1.8.2 Linear expansion in solids

When we want to talk about expansion in solid materials, we have to consider the two major kinds of materials: isotropic and anisotropic. Isotropic materials are materials that have identical values of properties in all directions, for example glass and metals, whereas anisotropic materials are materials in which properties change depending on the direction along the object. Common examples of anisotropic materials are wood and composite materials.

The change in any linear dimension of a solid (length, width, thickness) is called linear expansion. If the linear dimension is 'L', then changes in temperature (ΔT) cause changes in length (ΔL) such that

$$\Delta L = \alpha(\Delta T) \cdot L \tag{1.8}$$

where α is called the coefficient of linear expansion and is mathematically given as:

$$\frac{\Delta L}{L}\frac{1}{\Delta T} = \alpha. \tag{1.9}$$

1.8.3 Expansion in isotropic materials

As we discussed above, an isotropic material is a material having identical values of a property in all directions. Let's look at the various types of isotropic materials:

1. **Isotropic solids:** For an isotropic solid, the percent change of length for a given temperature change is the same for each of its dimensions. In three dimensions for any ΔT, ($\frac{\Delta L}{L}$) is the same in all directions (i.e., length, thickness, face/body diagonal, etc). Hence, for area and for volume for isotropic solids we can write:

$$\Delta A = 2\alpha(\Delta T) \cdot A \tag{1.10}$$

$$\Delta V = 3\alpha(\Delta T) \cdot V \tag{1.11}$$

The volume equation can be rewritten using the coefficient of volume expansion β where $\beta = 3\alpha$ as,

$$\Delta V = \beta(\Delta T) \cdot V \tag{1.12}$$

$$\beta = \frac{\Delta V}{V}\frac{1}{\Delta T}. \tag{1.13}$$

2. **Isotropic liquids:** For isotropic liquids, β is relatively independent of temperature. Liquids usually expand with increasing temperature, with typical β values around $[200 - 1000] \times 10^{-6}$/°C. This is an order of magnitude larger than the β of solids.
3. **Isotropic gases:** For isotropic gases β is strongly dependent on temperature. For an ideal gas $\beta = \frac{1}{T}$ when T is expressed in Kelvin. For a gas at room temperature and constant pressure, β is approximately 3300×10^{-6}/°C. This is an order of magnitude larger than β for liquids.

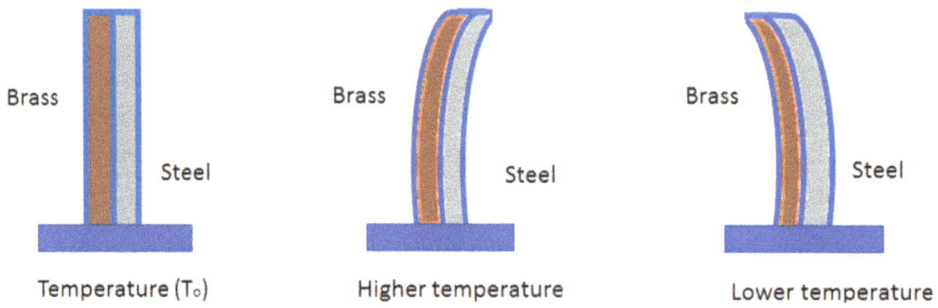

Figure 1.6. Bimetallic strip.

1.8.4 Bimetallic strip

A bimetallic strip consists of two flat strips made of different metals joined together. The common metals that are used to make a bimetallic strip are steel and copper, which expand at different rates as they are heated. The different expansion coefficients of each metal forces the strip to bend one way when it is heated above its initial temperature and forces the strip to bend the other way when it is cooled below its initial temperature. The metal with the higher coefficient of thermal expansion is on the outer side of the curve when the strip is heated and on the inner side when cooled, as seen in figure 1.6.

One common use of bimetallic strips is in the thermostats used to control automatic room temperatures. As a room warms up or cools down, the bimetallic strip in the thermostat bends one way or the other. If it bends far enough, it comes into contact with a button that turns an air conditioner or a furnace on or off. Bimetallic strips are also used in a wide range of mechanical and electrical devices, for example, heat engines, clocks, etc.

Worked example 1.6. For steel α is 1.1×10^{-5} 1/°C. Estimate the total variation in length of a 1 km steel bridge between a cold winter night and a hot summer day. Assume that the temperature the difference between a cold winter night at a hot summer day is about 45 °C.

Solution: In order to estimate the total variation in length of a 1 km steel bridge between a cold winter night and a hot summer day, we can use equation (1.9).

$$\begin{aligned}\Delta L &= \alpha(\Delta T) \cdot L \\ &= (1.1 \times 10^{-5} 1/°\text{C})(45\ °\text{C})(1000\ \text{m}) \\ &= 0.495\ \text{m}.\end{aligned}$$

Please note here that I have left change in temperature (ΔT) in (°C) because the α for the steel in this example is given in (°C).

Worked example 1.7. A brass cube ($\alpha_{\text{Brass}} = 1.9 \times 10^{-5}$ 1/°C) has an edge length of 20 cm at room temperature (20 °C). The cube is then heated to 75 °C.

a. What is the increase in the surface area?
b. What is the increase in the volume?

Solution: We have a brass cube that has length 20 cm which is heated to 75 °C. This means the change in temperature (ΔT) = 55° C.

a. The increase in the surface area.

Area of one face = length × width = 0.20 m × 0.20 m = 0.04 m^2.

A cube has six faces, which implies the area of the cube will be 6 × 0.04 m^2 = 0.24 m^2.

We can therefore use equation (1.10) to find the increase in area as:

$$\begin{aligned}\Delta A &= 2\alpha(\Delta T) \cdot A \\ &= 2(1.9 \times 10^{-5})(55)(0.24) \\ &= 0.0005 \text{ m}^2.\end{aligned}$$

b. The increase in the volume.

The volume of the cube is given as:

$$V = 0.20 \times 0.20 \times 0.20 = 0.008 \text{ m}^3$$

Therefore, we can use equation (1.12) to get:

$$\begin{aligned}\Delta V &= 3\alpha(\Delta T) \cdot V \\ &= 3(1.9 \times 10^{-5})(55)(0.008) \\ &= 0.000011 \text{ m}^3.\end{aligned}$$

1.9 Pressure

Pressure is the compressive force per unit area and is defined as a normal force exerted by a fluid per unit area. Suppose we have a perfect fluid at rest in a container. Assume we put a small object with surface area (ΔA) inside the fluid. The fluid exerts a force (ΔF) on one side of this area but according to Newton's third law, an equal and opposite force ($-\Delta F$) is exerted by the fluid on the other side of this area. The pressure (P) at this area will be:

$$P = \frac{\text{Normal Force}}{\text{Area}} = \frac{\Delta \vec{F}}{\Delta A} = \lim_{\Delta A \to 0} = \frac{\Delta \vec{F}}{\Delta A} = \frac{d\vec{F}}{dA} \frac{\text{N}}{\text{m}^2} = \text{Pa} \tag{1.14}$$

Note since pressure is force per unit area, it seems like it should be a vector quantity, but it is not. Pressure at any point in a fluid is the same in all directions, so it has a magnitude at every point but not a specific direction, thus it is a scalar quantity.

The force exerted by the fluid at rest on any area must always be at right angles to the area, as seen in figure 1.7. When the force is not perpendicular, then it has a component along the surface, which results in an equal and opposite force to be exerted on the fluid that causes the fluid to move. This negates the fluid-static condition since the fluid starts to move and we will have the condition of fluid-dynamics.

1.9.1 SI unit of pressure

The SI pressure unit pascal (Pa) is too small for pressures encountered in real life, so its multiples kilopascal, 1 kPa (10^3 Pa) and megapascal, 1 MPa (10^6 Pa) are used in practice.

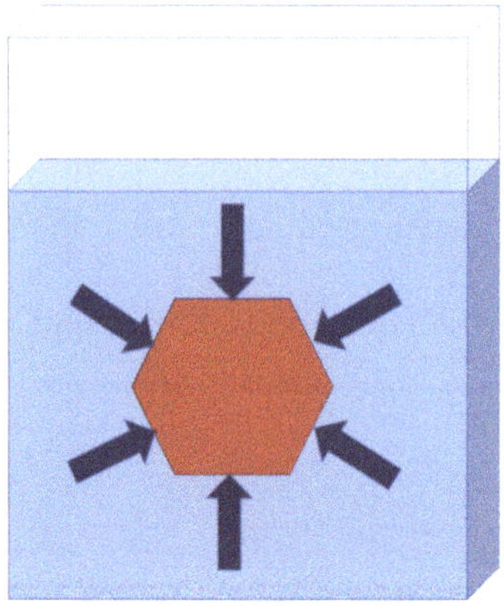

Figure 1.7. Pressure is always perpendicular to the surface of an object.

Usually the pressure unit used is the standard atmosphere ($P_{atm} = 1.01325 \times 10^5$Pa = 101.325 kPa). P_{atm} is defined as the pressure produced by a column of mercury 760 mm in length at 0 °C ($\rho_{Hg} = 13\ 595$ kg m^{-3}) under standard acceleration due to gravity ($g = 9.807$ m s^{-2}). The atmospheric pressure is the weight of the air above a location per unit surface area.

Atmospheric pressure changes with elevation and weather conditions. The standard atmospheric pressure (P_{atm}) decreases with elevation; basically it changes from 1.013 25 $\times$ 10^{5}Pa at sea-level to 89.88 kPa at 1000 m, to 79.50 kPa at 2000 m to 5.53 kPa at 20 000 m. At the peak of Mount Everest $\approx$ 8800 m, P_{atm} reduces to $\approx$26.5 kPa. One effect of high altitudes is that cooking takes longer. Water boils at a lower temperature at higher altitudes, so foods prepared by boiling or simmering cook at a lower temperature, and hence take longer to cook. Another effect is that air at higher altitudes is colder, less dense, and it contains fewer oxygen molecules. This means we need to take more breaths in order to get the same amount of oxygen as we would require at lower altitudes.

1.9.2 Variation of pressure in with depth

People who scuba dive know that the pressure increases as we dive to greater depths. The deeper you go, the higher the pressure. The increasing water pressure with depth limits how deep a submarine can go. To calculate the variation of pressure with depth, let's consider a cylinder of mass (m) and cross sectional area (A) and height $y_2 - y_1 = \Delta y$ as shown in figure 1.8.

The mass of this object is given as:

$$dm = \rho dV = \rho(dA)(dy) = \rho dAdy$$

where ρ is the density. The weight of the object will be given as is:

$$W = (dm)g = \rho dVg = \rho(dA)(dy)g.$$

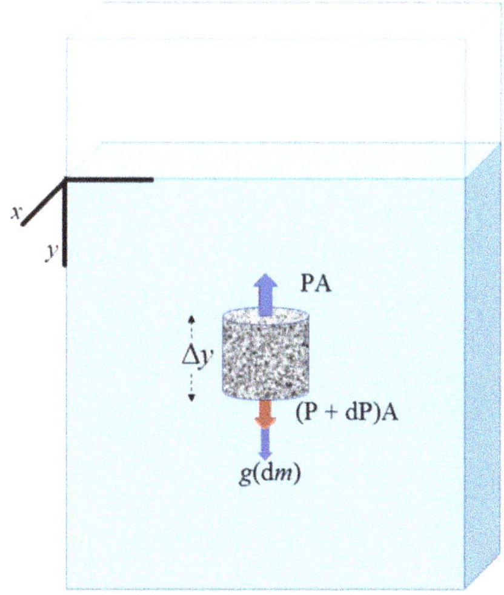

Figure 1.8. Forces acting on the object in terms of pressures.

Using figure 1.8 we can write Newton's Force equation[7] for this situation, given $P_{\text{up}} \equiv P$, $P_{\text{down}} \equiv P + \Delta P$ and $m = \rho dAdy$ we get:

$$\begin{aligned} \Sigma F_y &= F_{\text{up}} - F_{\text{down}} - mg \\ 0 &= PA - (P + dP)A - (\rho Ady)g \end{aligned}$$

Which simplifies to:

$$\frac{dP}{dy} = -\rho g \tag{1.15}$$

Equation (1.15) determines how pressure varies with elevation above a particular reference level in static equilibrium. When dy is positive, elevation increases and pressure decreases. From this equation we see that pressure in a fluid increases linearly with depth. The factor (ρg) is often called '*weight density*'[8] of the fluid and for water it is numerically equal to 9800 N m^{-3}.

If we assume P_1 to be the pressure at elevation y_1 and P_2 to be the pressure at elevation y_2, then we can integrate equation (1.15) to get:

$$P_2 - P_1 = -\rho g(y_2 - y_1) = -\rho g \Delta y \tag{1.16}$$

Assuming that in this case the density of the fluid and g remain constant and do not change, equation (1.16) gives the pressure change for a homogenous, incompressible liquid. For a given fluid, the vertical distance ($y_2 - y_1 = \Delta y$) is sometimes used as a measure of pressure, and it is called the '*pressure head*'. Pressure head is the height of

[7] Newton's Force equation results from Newton's second law of motion which states that the sum of the forces acting on an object is numerically equal to the mass of the object times the acceleration produced due to the unbalanced force acting on it ($\Sigma F = ma$, when $a = 0$, $\Sigma F = 0$).

[8] Weight density is defined as the weight of a substance per unit volume and has units N m^{-3} or lbf ft^{-3}.

a fluid (liquid or gas) column that corresponds to a particular pressure exerted by the fluid column on the base of the container.

$$y_2 - y_1 = \Delta y = H = \frac{\Delta P}{\rho g}$$

where H is the height of the fluid.

Pressure in a fluid at rest is independent of the shape or cross section of the container. It changes with the vertical distance, but remains constant in other directions. Therefore, the pressure is the same at all points on a horizontal plane in a given fluid, as seen in figure 1.9.

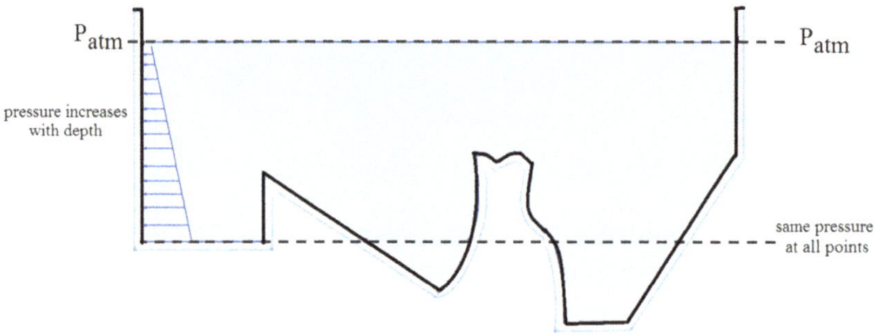

Figure 1.9. The pressure is the same at all points on a horizontal plane in a given fluid regardless of geometry, provided that the points are interconnected by the same fluid.

1.9.3 Commonly used pressures

Three most commonly used pressures used for calculations in thermodynamics are shown in figure 1.10, and they include:

1. **Absolute pressure:** The actual pressure at a given position is called the absolute pressure (P_{abs}). Absolute pressure is measured relative to the absolute vacuum, which is also called absolute zero pressure.

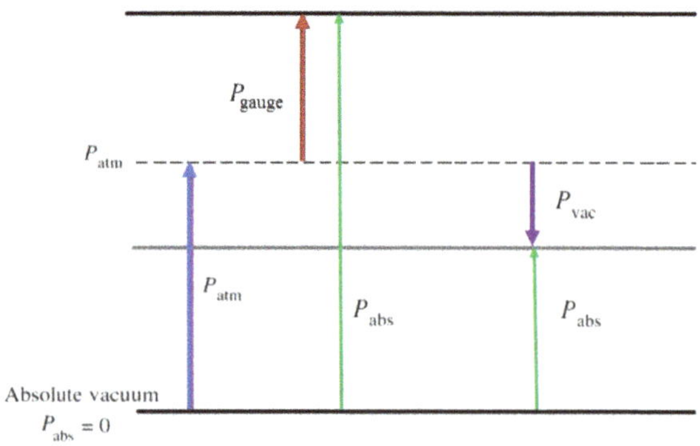

Figure 1.10. Absolute, gage, and vacuum pressures.

2. **Gauge pressure:** Most pressure-measuring devices are calibrated to read zero in the atmosphere and they actually read the difference between the absolute pressure and the local atmospheric pressure. This difference is called the gauge pressure (P_{gauge}). The gauge used to measure air pressure in a car tire reads the gauge pressure. The reading of 32 psi in a tire indicates a pressure of 32 psi above the atmospheric pressure.
3. **Vacuum pressure:** Pressures below atmospheric pressure are called vacuum pressures (P_{vac}) and are measured by vacuum gauges that indicate the difference between the atmospheric pressure and the absolute pressure.

Absolute, gauge, and vacuum pressures are all positive quantities and are related to each other as,

$$\begin{aligned} P_{gauge} &= P_{abs} - P_{atm} \quad \text{(for pressures above } P_{atm}) \\ P_{vac} &= -P_{abs} + P_{atm} \quad \text{(for pressures below } P_{atm}) \end{aligned} \tag{1.17}$$

Worked example 1.8. A vacuum gauge connected to a chamber reads 8 psi at a location where the atmospheric pressure is 1 atm (14.696 psi). Determine the absolute pressure in the chamber.

Solution: Use equation (1.17) to find the absolute pressure,

$$\begin{aligned} P_{vac} &= -P_{abs} + P_{atm} \\ 8 \text{ psi} &= -P_{abs} + 14.696 \text{ psi} \\ \Longrightarrow P_{abs} &= 6.696 \text{ psi} \end{aligned}$$

Worked example 1.9. The hydraulic lift, as shown in figure 1.11, in a car repair shop has an output diameter of 10 m and is to lift cars up to 2000 kg. Determine the fluid gage pressure that must be maintained in the reservoir. Assume the weight of the piston of the lift is negligible.

Solution: Pressure is defined as force per unit area, thus the gage pressure required is simply the ratio of the weight of the car to the area of the lift. Given diameter = 10 m, implies radius = 5 m.

$$P_{gauge} = \frac{mg}{\text{Area}} = \frac{mg}{\pi r^2} = \frac{(2000 \text{ kg})(9.8 \text{ m s}^{-2})}{\pi(5)^2} = 249 \text{ Pa}.$$

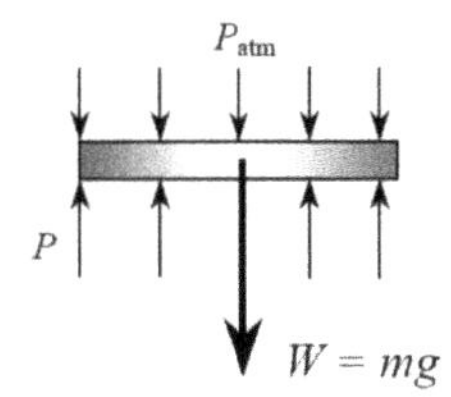

Figure 1.11. Hydraulic lift—example.

1.9.4 Pascal's principle

One consequence of the fluid pressure remaining constant in the horizontal direction is that the pressure applied to a confined fluid increases the pressure throughout by the same amount. This is called **Pascal's principle**, which states that pressure applied to a fluid in a closed container is transmitted equally to every point of the fluid and to the walls of the container. Blaise Pascal (1623–1662 CE) knew that the force applied by a fluid is proportional to the surface area. He experimented with two hydraulic cylinders of different areas, and realized that when they are connected, the larger cylinder can be used to exert a proportionally greater force than that applied to the smaller cylinder. This is called the **Pascal's machine** and it uses the concept of *hydraulic lift.*

A hydraulic lift works by using an incompressible liquid to multiply the effect of the force applied to lift something very large or very heavy by using a much smaller force in comparison. A hydraulic lift system contains two pistons connected by a tube, all filled with an incompressible fluid like oil or sometimes water. Since the fluid is incompressible, we know two things; one is that the pressure is constant and two, very little energy is lost to friction. This means that the force is directly proportional to the surface area of the piston.

If the pistons have the same diameter, then the force applied on one piston to push it down by a certain distance is the same as the force that lifts the second piston up by the same distance. However, if one of the pistons has a larger surface area than the other piston, the effect of the force can be multiplied by the same factor as the increase in area of the piston. For example, in a particular hydraulic system one of the pistons is ten times larger than the other, then moving the smaller piston over ten meters with a certain amount of force will allow the bigger piston to move by one meter. Hence, the same input force can be used to lift bigger objects.

Pressure on the fluid at the smaller piston will be given as:

$$P_{\text{in}} = \frac{F_{\text{in}}}{A_{\text{in}}}$$

Pascal's principle says that:

$$P_{\text{in}} = P_{\text{out}}$$

$$F_{\text{in}} = F_{\text{out}} \frac{A_{\text{in}}}{A_{\text{out}}}$$

$$F_{\text{in}} = (Mg) \frac{A_{\text{in}}}{A_{\text{out}}} \tag{1.18}$$

The area ratio $\left(\frac{A_{\text{out}}}{A_{\text{in}}}\right)$ is called the '*ideal mechanical advantage*' of the hydraulic lift. Usually, $\left(\frac{A_{\text{in}}}{A_{\text{out}}}\right)$ is less than 1 and this implies that F_{applied} can be much smaller than the weight (Mg) to be lifted.

Downward movement of the smaller piston through a distance (d_{in}) displaces a volume of fluid equal to, $V = A_{\text{in}} d_{\text{in}}$. When the fluid is incompressible this volume must be equal to the volume of fluid displaced by the upward motion of the larger piston, i.e., $V_{up} = V_{down}$. Let's suppose we need to lift a car up; then in this case we can assume $V_{down} = A_{in}d_{in}$ and $V_{up} = A_{out}d_{out}$. From these we can write:

$$V_{down} = V_{up}$$

$$d_{out} = (d_{in})\frac{A_{in}}{A_{out}} \tag{1.19}$$

If $\left(\frac{A_{in}}{A_{out}}\right)$ is a small number then the distance moved by the larger piston is much smaller than the distance the applied force moves the smaller piston. The price we pay for gaining the ability to lift a large load is losing the ability to move it very far.

$$\begin{aligned} \frac{F_{in}}{F_{out}} &= \frac{d_{out}}{d_{in}} \\ F_{in}d_{in} &= F_{out}d_{out} \\ W_{in} &= W_{out} \end{aligned} \tag{1.20}$$

Since, [work ≡ force × displacement]. This means, work done by external force on the smaller piston is numerically equal to the work done by the fluid on the larger piston. So if we ignore friction and other dissipative forces there is not a net gain or loss in energy in this hydraulic system.

Worked example 1.10.

1. Suppose the small piston of the hydraulic lever as shown in figure 1.12 has a diameter of 4 cm and the large piston of the hydraulic lever has a diameter of 50 cm. What weight on the small piston will support a 20 kN (2039 kilograms force (kgf)) weight on the large piston?

   Solution: Given the diameters, we calculate radii for the small piston as 2 cm and the large piston as 25 cm. We can use equation (1.18) to get,

$$\begin{aligned} F_{in} &= (Mg)\frac{A_{in}}{A_{out}} \\ &= (Mg)\frac{\pi r_{in}^2}{\pi r_{out}^2} = (20\text{ kN})\left(\frac{2^2}{25^2}\right) = 0.128\text{ kN} = 128\text{ N}. \end{aligned}$$

Figure 1.12. Hydraulic lever.

2. Through what distance must the small piston move to raise the weight by 2 cm?
Solution: Use equation (1.20),

$$F_{\text{small}} h_{\text{small}} = F_{\text{large}} h_{\text{large}}$$

$$h_{\text{small}} = \frac{F_{\text{large}}}{F_{\text{small}}}(h_{\text{large}}) = \frac{20000}{128}(2\text{ cm}) = 312.5\text{ cm}.$$

1.9.5 Pressure in solids

For solids, pressure is equivalent to *normal stress*, (σ) which is the force acting perpendicular to the surface per unit area,

$$\sigma = \frac{\text{Normal Force}}{\text{Area}} = \frac{N}{m^2} = \text{Pa}. \tag{1.21}$$

Which can be rewritten as,

$$\sigma \cdot dA\hat{n} = \vec{F} \tag{1.22}$$

where $\hat{n}$ is the unit vector along the outward drawn normal. The force exerted by an object sitting on a table is equal to the weight of the object regardless of the area in contact. However, pressure depends on the surface area in contact. Different orientations will have different areas in contact with the surface and therefore exert different pressures. More surface contact will give lower pressure and less surface area in contact will result in higher pressure.

Worked example 1.11. How much pressure will a 100 pound person with a total foot imprint area of two feet on the ground equaling 25 in^2 exert on the floor?
Solution:

$$\sigma = \frac{\text{Normal Force}}{\text{Area}} = \frac{100\text{ lbf}}{25\text{ in}^2} = 4\text{ psi} \tag{1.23}$$

Now if the person stands on one foot, the pressure doubles to 8 psi.

If the person gains weight, they are likely to have foot pain because of the increased pressure. This also explains how a person can walk on fresh snow without sinking by wearing large snow snowshoes.

Concept questions 1.2

1. Why does glass sometimes break when you pour boiling liquid into it too quickly?
2. A metal washer has a hole in the middle. As the washer is heated, does the hole in the center expand, contract or stay the same? Explain in your own words.
3. Will putting boiling water on a frozen windshield of your car cause the windshield to shatter?

4. What is the difference between force and pressure?
5. What are the different types of pressures?
6. What are five units of pressure?
7. Which of the following is not a unit of pressure?
   (a) Pascal
   (b) Torr
   (c) Dynes
   (d) Atm

Exercises 1.2

1. A copper flask with a volume 150 cm^3 is filled to the brink with milk. If the temperature of the system is increased from 10 °C to 22 °C, how much milk will spill over from the flask? Hint: $\alpha_{Cu} = 1.7 \times 10^{-5}$ 1/°C and $\alpha_{milk} = 2.1 \times 10^{-4}$ 1/°C.
2. A brass cube has an edge of 10 cm at room temperature (about 20 °C). Assuming $\alpha_{Brass} = 19 \times 10^{-6}$ 1/°C, find,
   Change in its surface area when it is heated to 50 °C and 75 °C.
   Change in its volume when it is heated to 50 °C and 75 °C.
   Change in its surface area when it is cooled to 0 °C.
   Change in its volume when it is cooled to 0 °C.
3. When the temperature of a Copper penny is raised by 100 °C, its diameter increases by 0.18%. Find the percent increase in the following:
   Area of a face
   The thickness
   The volume
   The mass of the penny
4. A vacuum gauge connected to a chamber reads 35 kPa at a location where the atmospheric pressure is 92 kPa. Determine the absolute pressure in the chamber.
5. Determine the atmospheric pressure at a location where the barometric reading is 750 mm Hg. Take the density of mercury to be 13 600 kg m^{-3}.

Further reading

- Cengel Y and Boles M 2006 *Thermodynamics: An Engineering Approach* 5th edn (New York: McGraw-Hill Higher Education)
- Ohanian H 1985 *Physics* (New York: W.W. Norton & Company)
- Resnick R, Halliday D and Krane K 2002 *Physics* vol 1 5th edn (New York: Wiley)
- Schroeder D 2000 *Thermal Physics* (Reading, MA: Addison-Wesley Longman)
- Zain S 2019 *Techniques of Classical Mechanics: From Lagrangian to Newtonian Mechanics* (Bristol: IOP Publishing)

Chapter 2

Basic ideas of thermodynamics

2.1 Scope of thermodynamics

As we saw in chapter 1, thermodynamics has its roots in the early 17th century and was developed with the primary goal of improving the efficiency of steam engines. These machines converted heat ('*thermo*') to mechanical energy ('*dynamics*'). These days, however, thermodynamics is a broad subject and it describes many things, from car engines and refrigerators, to propulsion systems for rockets and aircraft, to coal and nuclear power plants.

At the heart of it, classical thermodynamics is an experimental science, and is founded on formalisms and generalizations based on experimentation. Thermodynamics allows us to derive relationships between quantities, such as heat, work, specific heat capacities, and coefficients of expansion, and it especially allows us to calculate the change of these quantities under the change in temperature.

2.1.1 What is heat?

The biggest question scientists worried about at the beginning of thermodynamics was, what exactly is '**heat**'? Initially, scientists hypothesized that 'heat' was a kind of fluid whose total amount is always conserved. The '*heat-fluid theory*' developed from this idea and it stated that a fluid called '***caloric***' or simply '*heat*' flows from one body to another which results in the 'heating' of the object.

French physicist, Sadi Carnot used the 'heat-fluid theory' in 1824 CE to explain comparatively clearly the limitations of transformation of heat into work and vice versa; this is now called the Second law of thermodynamics. About two decades later, in 1842 CE, R J Mayer discovered the equivalence of heat and mechanical work and made the first declaration of the principle of the conservation of energy, which is now called the First law of thermodynamics.

doi:10.1088/978-0-7503-3083-1ch2

Today we know that the actual basis for the equivalence of heat and dynamical energy is found in the kinetic theory of matter[1]. The kinetic theory of matter reduces thermal phenomena to the motion of atoms and molecules that make up the system. From this point of view, we consider properties of a large ensemble of particles in the study of 'heat' and ignore the microscopic details of individual particles. This special branch of study is called '*statistical mechanics*', and we will talk about it in detail in chapter 11.

2.2 Thermodynamic processes

The term 'sytem' initially introduced in chapter 1, is a fundamental term in thermodynamics and in order to study thermodynamics we have to understand what we mean when we say 'thermodynamical systems'. Basically, the principles of thermodynamics tell us the important relations we need in order to specify *all* the relevant properties of a *system*. In thermodynamics, a system refers to a certain portion of the Universe within some closed surface called the '*boundary*' of the system.

Any other system that can exchange properties, like energy or mass, with our system under observation is called the '*surroundings*' of the system. Many thermodynamic problems involve the exchanges of energy between the system and its surroundings. The sum of the system and surrounding make up our Universe.

2.2.1 What is a process?

If the state of a system changes, then the system is considered to be undergoing a '*process*' or a '*transformation*'. A series of states through which the system goes defines the '*path of the process*'. When a process occurs such that the system remains infinitesimally close to equilibrium states at all times, it is called a '*quasi-equilibrium*' or '*quasi-static*' process. *Quasi* means 'almost' and *static* means unchanging, which means that the thermodynamic properties of the system do not change with time. A quasi-equilibrium process is an idealized process and even though it is not a true representation of an actual process, it is still a good representation of a real process. A quasi-static process is a sufficiently slow process because the system requires time to adjust internally so that properties in one part of the system do not change any faster than in the other parts. Scientists and engineers are always interested in studying and employing quasi-static processes for two reasons: one, they are easy to analyze; and two, work-producing devices deliver the most work when they operate on quasi-static processes.

---

[1] The kinetic theory of matter applies the laws of mechanics to the individual molecules of a system. This theory has many applications, including allowing us to,

1. Calculate the numerical value of the specific heat capacity of a gas.
2. Understand the properties of gases in terms of the law of force between individual molecules.

A process is called '*cyclic*' if the properties of the system return to their original values at the end of the path[2]. That is, a process where the initial and final states are identical is called a cyclic process.

The prefix '*iso-*' is used when a particular property in a process remains constant. An *iso*-thermal process is a process during which the temperature (T) of the system remains constant, an *iso*-baric process is a process during which the pressure (p) remains constant and an *iso*-choric process is a process during which the specific volume (V) remains constant. Here we will discuss each process very briefly,

1. **Isochoric transformation/process:** The term isochoric comes from two Greek words *iso*-(equal) and *choros*-(space). It is a thermodynamic process during which the volume of the closed system undergoing the process remains constant. In an isochoric transformation the system performs no external work and any thermal energy transferred to the system externally absorbs in the form of internal energy. For an isochoric transformation,

$$dV = 0 \Longrightarrow V = \text{constant} \tag{2.1}$$

2. **Isobaric transformation/process:** The term isobaric comes from the Greek *iso*-(equal) and *baros*-(weight). In an isobaric process the pressure stays constant,

$$dp = 0 \Longrightarrow p = \text{constant} \tag{2.2}$$

3. **Isothermal transformation/process:** The term isothermal comes from two Greek words *iso*-(equal) and *thermos*-(hot), and its first known use was in 1826 CE. An isothermal process is a thermodynamic transformation at constant temperature, that is, it is a variation of the state of a physical system during which the temperature of the system does not change with time,

$$dT = 0 \Longrightarrow T = \text{constant} \tag{2.3}$$

For an isothermal process, the system under consideration is in contact with an outside thermal reservoir (called a heat bath or a source), and the change occurs slowly enough to allow thermal equilibrium to be maintained through the transition.

4. **Adiabatic transformation/process:** The word adiabatic comes from the Greek word *adiabatos*, which means '*not to be passed*'. An adiabatic transformation occurs without the transfer of heat or the transfer of mass between a thermodynamic system and its surroundings. There are two ways a process can be adiabatic; one, when the system is thermally isolated from its environment and its boundary is a thermal insulator and two, when the system and the boundary are at the same temperature, so there is

[2] Note that even if a system returns to its original state upon completing a cycle, the state of the surroundings may change because of the change in the system.

no temperature difference and hence no driving force for heat transfer. In an adiabatic process,

$$Q = 0 \tag{2.4}$$

where Q is the heat in a system.

Adiabatic expansion causes a drop in temperature of a gas, whereas adiabatic compression causes a rise in temperature of the gas. For a reversible process, an adiabatic transformation is identical to an isentropic[3] transformation.

2.2.2 The steady-flow process

When variables, such as pressure, density, and flow velocity are constant over time the flow is called a *steady flow*, which happens at low speeds. Many devices we use in everyday life operate continuously for long periods of time under the same conditions. These devices are called '*steady-flow devices*'[4], and include but are not limited to devices such as turbines, pumps, power plants and/or refrigeration systems. The process that these devices go through is called the '*steady-flow process*', which is defined as a process in which a fluid flows steadily through a control volume. The fluid properties may change point to point within the control volume but they remain the same during the entire process. This means during a steady-flow process, the volume (V), the mass (m), and the total energy content (E) of the control volume remain constant.

2.2.2.1 Continuity equation

Conservation of mass[5] implies that for a fluid in steady motion, the total rate of mass flow across any two cross-sectional areas in a pipe must be the same. Conservation of mass in a steady-flow process, where the flow rate does not change over time, implies that at any point the inflow rate must equal outflow rate. This statement is called the equation of continuity or the continuity equation and is given mathematically as,

$$v_1 A_1 = v_2 A_2 \tag{2.5}$$

where v_1 is the speed at cross-section A_1 and v_2 is the speed at cross-section A_2. Common applications of equation of continuity are pipes with flowing fluids such as in water pipes, power plants, etc.

2.2.2.2 Bernoulli's principle

Bernoulli's principle is the statement of conservation of energy in fluids. It states that for a non-viscous, in-compressible fluid in a steady flow, the sum of pressure, potential and kinetic energies per unit volume are constant at any point. Physicists visualize flow using figures called '*streamlines*', which show the directions of the flow at any particular

---

[3] Isentropic process is a process in which the entropy of the system remains unchanged.

[4] The terms steady and uniform are used frequently in thermodynamics, so it is important to have a clear understanding of their meanings. The term steady implies no change with time and the term uniform implies no change with location over a specified region.

[5] The law of conservation of mass states that mass can neither be created or destroyed.

point in the flow. In a steady flow condition, the pattern of streamlines does not change with time, as can be seen in figure 2.1.

Bernoulli's principle relates the speed of a fluid to the pressure of the fluid and it states that at points along a horizontal streamline, higher pressure regions have lower fluid speed and lower pressure regions have higher fluid speed. In other words, Bernoulli's principle says that in a horizontal pipe when diameter decreases, the fluid will move faster and that will result in lower pressure. This is called the '*Bernoulli effect*', and is shown in figure 2.2. The Bernoulli effect seems counter intuitive since we generally associate higher speeds with higher pressures. But think about it this way, as the velocity increases through the constriction the kinetic energy increases. This increase in kinetic energy occurs at the expense of pressure energy, hence pressure goes down. The Bernoulli equation is mathematically given as:

$$P_1 + \frac{1}{2}\rho v_1^2 + \rho g h_1 = P_2 + \frac{1}{2}\rho v_2^2 + \rho g h_2 = \text{constant} \tag{2.6}$$

where P_1 is the pressure energy term; $(\frac{1}{2}\rho v_1^2)$ represents kinetic energy per unit volume and $(\rho g h_1)$ is potential energy per unit volume. This equation can also be written as,

$$\frac{P}{\rho} + hg + \frac{v^2}{2} = \text{constant} \tag{2.7}$$

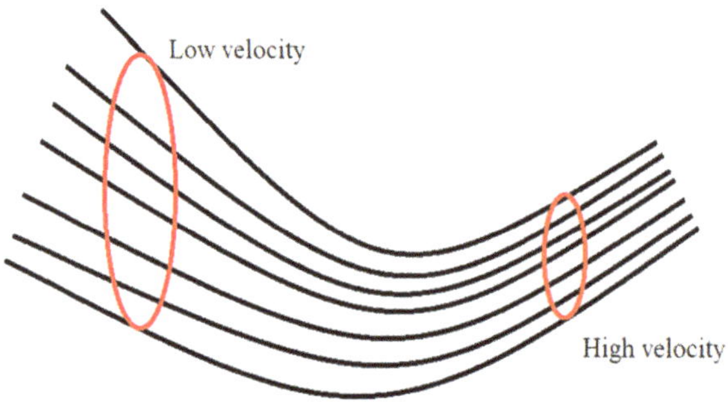

Figure 2.1. Wide streamlines versus narrow streamlines.

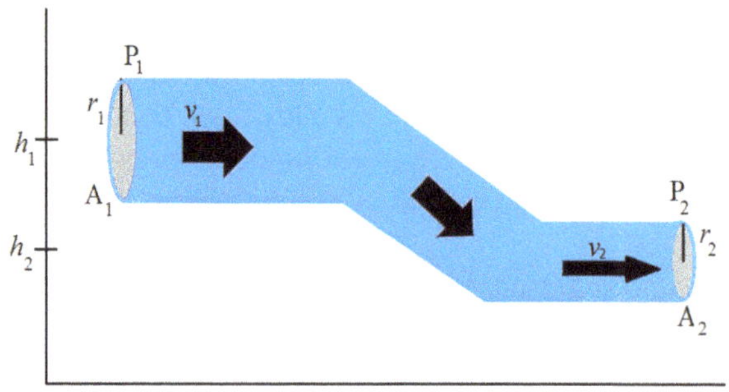

Figure 2.2. Bernoulli's equation.

where the first term $\left(\frac{P}{\rho}\right)$ expresses the potential energy per mass of liquid due to pressure and is called the *pressure head.* The second term (hg) is the potential energy per mass of liquid due to gravity and is called the *elevation head.* The term $\left(\frac{v^2}{2}\right)$ is the kinetic energy per mass of liquid. Some important features of Bernoulli's equation include,

1. **Static Pressure:** Fluid statics is a special case of fluid dynamics. In equation (2.6) when fluids are not moving we can put: $v_1 = v_2 = 0$.

$$\begin{aligned} p_1 + \rho g h_1 &= p_2 + \rho g h_2 \\ \Delta p &= \rho g \Delta h \end{aligned} \tag{2.8}$$

   where Δp is called the *static pressure.*

2. **Dynamic Pressure:** Suppose the fluid flows horizontally. We know for a horizontally moving fluid heights are equal, $h_1 = h_2$ and we can write equation (2.6) as:

$$p_1 - p_2 = \frac{1}{2}\rho g(v_2^2 - v_1^2). \tag{2.9}$$

   When the speed is large, the pressure must be small and vice versa. It is worth noting here that the term $\left(\frac{1}{2}\rho v^2\right)$ has the dimension of pressure and hence we call it *dynamic pressure.*

2.3 State properties of a system

The state of a system is defined by the values of certain experimentally measurable quantities called the '*state variables*' or simply the '*properties*'. It is important to note that properties describe states only when the system is in equilibrium. For fluid systems, typical properties are the temperature (T) of the system, the pressure (p) exerted by it, and the volume (V) it occupies. The properties of a fluid system are generally uniform throughout the system. However, in cases where the properties of a system vary from point to point, we analyze the system by subdividing it, either conceptually or in practice, into smaller systems in which the properties are assumed to be constant.

Complex systems require specification of more unusual and complex properties. For example, the state of an electric battery requires the specification of the amount of electric charge it contains. Similarly, a magnetic body requires the specification of the magnetization of the magnetic body.

2.3.1 Equation of state

Systems in thermodynamics are described by the properties of the system such as, pressure (defined as force per unit area), volume, and temperature of the system. The systems where the relevant properties are pressure, volume and temperature are called pVT-systems. For a given system T, V, and p are not always independent of one another but are connected. In this case the equation of state is given as,

$$\mathcal{F}(p, V, T) = 0. \tag{2.10}$$

The form of the equation of state depends on the particular properties of the substance in our system and as of yet, there is not one unique equation of state that accurately describes properties of all substances (solids, liquids and gases) under all conditions. The best known equation of state is called '*the ideal gas law*', which can be written as $\mathcal{F}(p, V, T) = pV - nRT = 0$, where p is the pressure, V is the volume and T is the temperature. This law, discussed in detail in chapter 4 is quite accurate for gases at low pressures and moderately high temperatures but fails at higher pressures and lower temperatures.

In engineering and physics problems, equations of state are essential in the pVT calculations for mechanical processes and in the modeling of various states of systems. In the equation of state, when two variables are known, it is nearly always possible to solve for the third with respect to the others. Hence, the state of the system is completely determined by knowing any two of the three quantities p, V, or T. For example, in the (V, p) representation, shown in figure 2.3 a point defines a particular state of the system. The points representing states of equal temperature lie on a curve called an *isotherm*, as shown in figure 2.3.

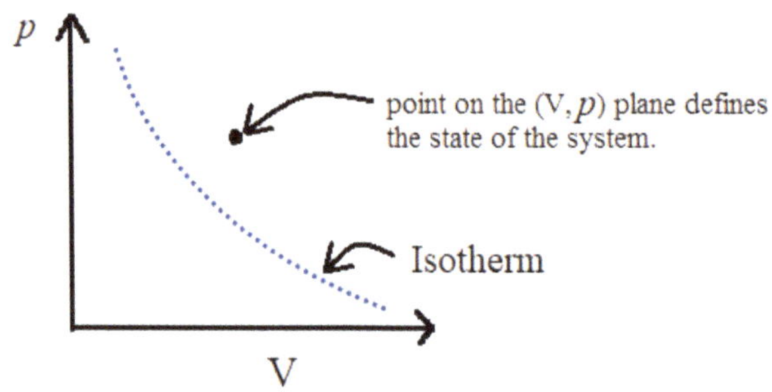

Figure 2.3. A point on this graph defines a state of the system.

2.3.2 Types of properties of a system

Physical properties are properties that can be measured or observed without changing the chemical composition of a system. Properties of a system are divided into two main types: intensive or extensive properties.

1. **Intensive properties:** Intensive properties do not depend on the quantity of matter in the system. Intensive properties are used to identify various substances because they do not change with the amount of matter present. Examples of intensive properties include temperature, density, melting point, etc. Think of intensive properties as properties that can exist at a particular point in space. For example, temperature is measured by the help of a thermometer to the various points. If you are in a room you can measure the temperatures at different locations by moving the thermometer and taking a reading. Similarly, you can measure the atmospheric pressure by means of a pressure sensor at different altitudes.
2. **Extensive properties:** Extensive properties are the properties of a system that depend on the size of the system. A good example is the volume of a system.

For a solid cube, doubling the length of all edges implies the volume increases by factor of eight, since volume = edge^3. Doubling the length of a cube side implies that its mass must also increase, hence mass is also an extensive property of the system.

Extensive properties cannot be used to identify a particular substance because they can change. Extensive properties are additive, which means that if the system is divided into a number of sub-systems, the value of the property for the whole system will equal the sum of the property values for the individual sub-systems. For example, if a pizza is divided into eight slices the total mass of the pizza will remain the sum of eight slices and each slice (sub-system) will be 1/8th of the entire pizza. For detailed properties of extensive and intensive properties, please refer to appendix D.

2.3.3 Specific and molal specific properties

State properties also have two related types of properties: specific and molal specific properties.

1. **Specific properties:** A specific property is obtained by dividing an extensive property of a system by the mass of the system.

$$\text{Specific properties} \equiv \frac{\text{Value of extensive property}}{\text{Mass of the system}} \tag{2.11}$$

   For example, specific heat capacity (c_p) which is proportional to the size of the system is a specific property of a system. We obtain specific heat capacity (c_p) by dividing the extensive property, heat capacity (C_p) by the mass of the system. We will discuss this in more detail in chapter 3.

2. **Molal specific value:** The molal specific value is the ratio of an extensive property to the number of moles of a system.

$$\text{Molal Specific Properties} \equiv \frac{\text{Value of extensive property}}{\text{Number of moles of the system}} \tag{2.12}$$

   For example, if n is the number of moles of a system the molal specific volume is:

$$v \equiv \frac{V}{n} \tag{2.13}$$

Mole—Happy Mole Day:

Yes, October 23 is called 'Mole Day'. YAY. Happy Mole Day[6]. But what exactly is a mole? The mole is one of the base units in the International System of Units (SI) with symbol *mol*. Mole is defined as *the amount of any substance that contains as many elementary entities, e.g., atoms, molecules, ions, electrons, or photons, as there are atoms in 12 g of carbon-12.*

[6] This has nothing to do with little creatures that burrow underground. Mole Day is an unofficial holiday celebrated among chemistry enthusiasts on October 23, between 6:02 AM and 6:02 PM, making the date: 6:02 10/23.

The number of molecules per mole is known as Avogadro's number (N_A) or Avogadro's constant, with numerical value approximately 6.022×10^{23} mol^{-1}. A mole is a number like a dozen, just a lot bigger. A dozen means 12; whereas a mole means 6.022×10^{23}. Just as a dozen could be a 12 of 'anything'; a mole will be 6.022×10^{23} many of anything. A mole of aluminum and a mole of gold both have 6.022×10^{23} atoms.

One mole of people has 6.022×10^{23} people in it and one mole of elephants has 6.022×10^{23} elephants. Suppose 1 man weighs 100 kg, a mole of people will weigh about 6.022×10^{25} kg. Compared to the mass of Earth (5.9×10^{24} kg), 1 mole of people will be an order of magnitude greater. As another example, if one elephant is assumed to weigh about 5000 kg, then a mole of elephants will weigh about 3.01×10^{27} kg, which is greater than the weight of the planet Jupiter, since the mass of Jupiter $\approx 1.89 \times 10^{27}$ kg!

Worked example 2.1. How many molecules?

1. 1 mole of iron (Fe).

$$1 \text{ mol} \times 6.022 \times 10^{23}\text{mol}^{-1} = 6.022 \times 10^{23}\text{molecules}$$

2. 0.5 mole of iron (Fe).

$$0.5 \text{ mol} \times 6.022 \times 10^{23}\text{mol}^{-1} = 3.011 \times 10^{23}\text{molecules}$$

3. 1 mole of water (H_2O).

$$1 \text{ mol} \times 6.022 \times 10^{23}\text{mol}^{-1} = 6.022 \times 10^{23}\text{molecules}$$

4. 0.5 mole of water (H_2O).

$$0.5 \text{ mol} \times 6.022 \times 10^{23}\text{mol}^{-1} = 3.011 \times 10^{23}\text{molecules}$$

Worked example 2.2. Find the molar masses of sodium hydroxide, NaOH and water, H_2O.

1. **Molar mass of sodium hydroxide, NaOH:** To find the molar mass of NaOH, start with the atomic mass unit[7] of NaOH found in a periodic table, sodium (Na) = 23 amu, oxygen (O) = 16 amu and hydrogen (H) = 1 amu. This means one mole of Na atoms would weigh 23 g, one mole of O_2 atoms would weigh 32 g and one mole of H_2 atoms would weigh 2 g. Hence we can find the molar mass of NaOH as:

$$\text{NaOH} = \text{Na} + \text{O} + \text{H} = 23 + 16 + 1 = 40\text{g mol}^{-1}$$

1 mole of NaOH is about 40 g.

[7] The atomic mass of an element is the average mass of the atoms of an element measured in atomic mass unit (amu). The atomic mass is a weighted average of all of the isotopes of that element, in which the mass of each isotope is multiplied by the abundance of that particular isotope. Atomic mass is also referred to as atomic weight, but the term '*mass*' is more accurate.

2. **Molar mass of water (H_2O):** We know H = 1 amu and O = 16 amu, which implies,

$$\text{water}(H_2O) = 2H + O = 2 + 16 = 18\text{g mol} - 1.$$

We can thus say 1 mole of water is about 18 g (16 g for the oxygen and 2 g for the hydrogen).

Worked example 2.3. Suppose you are given 10 g of nickel (Ni). How do you find the number of moles and the number of molecules?

1. **Number of moles:** The periodic table gives the atomic number of Ni = 28 and its atomic mass as 58.69. Hence its molar mass is 58.69 g mol^{-1}. So we calculate the number of moles as,

$$\frac{10\text{ g}}{58.69\text{ g mol}^{-1}} = 0.17\text{ mol}$$

2. **Number of molecules:** We can find a number of molecules as,

$$0.17\text{ mol} \times 6.022 \times 10^{23}\text{mol}^{-1} = 1.02 \times 10^{23}\text{molecules}$$

Concept questions 2.1

1. What is the main difference between intensive and extensive properties of a system. Explain with examples.
2. What is the definition of a mol? What is Avogadro's number?
3. What is the difference between a mol and amu?

Exercises 2.1

1. Suppose you put a water bottle at room temperature in a fridge to cool it. Would you model the water bottle as a closed system, an isolated system or an open system?
2. Calculate the number of atoms in 0.750 mole of iron (Fe).
3. What is the mass of 12×10^{23} carbon atoms?
4. Use the ideal gas law ($pV = mRT$) to determine the mass of air in a room with dimensions $(5 \times 5 \times 3)\text{m}^3$ at standard atmospheric conditions ($p = 1.01 \times 10^5$ Pa and 21 °C). For the ideal gas law, remember to convert °C to K and use $R = 287$ Pa m^3 kg^{-1} K^{-1}.

2.4 Energy

Let's start this section by asking a very simple question, what is classical thermodynamics? Very briefly, classical thermodynamics is the study of energy and its transformation. The law of conservation of energy states that energy is conserved; that means you cannot get '*something*' out of '*nothing*' and vice versa. The First and Second laws of thermodynamics are both related to energy, so understanding energy is fundamental to understanding thermodynamics.

Worked example 2.4. Pause for a second, and at this point, I would like you to look through your physics textbook or go to the Internet on any device you have handy and look up the term '*energy, the definition of*' and write the first definition you find that is not in-terms of work. Once you have done that, next try to find a definition of energy that does not relate energy to something else like activity, or power. *Write the definition of energy that is* **not** *in-terms of work, activity, or power* here:

Are you surprised by what you found?

It is very hard in literature to find a definition of energy that is not related to something else, like the ability to do work or power. The only thing we can confidently say is that no matter what this '*energy*' thing is, it is '**always**' conserved.

Now you must be asking, why does it work? Well, your guess is as good as mine! So for now we can call conservation of energy a 'miracle' of nature and the most fundamental law of physics. Remember, most physics is a *description* of nature, not necessarily an *explanation* of it. Fortunately, for us as a scientific community, this is enough to get all we need to get done!

2.5 The story of thermodynamics

Basically, what any conservation law including the law of conservation of energy says is that things do not change by themselves, for example, if you put ten ten-dollar bills in a room, two of them will not suddenly decide to 'walk-away'.

To further explain this idea, I will tell you a story taken from Richard Feynman, a Nobel Laureate and a physics professor at California Institute of Technology via the book 'Understanding Thermodynamics' by Van Ness (Reading, MA: Addison-Wesley Publishing Company, vol 1, 1963; vol II, 1964). I apologize for all the fibs, additions, and errors included in the story and I take full responsibility for these.

2.5.1 The story of 23 sugar cubes

Imagine 'mischievous Sami' a naughty little boy lives in a lovely house with his mother, father and a little sister, Eman. He has his own room with two windows, one facing west (call it window W), overlooking a swimming pool and the other facing south (call it window Q), as shown in figure 2.4.

Over summer vacations, while his parents are at work, their older brother, Hareem, babysits Sami and Eman, and to keep Sami out of his hair, Hareem gives him 23 sugar cubes to play with and warns Sami not to eat any of the sugar cubes. Hareem leaves Sami in his room with the sugar cubes to play with but looks in from time to time to see how he is doing. Every time Hareem looks in he counts the number of sugar cubes, and he always gets 23 sugar cubes.

However, a couple of hours later, when Hareem comes back to check, he finds only 21 sugar cubes. Oh no, is Sami in trouble? Before Hareem can yell at him, Sami points to an old chocolate box. Hareem heads toward the box, but before he can

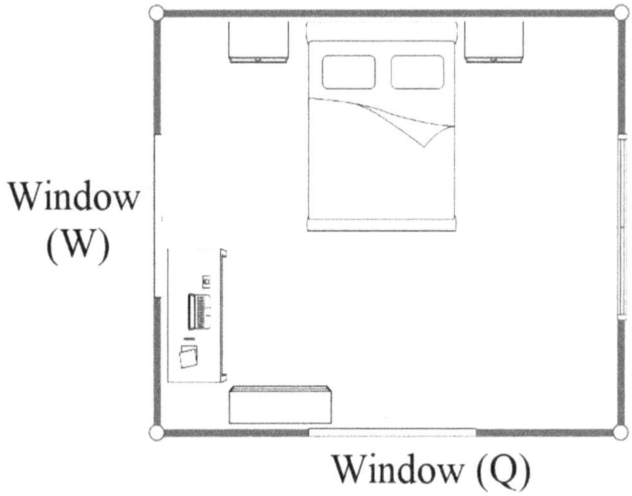

Figure 2.4. Room schematic for Sami's room.

open it Sami says, 'Please, do not open the box'. For once Hareem trusts Sami and leaves the room.

He comes back again after a little while, and discovers 23 sugar cubes again. This time however he decides to weigh the empty box and sugar cubes. He finds the empty box to be 100 g and one sugar cube to be 2 g. Being a physicist Hareem sets up a formula to check the number of sugar cubes without opening the box:

$$\text{No. on Floor} + \frac{\text{total mass of box} - 100\text{ g}}{2\text{ g}} = 23 \tag{2.14}$$

This formula works fine for a while but later in the afternoon the formula stops working. Two sugar cubes are missing again! Hareem notes that window (W) is now open and realizes Sami might have thrown the cubes out of the window (W). He looks out and sees the swimming pool under the window. He thinks to himself, if Sami has thrown the cubes out of the window they must have dissolved in the swimming pool by now. Luckily, Hareem puts together a device that can check the sugar content of the pool. He alters the above equation to account for this change:

$$\text{No. on Floor} + \frac{\text{mass of box} - 100\text{ g}}{2\text{ g}} + (\alpha \times S) = 23 \tag{2.15}$$

where α is a proportionality constant and S is the sucrosity of the pool. The formula works again! It once again accounts for a total of 23 sugar cubes, yay, all is well in the land again.

Hareem leaves the room and then realizes that instead of keeping track of everything all the time, he could make his life easier if he only deals with the changes in each term in the formula from one reading to the next. So he rewrites the above equation in terms of *changes in things* (Δ's) as given in equation (2.16).

$$\Delta(\text{No. on Floor}) + \frac{\Delta(\text{mass of box})}{2\text{ g}} + (\alpha \times \Delta S) = 0 \tag{2.16}$$

Let's look at this equation carefully. What this equation is saying is that in order for the sugar x to be '*conserved*', the sum of all the changes in the number of sugar cubes in all observations must be zero. Life is good again, until it's not.

The next time Hareem peaks in to check on Sami, equation (2.16) seems to have stopped working again. The sum now comes out to be (−4) instead of zero, which implies four sugar cubes are missing. Hareem looks around and notes that both windows are now open. He did not account for the sugar cubes that might have gone through window (Q) in equation (2.16).

He looks out the window (Q) but does not see any sugar cubes, however he sees a bunch of cats running away. Hmm, the pond was bad enough, but now how is he supposed to keep track of all the cats? He thinks for a while, then remembers that he once bought some detection devices. He runs to his room across the hall and gets two detectors; one for each window. The detectors count the sugar cubes as they fly past. This way he does not have to keep track of what goes on outside and he does not need to go keep track of what happens inside. He just has to keep track of what passes through the windows. Yay, that's good news, however he now has to revise his formula. He revises it to be:

$$\Delta(\text{No. on Floor}) + \frac{\Delta(\text{mass of box})}{2\text{ g}} + W + Q = 0 \tag{2.17}$$

where W is the number of sugar cubes passing through window (W) and Q is the number of sugar cubes passing through window (Q). Note we did not put (Δ) with W and Q. This is because W and Q just represent the number of objects passing through the windows, they do not represent the change in anything.

Suddenly, we have a very technical formula in equation (2.17). Keeping in mind we have actually ignored a lot of details, for example what the cats are up to, or what Sami is doing during this time. Instead we have focused our attention on a few things, namely sugar cubes, the room, and the two windows.

In the terminology of thermodynamics, we could call the room our *system*, the walls with the windows as the *boundary* of the system, and everything outside as the *surroundings*. Ideally, we should be able to ignore the surroundings, but in reality we really cannot ignore the surroundings completely and have to account for some interaction of our system with the immediate surroundings with the system. We can however make it so that in our formula we 'appear' to only deal with the system. We can now rewrite equation (2.17) as:

$$\Delta(\text{No. on Floor}) + \frac{\Delta(\text{mass of box})}{2\text{ g}} = -W - Q \tag{2.18}$$

The left-hand side (LHS) of equation (2.18) now represents the changes in the system. The terms on the right-hand side (RHS) show what *passes out* of the system into the surroundings. So these terms actually represent changes in the surroundings. This very crudely put is a conservation law, let's call it '*the law of conservation of sugar cubes*'. For any conservation law to be true, we must somehow include both the system and its surroundings. However, by insisting that we only take into account the immediate surroundings we are in fact adding a restriction to our definition of the conservation law of sugar cubes. We do not expect one of Sami's sugar cubes to suddenly appear on Mars even though Mars is a part of the surroundings. Why not? The law of conservation of sugar cubes does not exclude this from happening. However, it will still not happen, because it is just not '*reasonable*', whatever reasonable may mean! Let's just leave it at that for now. The point is that the system-and-its-boundary formula excludes this possibility.

You may ask, 'Why did I not just say that conservation laws exist between a system and its local surroundings?' We can, but then we would have to define what we mean by 'local'. We could say that local could be any part of the Universe with which the system 'interacts'. Now we have to define what we mean by 'interacts' and so on. However, as always, mathematics saves us from this issue and simplifies our life for us.

To explain more, let me finish our story. All is well. Hareem is happy that he can account for all 23 sugar cubes, and Sami is happy that he is not getting in trouble. However, over time the sugar cubes keep disappearing until the final sugar cube leaves the room and enters the surroundings. Is this game over? Well, maybe not. Seeing how upset Sami is, Hareem gives him more sugar cubes but his time he does not count them. Can he still play the same game? The simple answer is, yes, he can still play the game. The formula will still work. All he needs to do is to first reset the window detector counters and note down an initial observation.

2.5.2 The story of jelly beans

The next day, Sami's friend, Logan, comes over and sneaks in a bag of jelly beans. Sami immediately hides the bag of jelly beans in his chocolate box. Hareem does not see this happen, but he gets suspicious. He knows Sami has '*something*' in his chocolate box but he does not know what it is. Can he use his formula from yesterday? Let's see.

Unlike yesterday, this time Hareem does not know what Sami has, but he knows that it must be sweet and sticky because Sami loves anything sweet and sticky. So he calls the mass of this 'something' '*a lump*' with mass '*a*'. Then the formula from yesterday (equation (2.18)) will be rewritten as:

$$\frac{\Delta(\text{mass of box})}{a} = -W - Q \tag{2.19}$$

Hareem knows the mass of the box from yesterday and can figure out W and Q (number of things out of the windows W and Q, respectively) from the detection devices he set yesterday. So the only unknown in the above expression is the mass of the lump '*a*' and that can be easily solved algebraically. This time Hareem calls the conservation law 'the law of conservation of lumps', because I have always wanted to say this phrase out loud!

2.5.3 The story gets cryptic

Perhaps this story has been absurdly obvious so far. Let's see if we can make it more cryptic. The RHS of the equation (2.19) is basically a counting scheme, one that counts W and Q directly, measured by the detector. Since mathematically we know for any equation the LHS must equal the RHS, we can deduce that the LHS of equation (2.19) should also be some kind of counting scheme. The LHS of this equation does in fact 'count', but it does so indirectly. Here the number of counts is given as a change in some function, say f, that depends on the mass of the box. So we can rewrite equation (2.19) as:

$$\Delta[f(\text{mass of box})] = -W - Q \tag{2.20}$$

For equation (2.20), we know the exact the nature of the function (f), it is just the mass of box, written mathematically as $f = f$ (mass of box).

For complex systems where the function depends on other properties of the box, such as the electric charge of the box or if the mass of the box changes with temperature, the 'conservation law of lumps' will become both difficult and abstract. Luckily, we will not deal with those situations in this text.

2.5.4 Law of conservation of energy

The law of conservation of energy is inherently more difficult and abstract than our 'simple' law of conservation of lumps. As far as we know energy does not come in the Universe as nice uniform lumps. In fact, the law of conservation of energy does not represent a conservation of any particular thing. Now what can that possibly mean? Let's examine in detail by asking a question: How is energy conservation similar to the law of conservation of sugar cubes and the law of conservation of lumps?

Answer: Conservation laws (be they of sugar cubes, jelly beans, or energy) are similar in their mathematical formalism. They all include terms that account for changes in the system and the surroundings. It is interesting that the simplest expression for each conservation law actually deals with the same basic three things:

1. The changes that occur in a particular system.
2. Quantities that pass the through boundary of the system.
3. The 'local' surroundings.

We can now rewrite the above equation (2.20) for energy conservation as we did for the conservation of lumps. Think of Hareem and the 'lumps'. He never actually saw the lumps. Similarly, no one has ever seen 'energy' but like Hareem we can use the concept of energy and write the energy equation as:

$$\Delta[U] = -(\text{energy out by } W) - (\text{energy out by } Q) \tag{2.21}$$

where U is the energy of the system. U can be assumed to be some function of the measurable properties of the system, just as the number of unseen jelly beans were a function of the mass of the box.

Please note that we don't really know the functional relationships for the energy ahead of time. So the best we can do is to make some educated guesses. For example we say that it depends on the temperature, pressure, etc. We really do not know them all, so the best we can do is:

$$\text{Energy of system} = U(T,\ P, \text{etc}) \tag{2.22}$$

where U is the *internal energy function* and the conservation formula (2.21) becomes:

$$\Delta[U(T,\ P, \text{etc})] = -\text{energy out by } W - \text{energy out by } Q \tag{2.23}$$

which can be simplified to:

$$\Delta U = -W - Q \tag{2.24}$$

On the RHS of equation (2.24), when defining conservation of energy in thermodynamics, we call W and Q boundaries, instead of windows, by which energy moves in and out of the system. So we redefine them as work (W) and heat (Q). Both W and Q have special technical meanings we will discuss in section 2.6. But for now let's assume we can measure these quantities experimentally and know all about them.

Note from equation (2.24), it seems like after this long winding story, I have somehow derived the formula for the energy of a system, or explained what energy is or in fact that we know what we are talking about. Nothing could be further from the truth; nobody really knows what 'U' is. The best we can do is to try to guess what energy is by finding energy (U) as a function of other things.

2.5.5 Law of conservation of energy equation—decoded

Let's remind ourselves what we have done so far in this section. You may think I have derived the equation of conservation of energy in section 2.5.4. Look carefully, what I actually have done is:

.... just *written down* the equation of conservation of energy!

Which is just about the best anyone can do. No matter how many fancy diagrams are made and no matter how confused the issue is made by mathematical manipulations, if you look carefully, you will find at the end each author (including me) is just simply *writing down* the law of conservation of energy. You may also think I have '*explained*' the law of conservation of energy. Again, rest assured, I have not. I have only given you a plausible mathematical equation that works. This is a very important fact to understand and appreciate.

One other thing that might be bothering you is that both W and Q on the RHS of equation (2.24) have negative signs. Remember the reason we did this was because Sami was only able to throw the sugar cubes '*out*' of the windows. Had the sugar cubes been able to enter through the window we would have to put a positive sign in the equation (2.24).

In fact, historically, the conservation of energy has been written as:

$$\Delta U = -W + Q \tag{2.25}$$

This is an accident of history. The first application of thermodynamics were made to heat engines, which '*take in heat*' and '*put out work*'. Hence signs represent heat in and work out as positive quantities. You may define these quantities any way you prefer, but you must be consistent in your calculations. We will talk more about it in the upcoming chapters. Having said that, I tend to write the First law of thermodynamics as:

$$\Delta[U(T, P, \text{etc})] = W + Q \tag{2.26}$$

In my convention, Q is positive when heat enters the system and negative when heat leaves the system and work done on the gas (during compression) is positive and work done by the gas (during expansion) is negative.

Now that I have defined this equation and the variables used in it, the next thing to do is to figure out how to use this equation. Physicists and engineers use this equation to calculate Q or W or both when another equation cannot be found that connects Q and W. In order to do that we need to know the function $U(T, P, \text{etc})$. The question is, one, how do we get the numerical value for $U(T, P, \text{etc})$ with just one equation? and two, how do I exactly define what '*etc*' is in the expression $U(T, P, \text{etc})$? These are all good questions; let's try to address some of them below.

2.5.6 What exactly is *U*?

Well, I do not have answers to everything asked above, but perhaps the last question is easy to answer. How do I exactly define what '*etc*' is in the expression $U(T, p, \text{etc})$? The answer involves the notion of '*state*' that was discussed in section 2.3. We said that internal state of a system is fixed when its measurable properties do not change any more. So we can find the answer to this question if we can find the '*measurable properties*' and fix them, which in turn will fix the state of the system. The only way to do this is to find these measurable properties experimentally, which we will try to do here.

Let's suppose we are looking for variables, temperature (T) and pressure (p). How do we get the relationship between U and these variables? This is the second major complication in thermodynamics. What we find is that in any analysis we must always use our conservation of energy equation. That might seem incredible since the use of equation (2.26) was to find Q or W!

So how are we to find, U, W, and Q at the same time? The secret is that we do not find all required state properties at the '*same time*'. We play the game forward and backwards, but never at the same time. Think of what Hareem did when he used his equation backwards to determine the mass of the lump he had never seen. He used the equation to check the conservation of lumps and again used the same equation to find the number of sugar cubes in the pool or sugar cubes eaten by cats. The same thing holds true for the conservation of energy equation in which we never see the 'energy'.

So how do we do that? We do that by setting up an experiment in the lab that measures T and p, but then also measures Q and W at various values of T and p. From these we calculate $U = U(T, p)$ for many values of T and many values of p. We can put these values in a table, a graph or an equation and evaluate $\Delta U(T, p)$.

> '*The main point is that all of this is done experimentally. That is about the best we can do!*'.

Once we have established that, we can rewrite the energy conservation formula for pretty much any process involving the same kind of system. We also find that it works time after time, every time.

It works perfectly for a while, until one day it does not work. Oh no! What do we do now? Well, we do what Hareem did. He looked for other explanations, like an open window or cats running away. So just like Hareem we look around and take into account any changes in our system that could've happened while we were not looking, say by the change in elevation. Could that have changed the energy of the system? Sure enough, with a bit of experimentation we can come up with another component to the formula, called the '*potential energy function*' which fixes our formula. If we go through the whole process again we will find other functions, like kinetic energy function for when our system is moving, to make our equation (2.26):

$$\Delta[U(T, P, \text{etc})] + \Delta[\text{PE}(h)] + \Delta[\text{KE}(m, v)] = W + Q \tag{2.27}$$

where h is the height, m is the mass, and v is the velocity of the system.

Through this discussion we have determined that we can keep changing our conservation of energy formula as our system changes. Is that cheating? Are we doctoring our equation so that it works every time? Not quite. Yes, we are adding terms to account for new information as it comes along. However these terms will disappear when the system stops undergoing the changes that are causing them. For example, when the system stops moving, velocity becomes zero and hence the ΔKE term in equation (2.27) disappears; the same happens when the elevation of the system stops changing, the ΔPE term vanishes.

2.5.7 Ultimate test for conservation of energy

The ultimate test of our accounting scheme came with the beginning of the nuclear age. In this case 'energy' appeared to come out of nothing and nowhere. Wait, what? how? Does this make conservation of energy wrong? Isn't the law of conservation one of the fundamental laws?

Well, not all is lost. Einstein was the first to give an explanation for this newly discovered phenomenon and proved that this new found 'energy' does not come out of nothing and nowhere but rather comes from change in mass (Δm). So the addition of a new term of a nuclear energy function ($-c^2\Delta m$) fixes our system once again! So now our equation becomes:

$$\Delta[U(T, P, \text{etc})] + \Delta[PE(h)] + \Delta[KE(m, v)] - c^2\Delta m = W + Q \tag{2.28}$$

The minus sign in the equation is necessary because Δm is negative, the mass of the system decreases.

2.5.8 When all is said and done

When all is said and done, what do we have? We have an equation (equation (2.28)) which gives a '*mathematical expression*' to the law of conservation of energy. How else could this law be expressed besides mathematically? Every form of energy we have discussed so far is known '*only*' as a function of other variables and functions are just *pen-and-paper* constructs. I cannot show you a function that has any other substance except on paper. This is why I cannot show you a 'chunk of energy' but I can write an equation for it. This is why there are no energy measuring devices that you can buy that can be stuck into a system and read its energy. The whole thing is man-made, like the equator line on a map!

So at the end of the day, we have nothing more than a scheme with a set of rules, which involves changes in energy functions. Remember, this is set up this way because we have no way to record absolute values of our energy functions. However, the most remarkable thing about our scheme is that it is extremely general, it can be applied to all kinds of systems from the largest stars to the smallest particles. It is equally valid over all time intervals, from pico-seconds to eons as well as overall matter, both living and non-living.

Simply put conservation of energy just works! Or in other words, at least we have not yet found a system where conservation of energy does not work.

2.5.9 Energy—the definition?

After all this explanation, we still have to talk about energy as if we know what we are talking about. So here is my best attempt: Energy is a fundamental concept in physics which cannot be defined in terms of further fundamental quantities and we will define it in the tried and tested path of the masters, as:

'*energy is the capacity of a body to do work!*'.

Conservation of energy says, energy can be stored, it can be transferred, but can never be created nor destroyed. There are many forms of energy, like thermal, mechanical, kinetic, potential, electrical, magnetic, chemical, nuclear energies, and many more. For this text, we will mostly consider only the following three types of energy:

1. Potential energy (PE): Energy associated with the position of the system within a field. In a gravitational field, PE = mgh; where m is the mass of the object, g is the acceleration due to gravity and h is the height.
2. Kinetic energy (KE): Energy associated with the net movement of the system. KE is mathematically given as $\frac{1}{2}mv^2$, where v is the velocity of the object and m is the mass of the object.

3. Internal energy (IE): Internal energy is energy that matter stores in its structure and in the motion of its constituents, like electrons, protons or neutrons. IE is also denoted by U.

Worked example 2.5. A container filled with hot liquid is cooled while being stirred. Suppose the initial internal energy of the fluid is 1000 kJ and during the cooling process it loses 500 kJ of heat. During this time the stirrer does 100 kJ of work on the fluid. Neglecting the energy stored in the stirrer, find the final internal energy of the fluid.

Solution: The container itself does not move, so the changes in kinetic and potential energy are zero (Δ KE = Δ PE = 0). If we take the liquid water as the system, we know that this is a closed system since no mass crosses the boundary of the container, $\Delta m = 0$. The question also tells us that heat is '*lost*' from the system and the work is done '*on*' the system. This means we can write equation (2.28) as,

$$\begin{aligned}\Delta[U(T, P, etc)] + \Delta[PE(h)] + \Delta[KE(m, v)] - c^2\Delta m &= W + Q\\ \Delta U &= W_{\text{on}} - Q_{\text{out}}\\ U_{\text{final}} - U_{\text{initial}} &= W_{\text{on}} - Q_{\text{out}}\\ U_{\text{final}} - 1000\text{ kJ} &= 100\text{ kJ} - 500\text{ kJ}\\ \Longrightarrow U_{\text{final}} &= 600\text{ kJ}.\end{aligned}$$

2.6 Temperature, heat, and work

The most important concepts in thermodynamics are temperature, heat, and work. However, we sometimes confuse the meaning of each when we use them in the thermodynamical framework. The reason for this confusion is because the definitions of these terms in thermodynamics are a bit different from their definitions in every-day life. Let's begin the discussion by quickly defining these terms as they are used in thermodynamics.

- **Temperature:** In thermodynamics, temperature is a measure of any object's ability to spontaneously give up energy. We can think of temperature as an inherent property of the system which describes its tendency to spontaneously exchange energy with its surroundings or another object. It is worth keeping in mind that this is not the true definition of temperature, however it is a true statement about temperature.
- **Heat:** Heat is any *spontaneous* flow of energy from one object to another caused by a difference in temperature between the objects. We can also say, heat is a way of changing the energy of a system by a change in temperature. In thermodynamics we say 'heat' flows from a 'hot' body to a ' cold' body. For example, 'heat' will flow spontaneously from a hot heater into a cold room until temperature equilibrium is reached.
- **Work:** Work, in thermodynamics, is defined as a transfer of energy into or out of a system which is *not spontaneous*. With work, we usually identify some

'external agent' that is actively putting energy into the system. For example, an external agent could be a piston in a piston-and-cylinder system that can be pushed or pulled to affect the temperature of gas inside the cylinder. When the piston is pushed down, the energy in the system increases, which implies the temperature of the system increases. However, we will not say that the system is being '*heated*' because the flow of energy is not spontaneous and it is not caused by a difference in temperature.

2.6.1 Work versus heat—which is which?

As we said before, the definitions of heat and work are not easy to differentiate in thermodynamics because their usage in thermodynamics is different than how they are used in everyday life. For example, when you rub your hands together to warm them up, you are not adding ' heat' to your hands rather you are doing 'work'. So even though your hands get warm when you rub them together we don't call it 'adding heat' in thermodynamics. Similarly, while warming food in the microwave you don't add 'heat' to your food, the microwaves do 'work' on the food even though during this process the food is getting hot.

One thing we say is both heat and work refer to energy *in transit*. We can talk about the total energy inside a system, but it is meaningless to ask how much heat or how much work is '*in*' the system. We can only say how much heat '*entered*' a system or how much work '*was done* ***on***' a system. Think back to the beginning of this chapter and the sugar cube example, we did not worry about the actual number of sugar cubes in the room, we only worried about how many the sugar cubes were leaving through windows (W) and (Q).

2.6.2 First law of thermodynamics

Using the terminology set-up in sections 2.5 and 2.5.9, we will use symbols Q and W to represent the amount of energy that enters a system as heat and the work done during '*our period of interest*'. I will use U for the total energy inside the system and say that if energy leaves the system ΔU is negative and when energy enters the system ΔU is positive.

From the law of conservation of energy, equation (2.28), we say ($Q + W$) is the total energy that enters a system and changes the system's energy by an amount ΔU, as shown in figure 2.5. This is probably the best statement of the First law of thermodynamics that I can give you. Therefore, I shall write the First law of thermodynamics as,

$$\Delta U = Q + W \tag{2.29}$$

2.7 Defining 'work done' in thermodynamics

In thermodynamics the '*work done*' is also called 'compressional' or 'boundary work' or 'useful work'. In order to explain '*work done*' let me start by explaining the simplest but essential device in thermodynamics: a piston-and-cylinder system.

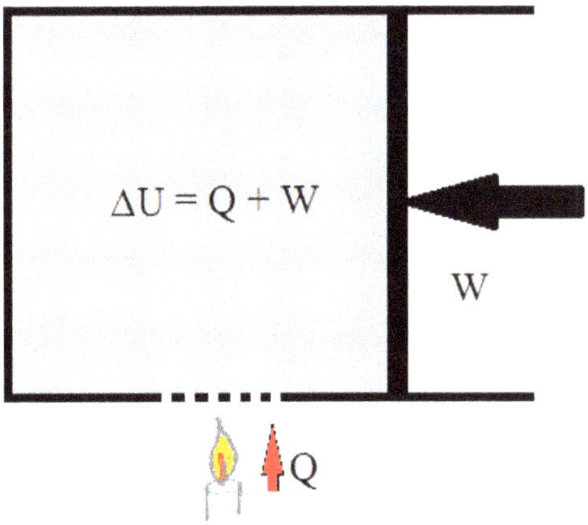

Figure 2.5. Total change of energy is the sum of heat (Q) added and work done (W) '***on***' the system.

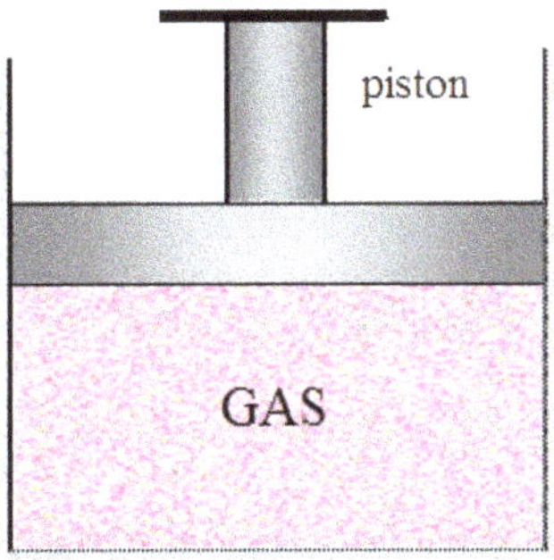

Figure 2.6. A simple thermodynamical device: piston-and-cylinder.

2.7.1 A simple device of thermodynamics

The piston-and-cylinder system, shown in figure 2.6, is the simplest device used in thermodynamics and is used in many applications in classical thermodynamics. It is a very important device that helped develop concepts of heat, internal energy, and work done in thermodynamics.

Let's begin with defining some initial conditions for the system under consideration. As we go through this example please keep in mind that I am defining an ideal system that does not exist in real life. The assumptions defined here never exist in real life, however, they help reduce inherent complications in our system. The assumptions include,

1. Surfaces between piston and cylinder are completely frictionless.
2. Our system has a perfect *adiabatic boundary* and both the piston and cylinder are made of perfectly insulating materials.

Pause for a second; what do you wish will happen under these ideal conditions? Ideally, we would like the gas in the cylinder to do the maximum amount of '*useful work*'. Useful work in this case will be that the piston moves up it and takes a mass

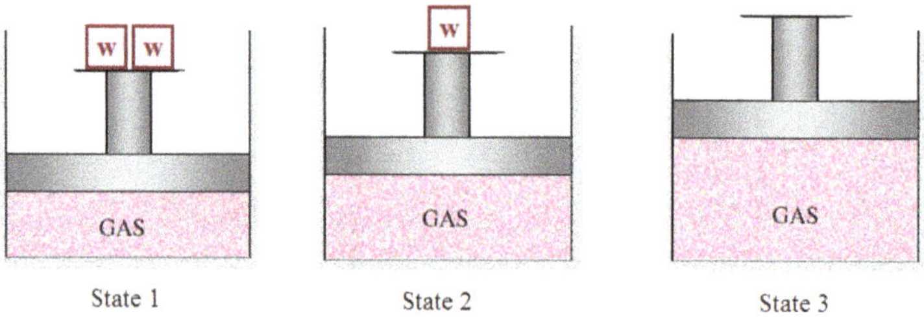

Figure 2.7. Piston-and-cylinder work.

resting on the piston up with it. To do this we will use the gas in the cylinder to move the piston, as shown in figure 2.7. There are some ways we could do that,

1. We could suddenly remove the weight on the piston. As a result the piston shoots up from state 1 to state 3, oscillates a little and then settles into a final equilibrium position. In this case, no useful work will be done because no mass on the piston was raised. What we have is exactly the same as work done when an empty elevator goes up in a building. Work was done by the elevator and electricity was used, but it was not '*useful work*'.
2. So what do we have to do to get '*useful work*' done? Clearly, the only way to do useful work is to lift the mass with the piston. One thing we can do is to divide the mass (W) on the piston into two equal parts ($w + w$) and remove each mass one after another. When we remove one of the masses ($w = \frac{1}{2}W$), we lift the piston to state 2 and do useful work against gravity. Next, we can remove the other mass ($\frac{1}{2}W$), which raises the piston to state 3; however, in this second lift no useful work will be done. In this case *useful work* was done during the first half of the process, however in the second half of the process '*no useful work*' was done.
3. Could we improve on step 2 and do even more *useful work*? Suppose in step 3 we use something granular (say sand) with small mass particles? We may now remove sand in smaller and smaller increments, which raises the piston and the remaining sand to a new equilibrium level with very little stabilizing oscillation. By removing sand this way we can go from state 1 to state 2 to state 3 until the final grain of sand has been raised to maximum piston position (state 5), as shown in figure 2.8. In this process, we have done *useful work* and have raised all of the mass (W) an average distance of roughly half the stroke of the piston[8] with very little stabilizing oscillation and hence very little loss of energy.

Essentially, our last two techniques used the mass on the piston to get '*useful work*' by dividing the mass (W) into progressively smaller divisions. If we extend this concept of division and imagine our grains of sand to be arbitrarily small in size we will get a limit on the amount of useful work we can extract from our system.

[8] A stroke refers to the full travel of the piston along the cylinder, in either direction up and down.

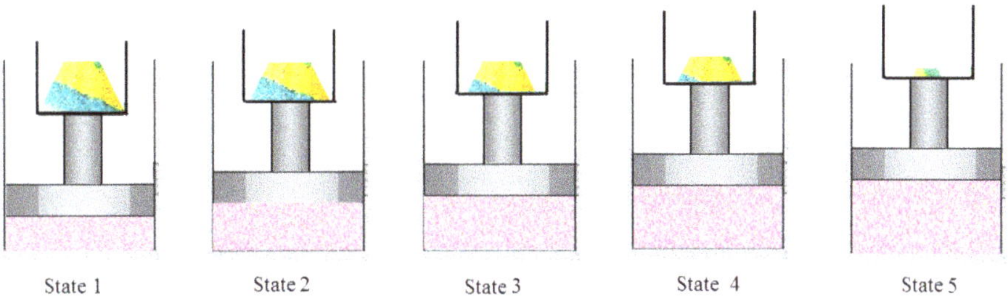

Figure 2.8. Piston-and-cylinder work with grains of sand.

This imaginary process of subtracting progressively smaller subdivisions of '*W*' can also work in reverse, that is, by adding sand back to the piston we can move the piston from state 5 to state 1. When this happens, we call the process a '*reversible process*'. The reversible process is very importance in thermodynamics. It not only represents the *limit* of what is physically possible, but it also provides a unique technique of mathematical analysis of the given problem. In thermodynamics sometimes the choices are:

1. Do calculations for a reversible process, or
2. Do no calculations at all.

2.7.2 Compressional or moving boundary work

The most important type of work in thermodynamics is the compression–expansion work or moving boundary work. Compression–expansion work is the work done on a system (often a gas) by compressing–expanding it in a piston-cylinder device. During this process, part of the boundary (the inner face of the piston) moves back and forth, compressing or expanding the volume occupied by the system. Which is why it is also called the '*pressure–volume work*'.

The amount of work done in classical mechanics equals the external force ($\vec{F}$) applied times the distance ($d\vec{x}$) an object is displaced.

$$W = \vec{F} \cdot d\vec{x} \tag{2.30}$$

This implies an external force moves the object in the direction of the applied force. This is a very simple formula that can be applied to a free body diagram of any solid object. However, if the system is complicated this formula inherently adds ambiguity to the system, by not clearly defining whether displacement ($d\vec{x}$) refers to the displacement of the center of mass or to the point of contact. In classical mechanics ($d\vec{x}$) in general is taken as the displacement of the center of mass.

Consider a typical cylinder-piston device, similar to the one introduced in the last section, with cross-sectional area of the piston (A) that encloses a particular gas, as shown in figure 2.9. Suppose the initial pressure of the gas is P_0 and the

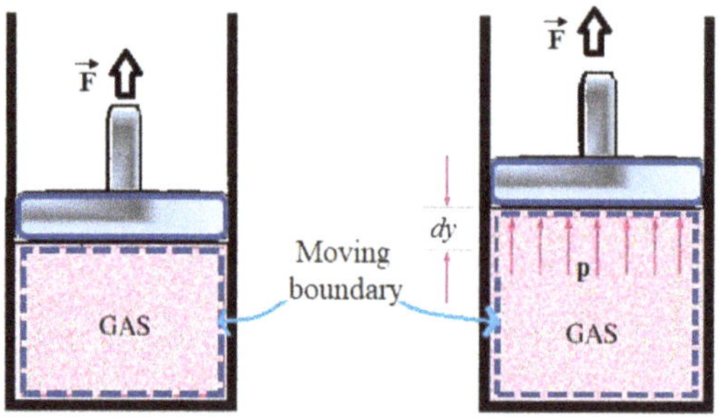

Figure 2.9. The work associated with a moving boundary is called boundary work.

total volume is V. If the piston is allowed to move a distance (dx) in a quasi-static equilibrium[9] manner, then the boundary work done during this process is,

$$W = \int dW = \int \vec{F} \cdot d\vec{x} = \int pA \cdot d\vec{x}$$
$$W = - \int pdV \tag{2.31}$$

Equation (2.31) says, the boundary work in the differential form is equal to the product of the absolute pressure (p) and the differential change in the volume dV of the system. This expression also explains why the moving boundary work is sometimes called the '*pdV work*'. Please keep in mind, this integral can only be evaluated if we know the functional relationship between p and V during the entire process.

In thermodynamics think of 'p' as the pressure a piston exerts '**on**' the gas. As we increase the external pressure we get a smaller volume. Similarly, as we decrease the external pressure we get a larger volume. Therefore, for a finite transformation from any initial state 1, with volume (V_1) to a particular final state 2, with volume (V_2), the total work done on the system will be given as,

$$W = \int_1^2 dW = -\int_{V_1}^{V_2} p\, dV \tag{2.32}$$

where the integral is taken over the entire transformation. Equation (2.32) implies if you have a tank of air at a pressure of 1 atm (1.013×10^5 N m^{-2}), and you want to reduce its volume by 1 liter (1/1000 m^3) you have to perform 100 J of work[10]. The boundary work on the gas is negative during an expansion process and positive during a compression process. Moving boundary work is the primary form of work involved in automobile engines, discussed in detail in chapter 8.

---

[9] We assume that as the gas is compressed it remains in equilibrium such that its pressure is uniform from one point to another, and hence is well-defined at all times. This is achieved by a reasonably slow motion of the piston to allow time for the gas continuously equilibrate to the changing conditions. The technical term for small or slow volume change in this sense is *quasi-static*.

[10] Please note, this is only true for a very large tank, so ΔV is small like dV.

2.7.3 The area under the process curve on a pV diagram

The state of the system can be represented on a (V, p) graph. When the state of the system is represented on a (V, p) graph, the work during a transformation can be represented by a curve that connects points (1) and (2). The shape of curve depends on the actual type of transformation under consideration. This integral and hence the work done can be represented geometrically by the shaded area in the graph shown in figure 2.10.

The quasi-static equilibrium expansion process is shown on a pV diagram in figure 2.11. Here, the differential area (dA) is equal to the differential work (pdV). The total area (A) under the curve from point (1) to (2) is obtained by adding these differential areas as,

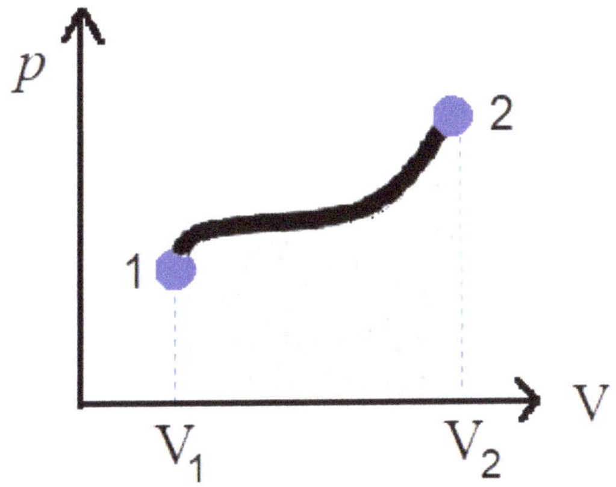

Figure 2.10. Work done during transformation.

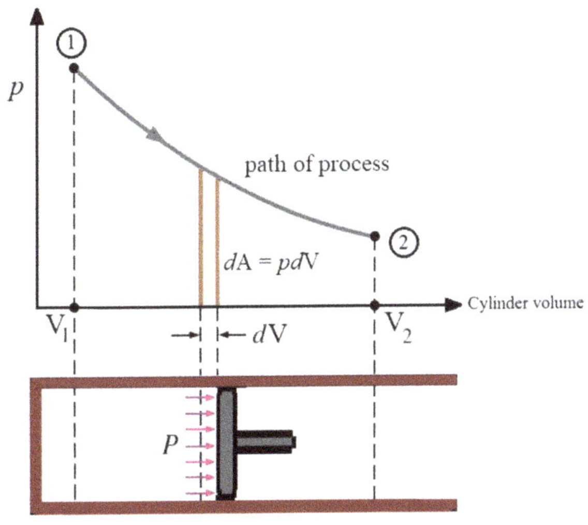

Figure 2.11. The area under the process curve on a pV diagram represents the boundary work.

$$A = \int_1^2 dA = -\int_{V_1}^{V_2} p\, dV \tag{2.33}$$

The area under the curve on a pV diagram numerically equals the work done during a quasi-equilibrium expansion or compression process of a closed system.

A gas can follow several different paths as it expands from state 1 to state 2 and each path has a different area under the curve, which implies that the work done will be different for each process.

Worked example 2.6. A gas is initially confined to a volume ($V_i = 10\ \text{m}^3$) in a cylinder by a frictionless piston held in place by stoppers. When the piston is released, it is forced outward by the internal pressure of the gas acting on the inside of the piston. A constant external atmospheric pressure (P_{atm}) acts on the piston from the outside. Calculate the work done on the gas and the work done by the gas if the final equilibrium position of the piston is $V_f = 20\ \text{m}^3$.

Solution: This is an expansion process for which the atmospheric pressure will not change during this volume change (P_{atm} = constant). Then the work done 'on' the gas:

$$\begin{aligned} W &= -\int_{V_i}^{V_f} p dV \\ &= -P_{atm}(V_f - V_i) = -[(1.01 \times 10^5\ \text{Pa})(20 - 10)]\ \text{m}^3 \\ &= -1.01 \times 10^6\ \text{J} \end{aligned}$$

From Newton's third law of motion:

Work done 'on' the gas = − work done 'by' the gas.

Therefore, work done 'by' the gas = $+1.01 \times 10^6$ J

2.7.4 Work done in cyclical transformations or processes

Cyclic transformations, also called *cycles*, are especially important in the study of thermodynamics. Cycles bring the system back to its initial state and therefore in these transformations the initial and final states of the system are the same.

On the (V, p) diagram seen in figure 2.12, the cycle is represented by points A, B, C, and D. The work done is equal to the area under the curve. In this case A and C are the points of minimum and maximum of our cycle, and A′ and C′ are the projections on the x-axis.

The work performed during the part of cycle ABC is '*positive*' and numerically equal to the area (ABCC′A′A) as shown in figure 2.13. The work performed during the part of cycle given by CDA will be '*negative*' and numerically equal to the area (CC′A′ADC) as shown in figure 2.13(b). The total amount of positive work will be equal to the difference between these two areas or in other words we can say, the total amount of positive work equals the area bounded by the cycle, as shown in figure 2.13(c). Please note in this case the work is positive because we are moving in a clock-wise direction. If we were moving in the opposite direction the work done would be negative.

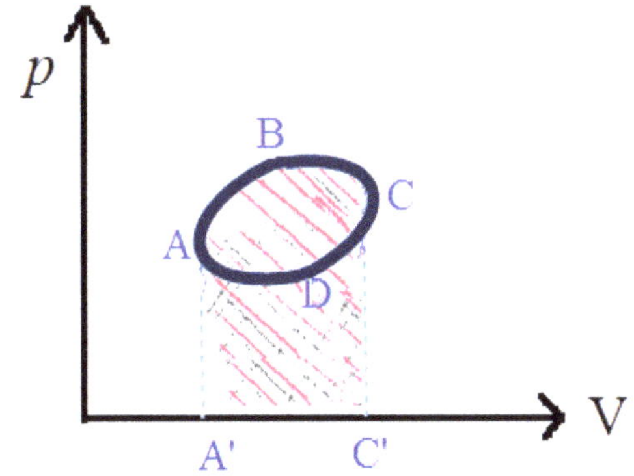

Figure 2.12. Cyclical transformation.

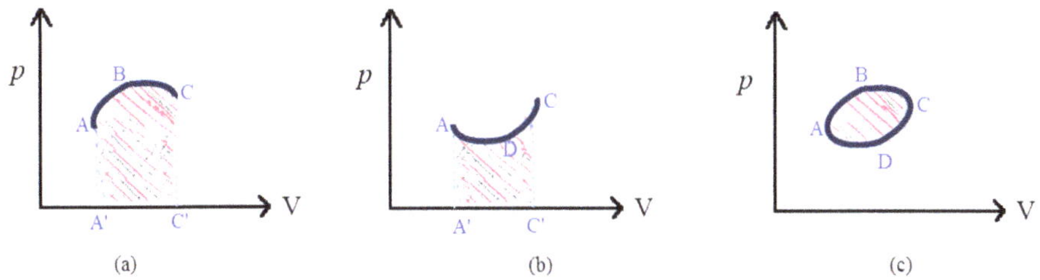

Figure 2.13. (a) Work performed during the part of cycle given by ABC. (b) Work performed during the part of cycle given by CDA. (c) Total Work performed during the cycle.

Worked example 2.7. Imagine a cylinder of helium with an initial volume of 1 L and an initial pressure of 1 atm. Somehow the helium is made to expand to a final volume of 3 L in such a way that its pressure rises in direct proportion to its volume.

1. Sketch a graph of pressure versus volume for this process.
2. Calculate the work done on the gas during the process, assuming that there is no 'other' type of work being done.
3. Describe what you might do to 'cause' the pressure rise as the helium expands.

Solution:

1. The graph of pressure versus volume for this process is given in figure 2.14.
2. The work done 'on' the gas during the process, which in this case is an expansion process, is given as the negative of value of the area under the graph, given by the shaded area in figure 2.14.

$$W = -\int P\Delta V = -[\text{Area under curve}] = \frac{1}{2}(2\text{ L})(2\text{ atm}) + (2\text{ L})(1\text{ atm})$$

$$= -(4 \times 10^{-3}\text{ m}^3)(1.01 \times 10^5\text{ Pa})$$

$$= -400\text{ J}.$$

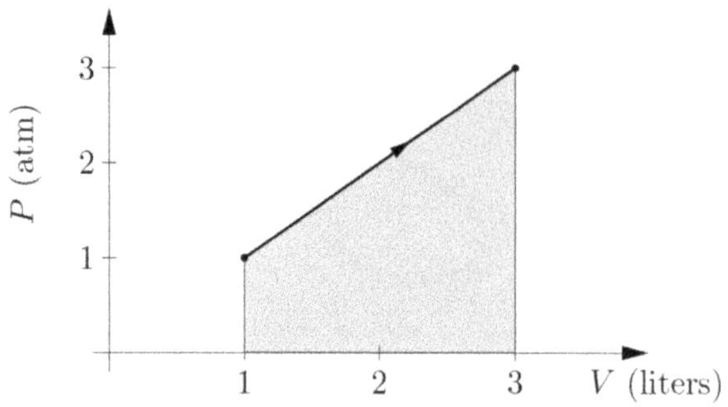

Figure 2.14. Work done during the process.

The negative sign indicates that 400 J of work is done *by* the gas on its surroundings.

3. To cause such an increase in pressure (and temperature) as the gas expands, we must provide heat, for instance, by holding a flame under the cylinder and letting the piston out slowly enough to allow the pressure to rise.

2.7.5 Boundary work for various processes

1. **Boundary work for isochoric processes:** As we saw in section 2.2, an isochoric process is a constant-volume process. This means that the volume does not change and since the volume does to change no work is done and the First law of thermodynamics can be written as,

$$\begin{aligned} \Delta U &= Q + W \\ \Longrightarrow \Delta U &= Q. \end{aligned} \tag{2.34}$$

2. **Boundary work for isobaric processes:** An isobaric process is a thermodynamic process in which the pressure remains constant: $\Delta p = 0$ and the First law of thermodynamics is written as,

$$\Delta U = Q + W. \tag{2.35}$$

Graphically, when pressure is constant (isobaric process), the work done on the gas is numerically calculated by finding the area under the curve of the (V, p) graph, as shown in figure 2.15 and is mathematically calculated by equation (2.36) as,

$$W = -P(V_f - V_i) \tag{2.36}$$

3. **Boundary work for isothermal processes:** The internal energy of a system depends on the temperature of the system. Under ideal conditions, a system that undergoes an isothermal process has a constant temperature, and thus it

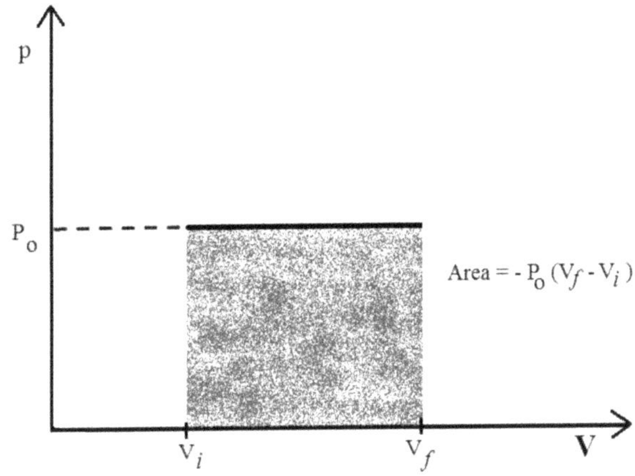

Figure 2.15. Work done when pressure is constant is given as the area under the graph of pV-graph.

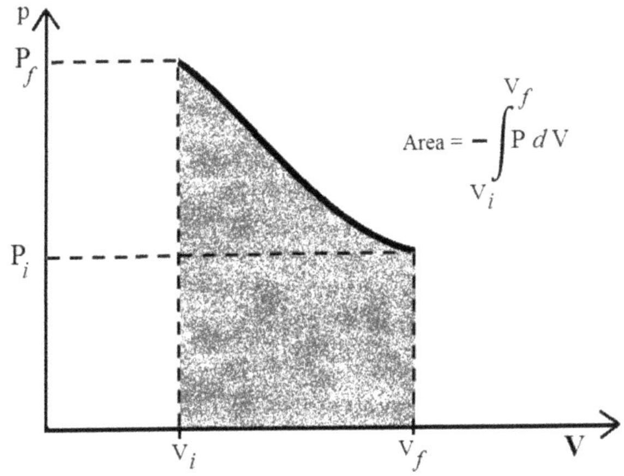

Figure 2.16. Work when pressure is not constant. Divide the process into tiny steps, calculate area under the graph for each step and then add all the areas.

will have constant internal energy. This implies that the change in internal energy (ΔU) is zero. So we can write the First law of thermodynamics as,

$$\begin{aligned} \Delta U = 0 &= Q + W \\ Q &= -W. \end{aligned} \tag{2.37}$$

4. **Boundary work when pressure not constant:** When the pressure is not constant, we divide the complete process in multiple steps and compute the area under the graph for each step. To find total work done on the gas we sum over the area of each step, graphically shown in figure 2.16 and mathematically calculated by equation (2.38),

$$W = -\int_{V_i}^{V_f} p(V)dV \tag{2.38}$$

2.8 Defining heat transfer

Usually, the state of equilibrium for a system is independent of how long it took the system to reach equilibrium. So in order to find the time or the rate of the processes we use completely different tools. The topic usually comes under the subject *transport theory* or *kinetics*. For more details about these processes of heat transfer please refer to appendix C. I will define them very briefly here.

2.8.1 Heat transfer processes

Processes of heat transfer can be classified, depending on the mechanism of heat transfer, into the following basic three kinds:

1. **Conduction:** Conduction is the transfer of heat by molecular contact. Heat transfers from the heater to the rest of the room when fast-moving molecules hit slow-moving molecules and give up their energy in the process.
2. **Convection:** Convection is the bulk motion of a gas or liquid, where the warmer material will expand. In the case of gases they will also rise upwards.
3. **Radiation:** Radiation is the emission of electromagnetic waves. Most emissions at room temperature are in the infrared region. They are also in the visible spectrum for very hot objects, such as the surface of the Sun.

Worked example 2.8. Give an example of a process in which no heat is added to a system, but its temperature increases.

Solution: A cylinder of compressed air at room temperature will decrease in pressure if you open the valve and let the air escape. The ideal gas law ($PV = nRT$) says temperature is directly proportional to pressure. This means as the pressure decreases the temperature of the air inside the cylinder will drop proportionally with the pressure decrease. Similarly, when you pump air into the cylinder from outside, the temperature of air in the cylinder rises proportionally with the pressure increase even though no heat is added to the cylinder.

Worked example 2.9. Give an example of a process in which heat is added to a system, but its temperature does not change.

Solution: When you boil water, on the stove or an open flame of any kind, energy is constantly flowing into the water due to heat applied, but the temperature stays at the boiling point (100°) and does not increase. The flowing energy goes into changing the water from liquid phase to gas phase; we will discuss phase change in chapter 10.

Worked example 2.10. Calculate the numerical value of the work done on the gas, assuming V_f is 2 m^3 and V_i is 1 m^3; p_f is 2 atm and p_i is 1 atm in each step (figure 2.17).

Solution: We have to find the numerical value of the work done on the gas in figure 2.17. The simplest technique is to tabulate the work done in each step as is done in table 2.1.

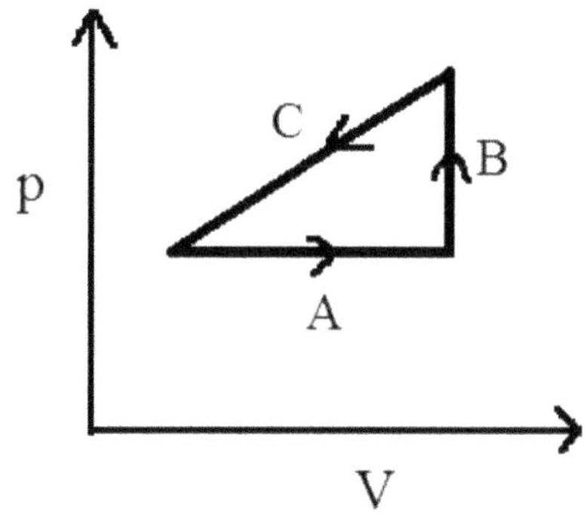

Figure 2.17. Example: Total work performed during the cycle.

Table 2.1. Summarizing all the steps for Example 2.10.

		Pressure	Volume	Work done
Step A	Expanding	$\Delta p = 0$	$\Delta V = V_2 - V_1$	$W_A = -\int pdV$
(Isobaric)	T decreasing			$W_A = -(1.01 \times 10^5 \text{Pa})(2 - 1)\text{m}^3$
	U decreasing			$W_A = -1.01 \times 10^5$ J
Step B	Constant V	$\Delta p = p_2 - p_1$	$\Delta V = 0$	$W_B = 0$
(Isochoric)	$p_1 \to p_2$			
Step C	Contracting	$\Delta p = p_1 - p_2$	$\Delta V = V_1 - V_2$	$W_C = \frac{1}{2}(b \times h) + (L \times W)$
	T increasing			$W_C = \frac{1}{2}(101\,000)(1) + (101\,000)(1)$
	U increasing			$W_C = 1.515 \times 10^5$ J

We can then add the work done in each step to calculate the work done on the gas as, $W_A + W_B + W_C = 0.505 \times 10^5$ J

2.9 Choosing an appropriate system

So far we have talked about energy and the conservation of energy from a general point of view. We have come up with a '*formalism*' for thermodynamics. However, this formalism has built-in limitations, the biggest limitation is that unlike the sugar cubes we talked about in section 2.5, energy is not actually a thing that can be observed. This means our thermodynamical formalism may or may not be appropriate for all systems under consideration and even though the law of conservation is always correct, the formalism we may have used to express it might not be correct if the thermodynamical system is not carefully and correctly selected.

In order to explain this better, let us consider a system made up of a block of steel (block 1) that is free to slide over another block of steel. Suppose a force ($\vec{F}_{\text{applied}}$) acts on block 1 as shown in figure 2.18 and apply our formalism to block 1 which is our system.

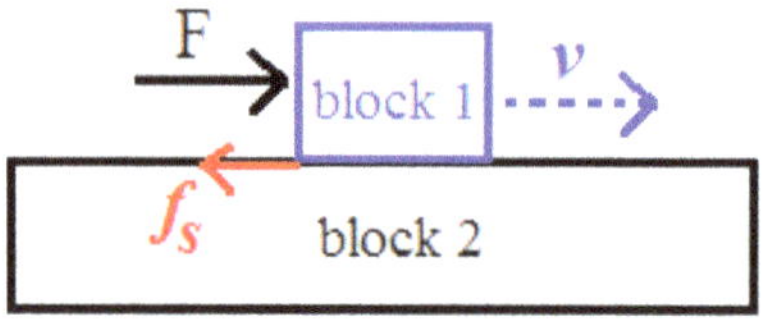

Figure 2.18. Experimental set-up.

When the block 1 is stationary, Newton's force equation is written as,

$$\begin{aligned} \Sigma\vec{F}_{\text{net}} &= 0 = \vec{F}_{\text{applied}} - \vec{f}_s \\ \vec{F}_{\text{applied}} &= \vec{f}_s \end{aligned} \tag{2.39}$$

where $\vec{f}_s$ is static friction. Since block 1 is at rest, the total work, defined as (force × displacement), is zero. From the First law of thermodynamics, equation (2.29), we have:

$$\begin{aligned} \Delta U &= Q + W \\ \Delta U &= Q \end{aligned} \tag{2.40}$$

Equation (2.40), shows that the change of internal energy of the block equals the heat transferred to the block from the surroundings. We know from experience that as the temperature of a sliding block increases, the internal energy of the blocks increases proportionally. Equation (2.40) shows us that heat is transferred to the block from the surroundings, which implies that the surroundings are at a higher temperature than the block. But how do we know this to be true? Are we wrong?

If 'Q' has the significance we've attached to it, then equation (2.40), must be incorrect, and if it is not then we must redefine 'Q'. Our error arose in choosing to make the system boundary the site of the transfer of energy, with the mechanism of this transformation being friction. Since friction always occurs at the boundary that separates the system from its surroundings, choosing our example the way we did was not a good choice.

This error leaves us with two choices, either abandon the formalism or select a new system. In situations like this, the usual choice is to select a new system and to make changes in our view of the system and redefine our system. Here when redefining the system, we should choose both blocks to be a part of the system, which means energy transformation as a result of friction will become part of the system. Both blocks experience a change in the internal energy and the First law of thermodynamics can be written as,

$$\begin{aligned} \Delta U_1 + \Delta U_2 &= Q + W \\ &= Q + (\vec{F} \cdot \Delta x) \end{aligned} \tag{2.41}$$

where $\vec{F}$ is the force that moves block 1 through distance Δx over block 2.

In this equation (2.41), Q is no longer the heat transferred between the block 1 and the surroundings, it is the heat transferred from the system (both block 1 and block 2) and the surroundings. This change solves our initial problem! Unfortunately, just based on thermodynamics we cannot evaluate ΔU_1, ΔU_2, or 'Q'. All thermodynamics tells us is what terms we need to take into account and what equations relating these terms we need. With this example, we note two very important points of thermodynamics:

1. Correct choice of the system is very important.
2. Many problems cannot be solved by thermodynamics alone.

Concept questions 2.2.

1. What is energy? Give at least ten types of energy.
2. What is the main source of energy of the human body?
3. What type of energy helps living things maintain a constant body temperature?
4. What is the difference between a reversible and a cyclic process?
5. What happens to the internal energy of a system when heat is added to a system? and what happens to the temperature of the system?
6. Why did we not use ΔQ or ΔW in the RHS of the First law, whereas we used ΔU on the LHS?
7. How do heat transfer and internal energy differ? In particular, which can be stored as such in a system and which cannot?
8. Suppose the internal energy of a system is increased by some process. How will you know if the increase is due to work done or due to the flow of heat? Hint: Is it a good idea to look at the system or its surroundings?

Exercises 2.2

1. Show that 1 Pa m^3 = 1 J, where Pa is pascal, the unit of pressure and J is joule, the SI unit of energy.
2. Suppose 200 J of heat leaves a system and 250 J of work is done on the system. What is the change in internal energy of the system? Did the internal energy increase or decrease during this time?
3. A sample of gas expands from 1 to 2 m^3 while its pressure increases from 1 to 2 Pa. How much work is done on the gas if its pressure changes with volume according to each of the three processes shown in the (V, p)-diagram in figure 2.19?
4. A sample of gas contracts from 2 m^3 to 1 m^3 while its pressure increases from 1 Pa to 2 Pa. How much work is done on the gas if its pressure changes with volume according to each of the three processes shown in the (V, p)-diagram in figure 2.20?
5. Suppose you metabolize 6×10^5 J of food energy while climbing 10 flights of stairs.
   (a) If your efficiency is 15 %, how much heat transfer to the surroundings has to occur to keep your temperature constant?
   (b) How does this relate to the fact that you warm up quickly when you exercise and how is this excess heat dissipated?

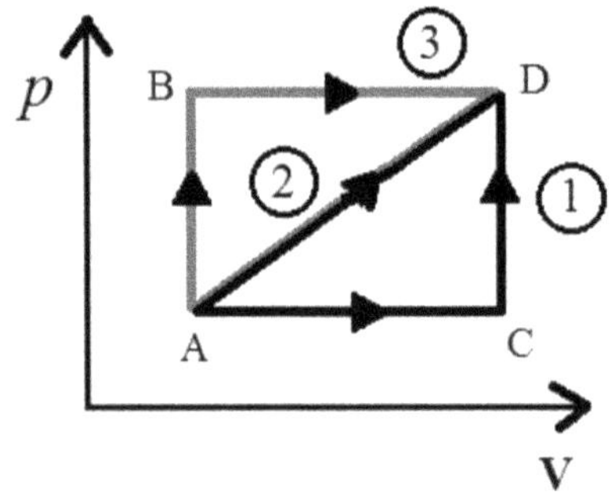

Figure 2.19. Exercise 2.2 number 3.

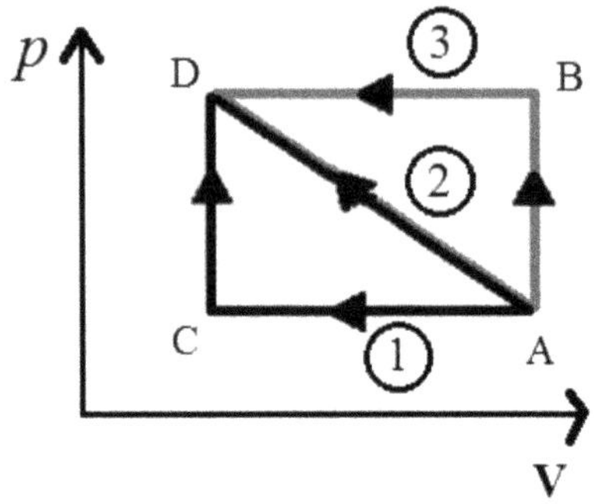

Figure 2.20. Exercise 2.2 number 4.

Further reading

- Cengel Y and Boles M 2006 *Thermodynamics: An Engineering Approach* 5th edn (New York: McGraw-Hill Higher Education)
- Fermi E 2012 *Thermodynamics* (New York: Dover Publications)
- Resnick R, Halliday D and Krane K 2002 *Physics* vol 1 5th edn (New York: Wiley)
- Schroeder D 2000 *Thermal Physics* (Reading, MA: Addison-Wesley Longman)
- Sears F and Salinger G 1975 *Thermodynamics, Kinetic Theory and Statistical Thermodynamics* 3rd edn (Reading, MA: Addison-Wesley Publishing Company)
- Van Ness H C 1969 *Understanding Thermodynamics* (New York: McGraw-Hill Book Company)
- Zain S 2019 *Techniques of Classical Mechanics: From Lagrangian to Newtonian Mechanics* (Bristol: IOP Publishing)

Chapter 3

State properties of systems

3.1 Thermodynamic state of a system

In this chapter we discuss the thermodynamic state of a system that was introduced in chapters 1 and 2 in detail. The thermodynamic state of a system is determined by specifying values of a set of '*measurable properties*' in the system that are sufficient to determine all other properties of a system, when the system is in '*equilibrium*'. In simple systems[1], these properties include pressure, volume and temperature of the system, whereas in complex systems, the specification of additional properties is required. In this text we mostly deal with '*simple systems*'. However, keep in mind that properties of a real system change from point to point. We have to analyze real systems by dividing them into multiple simple systems, where the properties of the divided systems are uniform. The subdivisions of a complex system can be either done conceptually or experimentally.

As defined in chapter 2, when the state of a system changes, it is said to be '*undergoing a process*' and the succession of states through which the system passes is called the '*path of the process*'. If, at the end of the process, the properties return to their original values, the system is said to have undergone a '*cyclic process*'. In thermodynamics, as the state of the system changes, we say the system passes through one equilibrium state to another equilibrium state.

Chapter 2, section 2.3.2, introduced us to intensive and extensive properties. We know the properties of all systems can either be '*intensive*' or '*extensive*' properties. Intensive properties, also called 'point properties', exist at all points in the system and include properties such as temperature or pressure. Intensive properties do not depend on the quantity of matter present in the system. Extensive properties, such as the volume of the system, on the other hand depend on the quantity of matter present.

[1] Simple systems are those whose properties are uniform throughout.

doi:10.1088/978-0-7503-3083-1ch3

3.1.1 The state of various systems

The state of a one-mass-particle system is said to be completely specified at any given time when both the position variables (x, y, z) and the velocity ($\dot{x}$, $\dot{y}$, $\dot{z}$) of the particle are known precisely at a given instant of time. For a system made of two mass-particles we need to specify six position variables (x, y, z)-coordinates and six velocity variables ($\dot{x}$, $\dot{y}$, $\dot{z}$) to completely satisfy the state of the system. For a system of N mass-particles, we require the knowledge of $6N$ variables (x, y, z, $\dot{x}$, $\dot{y}$, $\dot{z}$). This means as the number of particles increases in a system it becomes progressively difficult to keep track of each variable for each mass particle of the system at all times.

To simplify the situation in thermodynamics a concept called the '*state of the system*' is introduced. In the *state of the system* concept, rather than keeping track of all six individual variables for every single mass particle in the system, we analyze the average properties of system as a whole and introduce the concept of an '*ensemble*' of mass particles. In an '*ensemble*', instead of keeping track of detailed knowledge of the microscopic properties of each mass-particle, we keep track of the properties of the *ensemble* as a whole. We discuss *ensembles* in detail in chapter 11; here, I will just introduce the thermodynamic concept of the state of a system with a few examples.

1. **A system made of chemically defined homogeneous fluids and solids:** A homogeneous thermodynamic system is one whose chemical composition and physical properties are either the same throughout or change continuously from one point to another. A simple example of a homogeneous system whose properties remains the same is a sealed container of a gas or a glass of water sitting on a counter top. Another example of a homogeneous system is the atmosphere of Earth. Atmospheric air is a mixture of gases like nitrogen and oxygen. The composition of the vertical column of Earth's atmosphere continuously changes from point to point in both composition and physical properties under the influence of gravity.

   The following measurements can be made on a system composed of chemically defined homogenous fluids and solids.

   (a) **Temperature (T):** This is easy, leave a thermometer in thermal contact with the system, read the temperature (T).
   (b) **Volume (V):** Most thermodynamic properties are largely independent of shape, hence we mostly only talk about the volume (V) of the system.
   (c) **Pressure (p):** In most cases we assume *isotropic pressure* to define the state of a system which means only one value for the pressure is required for the complete definition of our state. This is true for both liquids and for gases.

2. **A heterogeneous system:** In order to define the state of a heterogeneous (also called a non-homogeneous) system, we have to divide the system into a number of homogeneous systems, whose number may be finite or infinite. The homogenous divisions are called '*phases*' of a heterogeneous system and

each phase in the system is separated by a thin physical boundary. This phase boundary is not imaginary or mathematical surface but a real layer that separates the phases. When we pass over the boundary, the physical properties and/or the chemical composition of the system change instantly into the properties of the next phase. A phase of a heterogeneous system that consists of several pure substances is called a '*solution*' or a '*mixture*'. A good example of a heterogeneous system is a glass of water and ice. This system has two homogeneous phases, water and ice, whose chemical composition is the same but physical properties of the two phases are very different.

3. **A system made of a homogeneous mixture of several chemical compounds:** In addition to temperature, volume, pressure and phases, we must also take into account the concentrations of different chemical compounds that make up the mixture when defining the state of a system composed of a homogeneous mixture of several chemical compounds.
4. **A system containing moving parts:** In thermodynamics, we mostly deal with systems where all parts are either at rest or are moving so slowly that their kinetic energies may be ignored. If this is not the case and the system is moving, the knowledge of static thermodynamic state is not enough to define the complete thermodynamic state and in order to define the moving state completely, we must additionally specify the velocities of various parts of the system. This is quite complicated so we will not deal with such systems in this text.

3.1.2 Equilibrium state

One of the basic postulates of thermodynamics, based on observations of real systems, is that all isolated systems evolve in such a way that they achieve a unique stable equilibrium state. States of equilibrium or '*equilibrium states*' are states that remain the same and do not change as long as external conditions remain unchanged. There are many different types of equilibriums corresponding to various physical variables. For example, a gas enclosed in a container with a constant volume will be in equilibrium when pressure is constant throughout and the temperature of the system equals the temperature of the surroundings. A system is said to be in equilibrium when all types of equilibria are simultaneously satisfied. The most common equilibrium conditions include:

1. **Thermal equilibrium:** A system is in thermal equilibrium when temperature throughout the system remains the same.
2. **Mechanical equilibrium:** A system is in the mechanical equilibrium state if there is no change of pressure within the system with respect to time at all points.
3. **Chemical equilibrium:** A system is in a chemical equilibrium when the chemical composition of the system remains the same over time.
4. **Phase equilibrium:** Phase equilibrium occurs when the masses of the phases in a system do not change with time.

3.1.3 Transformations in the state of a system

Transformations in thermodynamical systems progress from some initial to some final state through a series of continuous intermediate states. In the study of thermodynamic transformation, both the initial and final states as well as intermediate states are required to be in thermodynamic equilibrium.

1. **Reversible transformations:** Reversible transformations take our system from an initial state (at volume V_1) to a final state (at volume V_2) via many infinitesimally small intermediate steps such that the system can be returned to its initial state by just reversing the path taken by the system. All the states in a reversible transformation are '*equilibrium states*'. For a reversible process, the work done is given by:

$$W_{\text{reversible}} = -\int_{V_1}^{V_2} pdV \tag{3.1}$$

   Experimentally reversible transformations are brought about by changing external conditions so slowly that the system has time to adjust to the altered conditions. No process can be more efficient than a reversible process.
2. **A quasi-static transformation:** A quasi-static transformation is one that is performed sufficiently slowly so that the system is always in equilibrium. Thus at any stage of the process, the thermodynamic coordinates of the system exist and can, in principle, be computed. For such transformations, the work done on the system can be related to changes in these coordinates.
3. **Irreversible transformations:** A process that is not reversible is called an '*irreversible process*'. An irreversible process is a process for which the final state (state 2) will not return to initial state (state 1) by just reversing the process by which state 1 was first taken to state 2. All complex natural processes are irreversible and irreversibilities always lower the efficiencies of processes. Think of this as friction or drag in classical systems.

3.2 Heat capacity and specific heat

We can change the state of an object by transferring energy to it or by transferring energy out of it. This is done primarily in two ways, in the form of heat[2] or by doing work[3] on the object. During both of these situations the temperature (T) of the system changes. When heat is added to the system the temperature (T) of the system increases and when heat is removed from the system the temperature (T) of the system goes down. Similarly when the work is done '*on*' the system (say during a compression) the temperature of the system goes up and when the work is done '*by*' the system (during an expansion) the temperature of the system goes down.

[2] In a piston-cylinder system we can add heat two ways; one, adding heat while the piston is fixed, i.e., volume 'V' kept constant, or two, by adding heat while piston is allowed to move, where the pressure 'p' is constant.

[3] For example, rubbing two objects together exerts frictional force on both objects, which in turn increases the temperature of both objects.

3.2.1 Heat capacity

The name heat capacity is not quite correct and is a bit misleading. The name actually comes from the days of '*heat-fluid theory*' we introduced in chapter 2. In the heat-fluid theory the objects were thought to 'contain' a certain amount of a substance called '*heat fluid*' which was thought to be exchanged between hot and cold objects. Today we know that objects do not contain a certain amount of heat; objects gain or lose heat when they are in thermal contact with another object.

So what is heat capacity? To understand heat capacity think about of the amount of heat needed to cause a particular temperature change (ΔT). This means objects with a large heat capacity value require a large amount of heat for each degree change in temperature of an object, whereas objects with small heat capacities require very little heat for each degree change in temperature. Water is a good example of an object with a large heat capacity, whereas a piece of lead is a good example of an object with a small heat capacity.

I will reiterate here ***do not*** confuse the word 'capacity' as it is used in daily life with how it is used in thermodynamics. In thermodynamics, heat capacity is *not* the amount of heat a body can hold. It is simply the energy transferred to an object as heat when the temperature of the body increases or it is energy transferred out of the object as the temperature of the object decreases.

A very simple mathematical definition of heat capacity is that it is 'the ratio between the heat (Q) transferred to an object and the corresponding temperature change (ΔT) produced in the object'. Mathematically, heat capacity, symbol 'C' is given by the formula:

$$C \equiv \frac{Q}{\Delta T} \tag{3.2}$$

From the First law of thermodynamics ($\Delta U = W + Q$),

$$C = \frac{\Delta U - W}{\Delta T} \tag{3.3}$$

Another definition of heat capacity is that it is 'the amount of heat needed to raise the temperature of an object by one degree'. The most important thing about this definition is that it is *ambiguous*. Firstly, the amount of heat needed to raise the temperature of an object by 1° depends on two main things; one, the amount of the substance under consideration, and two, maybe more importantly, if you are additionally doing work on the object, and if you are doing work, how much work are you doing.

Another thing worth mentioning here is that heat capacity is always numerically '**positive**', just like speed[4], which implies that:

1. when Q is positive, then ΔT is positive as heat is added to a system.
2. when Q is negative, then ΔT is negative, implies as heat is removed from a system.

[4] Except for the case of black holes that have negative heat capacity.

To determine heat capacity of any object, we either, measure it, look it up in a reference for measured values or try to predict it theoretically. The last one is the most fun. For some objects we already know enough to predict the heat capacity.

Arguably a better name for heat capacity is the 'energy capacity' of an object, since it is the energy needed to raise the temperature of an object by one degree regardless of whether the energy actually enters as heat or work. However, it is a convention to call it heat capacity so I will call it heat capacity as well.

Worked example 3.1. The heat capacity of 1 kg of water is 4186 J K^{-1}. What is the temperature change of water if 505 J of heat is added to the system?

Solution: We can solve this problem by using equation (3.2).

$$\Delta T = \frac{Q}{C} = \frac{505}{4186} = 0.121\text{K}$$

3.2.2 Types of heat capacities

In practice there are two types of heat capacities that are more likely to occur.

1. **When no work is done on the system ($W = 0$):** When no work is done on the system, there is no change in volume ($\Delta V = 0$) and the heat capacity is called *heat capacity at constant volume* (C_V), mathematically given as:

$$C_V \equiv \left(\frac{\Delta U}{\Delta T}\right)_V = \left(\frac{\partial U}{\partial T}\right)_V \tag{3.4}$$

   where the subscript V denotes constant volume.

2. **When the work is done on the system is not zero ($W \neq 0$)**: It is a common observation that most objects expand when they are heated, which causes the object to do work on its surroundings. This implies this heat capacity is larger than the heat capacity at constant volume (C_V), discussed above, since we need additional heat to compensate for the energy lost as work. If the pressure surrounding the system is kept constant, then the total heat needed is *unambiguous* and we refer to the heat needed per degree as the *heat capacity at constant pressure* (C_p). Using the compression–expansion work equation (2.32) ($W = -\int p\, dV = -P\Delta V$) for constant pressure in equation (3.3) we get:

$$\begin{aligned} C_p &= \left(\frac{\Delta U + (P\Delta V)}{\Delta T}\right)_P \\ &= \left(\frac{\partial U}{\partial T}\right)_P + P\left(\frac{\partial V}{\partial T}\right)_P \end{aligned} \tag{3.5}$$

   It is worth noting that the first term on the RHS of equation (3.5) is not the same as the $\left[\left(\frac{\partial U}{\partial T}\right)_V\right]$ term given in equation (3.4). The last term $\left[P\left(\frac{\partial V}{\partial T}\right)_P\right]$ is the additional heat needed to compensate for the energy lost as work. This term is also directly proportional to the volume increase, which means that larger volume increases will result in this term being proportionally larger. For solids and liquids, $\left(\frac{\partial V}{\partial T}\right)$ is usually small and hence can be ignored. However, for gases it may not be ignored.

3.3 Specific heat capacity

It takes 4186 J to increase the temperature of 1 kg of water by 1 °C, so it should take twice that much (2 × 4186 J) to increase the temperature of 2 kg of water by 1 °C. This implies heat capacity not only varies with the type of substance but it also varies with the amount of substance. Calculating this value for every mass of a system becomes almost impossible to do. So we define a new quantity called '*specific heat capacity*', expressed with a small '*c*' and defined as specific heat per unit mass,

$$c \equiv \frac{C}{m} = \frac{Q}{m\Delta T} \tag{3.6}$$

Specific heat capacity is independent of the mass of the substance and hence we do not have to keep track of mass changes in systems and only have to worry about the type of substance. From equation (3.6), we can express 'Q' in terms of specific heat as:

$$Q = mc\Delta T \tag{3.7}$$

In general, water is used as a benchmark when comparing specific heat capacities, and we usually define the '***calorie***' as a unit defined as the amount of energy needed to raise the temperature of 1 gram of water by 1 °C. Mathematically, 1 calorie = 4.18 J and for 1 g of water, C_V = 1 cal °C^{-1} ≅ 4.2 J °C^{-1}.

It is important to keep in mind that both the heat capacity (C) and the specific heat capacity (c) of a material depend on temperature and therefore are not constant. However, at ordinary temperatures and ordinary temperature intervals, specific heats (c) can be assumed to be constant. For example, the specific heat of water varies by less than 1% in the temperature interval from 0 °C to 100 °C, so we can consider it a constant over that temperature interval.

To increase the temperature of an object from T_i to T_f, we need to provide heat to it. If m is the mass and c is the specific heat capacity, then we must divide the temperature into small portions of ΔT. Suppose c_n is constant, we can write the heat added to the system as,

$$Q = m\sum_{n=1}^{N} c_n(\Delta T)_n \tag{3.8}$$

In the limit where heat (Q) goes to zero and 'c' is a function of T, for small changes in temperature, we can write heat added to the system as,

$$Q = m\int_{T_i}^{T_f} c\, dT. \tag{3.9}$$

Please note that this situation is similar to that for the heat capacity. Equation (3.6) does not define specific heat capacity uniquely; we must specify the conditions under which the heat 'Q' is added to the substance. One condition is that the system remains at normal atmospheric pressure while we add heat. There are many other possibilities, each leading to a different value of 'c'. To get a unique value of 'c', we must specify other conditions, for example:

1. Constant pressure (c_p) which is the specific heat capacity at constant pressure (p).
2. Constant volume (c_V) which is the specific heat capacity at constant volume (V).

3.3.1 Specific heat capacities of various substances

Specific heat capacities at constant pressures of some common substances are summarized in table 3.1. As we see in table 3.1 the specific heat of hydrogen is nearly 14.3 (joule per g kelvin) at room temperature, which is much higher as compared to other substances. Hydrogen exists as a gas and gas molecules spread, which makes it harder for heat to travel through them.

Worked example 3.2. One liter of water is heated from 10 °C to boiling 100 °C. The specific heat capacity of water is 4.186 J g^{-1} °C^{-1}. Find the heat required for this process.

Solution: For water we need to convert 1 liter to kg,

$$1 \text{ liter} = 1 \text{ liter} \times \frac{1 \text{ kg}}{1 \text{ liter}} = 1 \text{ kg} = 1000 \text{ g}$$

Table 3.1. Specific heat capacities at constant pressures of some common substances.

Substance	c in J g^{-1} K^{-1}
Hydrogen	14.3
Water	4.186
Ice (−10 °C)	2.05
Zinc	0.387
Copper	0.386
Brass	0.380
Silver	0.233
Mercury	0.140
Lead	0.128
Gold	0.126
Glass	0.84
Granite	0.79

Then from equation (3.7),

$$\begin{aligned} Q &= mc\Delta T \\ &= (1000\ \text{g})(4.186\ \text{J g}^{-1}\,°\text{C}^{-1})(100\ °\text{C} - 10\ °\text{C}) \\ &= 4.186 \times 10^5\ \text{J} \end{aligned}$$

Hence, the heat required for this process is 419 kJ.

Worked example 3.3. 550 g of water at 32 °C is poured into a 210 g aluminum can with an initial temperature of 15 °C. Find the final temperature of the system assuming no heat is exchanged with the surroundings. Hint: Specific heats of water and aluminum are $c_w = 4.186$ J g^{-1} °C^{-1} and $c_{\text{Al}} = 0.900$ J g^{-1} °C^{-1} respectively.

Solution: We can solve this problem by using equation (3.7) and conservation of energy:

$$\begin{aligned} \Delta Q &= 0 \\ Q_w + Q_{\text{Al}} &= 0 \\ m_w c_w \Delta T + m_{\text{Al}} c_{\text{Al}} \Delta T &= 0 \end{aligned}$$

Knowing numerical values for $c_w = 4.186$ J g^{-1} °C^{-1} and $c_{\text{Al}} = 0.900$ J g^{-1} °C^{-1}, and the fact that $\Delta T = T_f - T_i$, we can use the actual numerical values given in the example to find that the final temperature of the system T_f is 30.7 °C.

3.3.2 Calorimetry

The word calorimetry comes from two words, the Latin word *calor* (heat) and the Greek word *metron* (measure). Calorimetry is the measurement of changes in state variables of a body due to heat transfer. The change in state could be due to physical changes, phase transitions, or chemical reactions.

Suppose a block of mass (m_b) with specific heat capacity (c_b) and initial temperature (T_b) is dropped into a calorimeter, basically a light-weight insulated flask containing water. If the water in the calorimeter has a mass (m_w), specific heat (c_w), and an initial temperature (T_w), we can find the final temperature of the block and water by assuming that the calorimeter is light enough that it can be ignored with respect to the system and that the calorimeter is completely isolated from its surroundings and no heat is exchanged between the calorimeter and its surroundings. There are two basic ideas involved in solving this problem:

1. The total energy of the system is conserved and the amount of heat lost by the block equals the heat gained by water.
2. Because the system is in thermal equilibrium the final temperature of the block and water are equal.

Mathematically, we write these conditions as,

$$Q_b + Q_w = 0. \tag{3.10}$$

where Q_b is the change in heat of the block and Q_w is the change in heat in water. Suppose the final temperature is 'T', and we write heat (Q) in terms of specific heat capacity and temperature. We can re-write equation (3.10) as,

$$m_b c_b(T - T_b) + m_w c_w(T - T_w) = 0. \tag{3.11}$$

Solving for the final temperature (T) we get:

$$T = \frac{m_b c_b T_b + m_w c_w T_w}{m_b c_b + m_w c_w} \tag{3.12}$$

which is the final temperature of the block and water.

Worked example 3.4. A half-kg block of metal with an initial temperature of 30 °C is dropped into a container holding 1.12 kg of water at 20 °C. If the final temperature of the block-water system is 20.4 °C, what is the specific heat capacity of the metal? Assume the container can be ignored and that no heat is exchanged with the surroundings.

Solution: Heat flows from the block to water since the block is at higher temperature. This means that Q of the block (Q_B) is negative and Q_w of the water is positive. If T is the final temperature of the system, then the amount of heat that flows out of the block will be given as,

$$Q_B = m_B c_B \Delta T = m_B c_B(T - T_B)$$

where c_B is the specific heat and m_B is the mass of the metal. The amount of heat that flows into the water is,

$$Q_w = m_w c_w \Delta T = m_w c_w(T - T_w).$$

where c_w is the specific heat and m_w is the mass of water. We can use conservation of energy to solve for the specific heat capacity as,

$$\begin{aligned} \Delta Q &= 0 \\ Q_w + Q_B &= 0 \\ m_w c_w(T - T_w) + m_B c_B(T - T_B) &= 0 \end{aligned}$$

Rearranging the above equation we can get:

$$c_B = \frac{m_w c_w(T - T_w)}{m_B(T - T_B)}$$

Putting the numerical values in the above equation we find that c_B numerically equals 391 J kg^{-1} K^{-1}. We can look up the value in a data table in a chemistry textbook and find that the block is probably made out of copper (Cu).

3.4 Latent heat of transformation

Can you put heat into a system without increasing its temperature? The answer is, yes, you can put heat into a system without increasing its temperature. It can be done at a '*phase transformation*'. In this section we talk about some things associated with phase transformations.

3.4.1 Phase transformation

A phase transformation occurs when a substance goes from one state to another, say from solid to liquid or a liquid to vapor or vice versa. At a phase transition, there is no change in temperature of a substance, $\Delta T = 0$. The equation for the heat capacity of the substance, equation (3.6), at a phase transformation then becomes,

$$C = \frac{Q}{\Delta T} = \frac{Q}{0} = \infty \tag{3.13}$$

which says that heat capacity during a phase transformation is infinite!

3.4.2 Latent heat

We see from equation (3.13) that during a phase transformation, heat capacity is infinite. However, one might still want to know how much heat is required to melt or boil a substance completely. The amount of heat required to melt or boil a substance divided by the mass of substance is called the '*latent heat of transformation*' (L). Mathematically given as,

$$L \equiv \frac{Q}{m}. \tag{3.14}$$

Just like we saw for the definition of heat capacity, this definition of latent heat of transformation is ambiguous as well, since any amount of work could be done during the process. By convention, however, we assume that,

1. Pressure is constant, usually taken to be 1 atm.
2. No other work is done besides the usual constant-pressure expansion or compression.

Latent heat for melting ice = 330 J g^{-1} or 80 cal g^{-1}. Latent heat for boiling water = 2260 J g^{-1} or 540 cal g^{-1}. To understand it better, remember, raising the temperature of water from 0°C → 100 °C requires 100 cal g^{-1} of energy.

Worked example 3.5. Your 100 g cup of tea is boiling-hot. About how much ice should you add to bring it down to a comfortable sipping temperature of 65 °C? Assume that the ice is initially at −15 °C. The specific heat capacity of ice is 0.5 cal g^{-1} °C^{-1} and take specific heat capacity to be the same as that of water.

Solution: The amount of heat we want to remove from the tea, given mass of tea = 100 g, c_{tea} = 1 cal g^{-1} °C^{-1}.

$$Q = m_{\text{tea}}c_{\text{tea}}(T - T_{\text{tea}}) = (100\ \text{g})(1\ \text{cal g}^{-1}\,°\text{C}^{-1})(65 - 100)\ °\text{C} = 3500\ \text{cal}$$

This heat enters the ice and brings it up to 0 °C, which then melts it. Then we bring the melted ice (now water) up to 65 °C. Suppose the mass that is required to do this is m.

This implies that the heat removed from the tea equals the heat added to the ice and the heat required to take the water 65 °C. We can write this mathematically as,

$$\begin{aligned} Q &= \text{ice to water} + \text{water from } 0^\circ \text{ to } 65^\circ + \text{latent heat} \\ &= m_i c_i (T - T_i) + m_w c_w (T - T_w) + mL \\ 7000 \text{ cal} &= m(0.5)(15^\circ) + m(1)(65^\circ) + m(80) \\ \Longrightarrow m &= 46 \text{ g} \end{aligned}$$

This means that we will need 46 g of ice to bring the tea down to a comfortable sipping temperature of 65 °C.

Concept questions 3.1.

1. Which of the following is *not* a state variable? (a) Work, (b) Internal energy, (c) Entropy, (d) Temperature, (e) Pressure.
2. Using the concept of heat capacity, describe which will warm up faster during the day, the water or the sand on the sea shore. What about at night?
3. Why is copper used to cover the bottom of many commercial pots and pans for stove-top heating?

Exercises 3.1.

1. To measure the heat capacity of an object, all we have to do is to put it in thermal contact with another object whose heat capacity we know. An example would be immersing a chunk of metal in boiling water (100 °C) and immediately transferring it to a Styrofoam cup with 250 g of water at 20 °C. After 60 seconds the temperature of the contents is 25 °C. Assume that during this time no significant energy is transferred between the cup and its surroundings. The heat capacity of the cup itself is negligible.
   (a) How much heat is gained by the water?
   (b) How much heat is lost by the metal?
   (c) What is the heat capacity of the chunk of metal?
   (d) If the mass of chunk is 100 g, what is its specific heat capacity?
2. Find the heat capacity of liquid water per molecule in terms of k_B, the Boltzmann constant.

3.5 Enthalpy

Keeping track of the compression–expansion work done during constant-pressure processes gets to be a pain after a while. There is a trick to making it easier. Instead of talking about the 'energy' content of a system, we can agree to use '*energy*' and '*add in*' the work needed to make 'room' for it by pushing the surroundings away

Figure 3.1. Magician pulling a rabbit out of thin air.

under constant pressure consideration. Think of this as magician trying to create a rabbit out of thin air (nothing) and placing it on a table on the stage (figure 3.1). What does she need to do to accomplish this? She needs,

1. Enough energy (U) to create the rabbit, and
2. Additional energy (P_{atm} V)[5] to push the atmosphere out of the way to make room for the rabbit on the table.

This additional energy needed is directly proportional to the pressure (P_{atm}) of the environment, and the total volume (V) of the system. Basically instead of (U) we will have ($U + P_{atm}$ V) which we call *enthalpy* (H) of a system and define as,

$$H \equiv U + P_{atm}V. \tag{3.15}$$

Hence, enthalpy is the '*total*' energy you would need to create the system out of nothing and to make room in the atmosphere for the system to inhabit.

Please note that this works in reverse as well where enthalpy will be,

1. Energy extracted during the annihilation of a system.
2. Plus, work ($P_{atm}V$) done by the atmosphere as it collapses to fill the vacuum left behind once the rabbit disappears.

3.5.1 The usefulness of enthalpy

Suppose a change takes place at constant pressure in our system, say we add some heat, or some chemicals react. This change will result in the change in energy,

[5] Same as the work done.

volume, and/or enthalpy of the system and so our definition of enthalpy must change and become,

$$H + \Delta H = (U + \Delta U) + P_{atm}(V + \Delta V) \tag{3.16}$$

which can also be rearranged as,

$$\begin{aligned} H + \Delta H &= (U + P_{atm}V) + (\Delta U + P_{atm}\Delta V) \\ \Longrightarrow H + \Delta H &= H + (\Delta U + P_{atm}\Delta V) \end{aligned}$$

Hence, the change in enthalpy during a constant-pressure process is,

$$\Delta H = \Delta U + P_{atm}\Delta V \tag{3.17}$$

Equation (3.17), tells us that enthalpy can increase for two reasons:

1. When the energy (ΔU) of the system increases.
2. When the system expands ($P_{atm}\Delta V$) and work is done on the atmosphere to make room for it.

From the First law of thermodynamics, we know that the change in energy in a system equals the heat (Q) added, plus compression/expansion (boundary) work ($W = -p\Delta V$), plus any other kind of work done on the system, hence $\Delta U = Q + W + W_{other} = Q - p\Delta V + W_{other}$. Put this value in equation (3.17), we get,

$$\Delta H = Q + W_{other} \tag{3.18}$$

Equation (3.18) makes it obvious that the change in enthalpy is caused '*only*' by heat plus other forms of work. Which means when no 'other' types of work are being done on the system; the change in enthalpy (ΔH) tells you directly the amount of heat added to the system.

3.5.2 Heat capacity and enthalpy

Suppose the temperature of an object is raised at constant pressure. In this case the change in enthalpy per degree of temperature equals the heat capacity at constant pressure called '*enthalpy capacity*'. Enthalpy capacity of the object is given as,

$$\begin{aligned} \frac{\Delta H}{\Delta T} &= \left(\frac{\partial H}{\partial T}\right)_P \\ &= \left(\frac{\partial U}{\partial T}\right)_P + p\left(\frac{\partial V}{\partial T}\right)_P \\ &\equiv C_P \end{aligned} \tag{3.19}$$

where C_P is the heat capacity at constant pressure.

3.5.3 Enthalpy of formation

Enthalpy of formation is a very important concept and to understand it, let's talk about how liquid water (H_2O) is formed. Chemistry textbooks tell us that for each mole of water produced, the change in enthalpy $\Delta H = -286$ kJ. This quantity is called the '*enthalpy of formation*', because each mole is being formed out of elementary constituents at 1 atm and room temperature (298 K) conditions.

Conversely when you simply burn a mole of hydrogen, you get 286 KJ of energy out. Most of this energy comes from the thermal and chemical energies of the molecules themselves, but a small amount also comes from work done by the atmosphere as it collapses to fill the space left behind by the consumed hydrogen.

There are two kinds of chemical reactions, endothermic and exothermic reactions. Both can be defined in terms of change in enthalpy of a system. When ΔH is positive, heat is absorbed by the system and the reaction is '*endothermic*'. This happens because the products of the reaction have a greater enthalpy than the reactants and heat is absorbed by the system. When ΔH is negative, the reaction is called *exothermic*, that is, heat is generated in the reaction and the overall enthalpy decreases. Data for enthalpy of formation for some common compounds are given in table 3.2.

Worked example 3.6. What is the change in enthalpy for the following reaction?

$$2Mg + O_2 \rightarrow 2MgO \tag{3.20}$$

Solution: The change in enthalpy (ΔH) is calculated by:

$$\Delta H = H_{\text{products}} - H_{\text{reactants}}$$
$$\Delta H_{Mg} = 0 \text{ kJ}$$
$$\Delta H_{O_2} = 0 \text{ kJ}$$
$$\Delta H_{MgO} = -601 \text{ kJ}$$

Table 3.2. Enthalpy of formation.

Substance	ΔH (kJ mol^{-1})
H_2O (l)	−285.83
H_2O (g)	−241.82
H_2CO_3 (aq)	−699.65
CO (g)	−110.53
CO_2 (aq)	−413.82
CO_2 (g)	−393.51
O_2 (g)	0
CH_4 (g)	−74.81
C_2H_2 (g)	+226.73
C_2H_4 (g)	+52.26
C_2H_6 (g)	−84.68

Then we can say:

$$\Delta H = 2\Delta H_{MgO} - (2\Delta H_{Mg} + \Delta H_{O_2})$$
$$\Delta H = 2(-601\text{ kJ}) - (2(0\text{ kJ}) + (0\text{ kJ}))$$
$$\Delta H = -1202\text{ kJ mol}^{-1}$$

Worked example 3.7. What is the change in enthalpy for the following reaction? Given that the enthalpies of formation for the compounds are (ΔH), $C_4H_{10} = -126$ kJ, $O_2 = 0$, $CO_2 = -394$ kJ, and $H_2O = -286$ kJ.

$$C_4H_{10} + O_2 \rightarrow CO_2 + H_2O \tag{3.21}$$

Solution: The change in enthalpy (ΔH) is calculated by:

$$\Delta H = H_{\text{products}} - H_{\text{reactants}}$$

Enthalpies of formation for the compounds are as follows:

$$\Delta H_{C_4H_{10}} = -126\text{ kJ}$$
$$\Delta H_{O_2} = 0\text{ kJ}$$
$$\Delta H_{CO_2} = -394\text{ kJ}$$
$$\Delta H_{H_2O} = -286\text{ kJ}$$

We first have to balance the equation before we can calculate the change in enthalpy.

$$2C_4H_{10} + 13O_2 \rightarrow 8CO_2 + 10H_2O \tag{3.22}$$

Then we can say:

$$\Delta H = 8\Delta H_{CO_2} + 10\Delta H_{H_2O} - (2\Delta H_{C_4H_{10}} + 13\Delta H_{O_2})$$
$$\Delta H = 8(-394) + 10(-286) - (2(-126) + (0))$$
$$\Delta H = -5760\text{ kJ mol}^{-1}$$

Worked example 3.8. Given that the enthalpy of vaporization for water is:

$$H_2O(l) \rightarrow H_2O(g);\ \Delta H_{\text{vap}} = +44.0\text{ kJ mol}^{-1} \tag{3.23}$$

Calculate H for each of the following processes:

1. Evaporating 4 moles of water.
   Solution: Enthalpy for one mole of water = ΔH_{vap} = +44.0 kJ mol^{-1}.
   Then, enthalpy for 4 moles of water = ΔH_{vap} = +4(44.0) = +176 kJ mol^{-1}.
2. Evaporating 4 g of water.
   Solution: Convert the mass of water to moles of water:

$$4\text{ g} \times \frac{1\text{ mol }H_2O}{18\text{ g}} = 0.22\text{ mol}$$

Hence the change in enthalpy in evaporating 4 g of water:
$\Delta H = 0.22\ (+44.0\ \text{kJ}) = +9.78\ \text{kJ}$

3. Condensation of 4 g of water vapor to liquid water.
Solution: Condensation is opposite to evaporation. Hence ΔH for condensation = 0.22 (−44.0 kJ) = −9.78 kJ

3.6 Determining energy (ΔU) and enthalpy (ΔH)

In this section we will develop ΔU and ΔH starting from the First law of thermodynamics.

1. **Under constant volume conditions:**
The First law of thermodynamics from chapter 2, is stated as,

$$\Delta U = Q + W \tag{3.24}$$

For constant volume processes, $dV = 0 \Longrightarrow$ work done $W = -\int dV = 0$. The First law then becomes:

$$\Delta U = Q_V \tag{3.25}$$

We find the change in enthalpy by determining heat at constant volume (Q_V) experimentally, by measuring the heat flow between the system and its surroundings at a constant volume.

2. **Under reversible and constant pressure conditions:**
Under reversible and constant pressure conditions, the First law of thermodynamics from chapter 2, will be written as,

$$dU = dq - p_{\text{ext}}dV$$

$$\int dU = \int dq_p - \int pdV$$

$$U|_{\text{i}}^{f} = q_p - pV|_{V_{\text{f}},\, p_{\text{f}}}^{V_{\text{i}},p_{\text{i}}}$$

$$U_{\text{f}} - U_{\text{i}} = q_p - (p_{\text{f}}\, V_{\text{f}} - p_{\text{i}}\, V_{\text{i}})$$

Rearranging the last expression we get,

$$(U_{\text{f}} + p_{\text{f}}\, V_{\text{f}}) - (U_{\text{i}} + p_{\text{i}}\, V_{\text{i}}) = q_p. \tag{3.26}$$

Since pressure (p), volume (V), and energy (U) are all state functions, the term ($U + pV$) is also a state function. We call this new state function enthalpy, H, and define it as follows:

$$H \equiv U + pV \tag{3.27}$$

Enthalpy (H) has units of Joules.

$$H_{\text{f}} - H_{\text{i}} = q_p \tag{3.28}$$

$$\Delta H = q_{\text{i}}\, p \tag{3.29}$$

For a process involving only pressure–volume (pV) work, ΔH can be determined by measuring the heat flow between the system and its surrounding. It is worth noting here that chemical reactions are mostly carried out at constant pressure rather than at constant volume.

Concept questions 3.2.

1. Where is the energy going in a process where there is a state change yet no temperature change?
2. Which freezes first, hot water in an ice cube tray or cold water in the same-sized tray?
3. Under what condition (adiabatic, isothermal, isobaric, or isometric) is enthalpy equal to heat?

Exercises 3.2.

1. A cube of copper of mass m_c = 75 g is placed in an oven at a temperature T = 312 °C until it comes to thermal equilibrium. The cube is then dropped quickly into an insulated beaker containing a quantity of water of mass m_w = 220 g. The heat capacity of the beaker alone is C_B = 0.190 J K^{-1}. Initially water and beaker are at T_i = 12 °C. What is the final equilibrium of the system? The system includes copper cube, water, and beaker. Hint: Specific heat of copper, 0.387 J g^{-1} K^{-1} and specific heat of water 4.186 J g^{-1} K^{-1}.
2. Use values given in table 3.2 to calculate the enthalpy change in the following reactions:
   (a) $2C_2H_6$ (g) + $7O_2$ (g) → $6H_2O$ (g) + $4CO_2$ (g)
   (b) CH_4 (g) + O_2 (g) → H_2O (g) + CO (g)
   (c) C_2H_2 (g) + O_2 (g) → H_2O (g) + CO_2 (g)

Further reading

- Cengel Y and Boles M 2006 *Thermodynamics: An Engineering Approach* 5th edn (New York: McGraw-Hill Higher Education)
- Fermi E 2012 *Thermodynamics* (New York: Dover Publications)
- Resnick R, Halliday D and Krane K 2002 *Physics* vol 1 5th edn (New York: Wiley)
- Schroeder D 2000 *Thermal Physics* (Reading, MA: Addison-Wesley Longman)

IOP Publishing

Thermodynamics and Statistical Mechanics
An introduction for physicists and engineers
Samya Zain

Chapter 4

The ideal gas

Matter can exist in three distinct physical states or forms, namely solid, liquid, or gas. Matter in each state is made up of the same kinds of particles but the behaviors of these particles differ in the three states. Solids and liquids are called 'condensed phases' because the particles are close together. Gases can be defined as 'fluids' that are able to expand freely and fill any available space, regardless of initial conditions.

Particles in a solid are fixed in place usually in a regular pattern but are free to vibrate in place. Particles in a liquid state are close together but have no regular arrangement. They are free to vibrate and move about inside the liquid. Particles in a gas are well separated from each other, have no regular arrangement and can move freely at relatively high speeds. In this chapter we will focus on the gaseous state of matter.

4.1 Experimental gas laws

Before we talk about ideal gases and the ideal gas law, it is useful to talk about some laws about gases that were developed by direct experimentations. Four names in particular are associated with experimental gas laws, including Robert Boyle (1627–1691 CE), Jacques Charles (1746-1823 CE), J L Gay-Lussac (1778–1850 CE), and Amedeo Avogadro (1776–1856 CE). Let's discuss them each briefly here:

1. **Boyle's law:** Robert Boyle (1627–1691 CE) showed experimentally that in a fixed amount of gas, maintained at a constant temperature, the pressure (p) and the volume (V) of the gas are inversely proportional to each other ($V \propto 1/p$). Mathematically, Boyle's experimental law is written as:

$$P_1V_1 = P_2V_2 \quad \text{for constant } N \text{ and } T \tag{4.1}$$

   where the subscripts 1 and 2 represent the states 1 and 2 of the system.
2. **Charles' law:** Jacques Charles (1746–1823 CE), wrote Charles' law, and he experimentally showed that when the pressure is kept constant, the volume is

doi:10.1088/978-0-7503-3083-1ch4

found to be proportional to the temperature of a system ($V \propto T$), mathematically written as,

$$\frac{V_1}{T_1} = \frac{V_2}{T_2} \quad \text{for constant } N \text{ and } p \tag{4.2}$$

3. **Gay-Lussac's law:** Gay-Lussac (1778–1850 CE) wrote Gay-Lussac's law which states when the volume is kept constant the pressure of the gas is proportional to temperature ($p \propto T$), where temperature is measured on the Kelvin scale.

$$\frac{P_1}{T_1} = \frac{P_2}{T_2} \quad \text{for constant } V \tag{4.3}$$

4. **Avogadro's law:** Avogadro showed in 1811 CE that under the same conditions of pressure and temperature, equal volumes of different gases contain the same number of molecules. In other words,

$$V \propto N \quad \text{for constant } p \text{ and } T \tag{4.4}$$

All of the above laws are combined in the ideal gas law, as seen in the next section.

4.2 The ideal gas

Real gases are considered to exhibit ideal gas behavior when the density of the real gas is low enough that molecules of the gas either do not interact with each other or when they do interact with each other they undergo elastic collisions, with no loss of kinetic energy.

4.2.1 Properties of an ideal gas

In an ideal gas the number of particles is considered to be very large. An ideal gas exhibits the following properties:

1. Ideal gases are generally made up of non-interacting identical particles[1].
2. Particles undergo random motion.
3. The size of the particles in an ideal gas is negligible when compared to the volume occupied by the entire gas.
4. The particles of an ideal gas obey Newton's laws of motion.
5. The particles of an ideal gas experience forces only during a collision.
   (a) Collisions are assumed to last for a negligible duration.
   (b) Collisions between the particles and collisions between particles and the walls of the container are completely elastic.

[1] Non-interacting in this case is used in the sense of not changing their internal state

4.2.2 Ideal gas law

The concept of ideal gas is very useful in the study of thermodynamics and statistical mechanics because ideal gases obey the ideal gas law. The ideal gas law is a simplified equation for the state for the ideal gas. In section 4.1 we defined experimental gas laws which can now be combined to give the ideal gas law. We know from Gay-Lussac's law the pressure of an ideal gas at constant volume is directly proportional to its temperature, (i.e., $p \propto T$), Charles' law says that the volume of an ideal gas is proportional to temperature (i.e., $V \propto T$), Boyle's law proves ($V \propto 1/p$) and Avogadro's Law states ($V \propto N$), where N is the number of gas molecules contained in the given volume. Combining the results from all the experimental gas laws we can write pressure as,

$$p \propto T\frac{N}{V} \tag{4.5}$$

which may also be written as the ideal gas law to be proportional to,

$$pV = kNT \tag{4.6}$$

where k is the constant of proportionality. k is experimentally found to be numerically equal to 1.38×10^{-23} J K^{-1}. Since the number of particles in an ideal gas is very large, an alternate way is to express the ideal gas law is in terms of 'n', the number of moles of gas rather than 'N', the number of molecules. N is mathematically given as n multiplied by Avogadro's number ($N = nN_A$) and the relation between k and R is ($Nk = nR$), so the ideal gas law can also be written as:

$$pV = kNT = nRT \tag{4.7}$$

where $R = 8.31$ J mol^{-1} K^{-1} is known as the '*universal gas constant*'. For air, $R = 287$ J kg^{-1} K^{-1}. Chemistry text books generally use this form of the ideal gas law, because chemists prefer not to deal with very large numbers like the number of molecules in a reasonable amount of gas.

Another form of the ideal gas law in terms of 'm', which is the grams of a gas whose molecular weight is M, is given by:

$$pV = \frac{m}{M}RT \tag{4.8}$$

Worked example 4.1. An insulated cylinder, maintained at 20 °C and 10 atm, is fitted with a piston and contains a gas with a volume of 20 L. Suppose the piston is lowered such that the volume of the gas decreases to 14 L while simultaneously the temperature is a raised to 25 °C. Assuming the gas in the cylinder acts like an ideal gas, find the final pressure of the gas.

Solution: In this case the number of molecules is not changing. So we can use the ideal gas law from equation (4.6) to give us:

$$\frac{P_1V_1}{kT_1} = N = \frac{P_2V_2}{kT_2} \tag{4.9}$$

We can simplify to find the final pressure of the gas as:

$$P_2 = P_1 \frac{V_1}{V_2} \frac{T_2}{T_1}$$

Remember, temperature (T) must be in Kelvin so we must convert; $T_1 = (273 + 20)$ K and $T_2 = (273 + 25)$ K and then put values in the above equation, $V_1 = 20$, $V_2 = 14$, $P_1 = 10$ atm, and we find $P_2 = 14.5$ atm.

Worked example 4.2. The universal gas constant (R) has units consistent with the ideal gas law, given by 4.7. These units are: (pressure) × (molar volume) / (temperature). Show that the units of R may also be:

$$\frac{\text{molar energy}}{\text{temperature}} \tag{4.10}$$

Solution: We know that pressure is force per unit area, hence the SI units of pressure will be N m^{-2} and volume is expressed in m^3. We start from the definition of the universal gas constant R and use dimensional analysis to get:

$$\frac{pV}{T} = \frac{\left(\frac{\text{N}}{\text{m}^2}\right)\text{m}^3}{T} = \frac{\text{N m}}{T} = \frac{\text{J}}{T} = \frac{\text{molar energy}}{\text{temperature}}$$

4.3 Microscopic model of ideal gas

The Zeroth law of thermodynamics states that at a very simple level, temperature is the state property that is considered to be equal between two objects after they have been in thermal contact with each other for a long enough period of time, called the relaxation time. When this happens we say two objects are in thermal equilibrium with each other.

In this section we find the relationship between temperature (T) and energy (U) when we consider a system made of an ideal gas. Suppose we have a container of length (L) with a piston of area (A) full of gas (dare I say, an ideal gas?) under ideal conditions, as shown in figure 4.1. Let's start by assuming that the container has only one particle of mass (m) in it and this particle bounces elastically off the walls. After some time, the particle bounces off one wall, its velocity component changes and

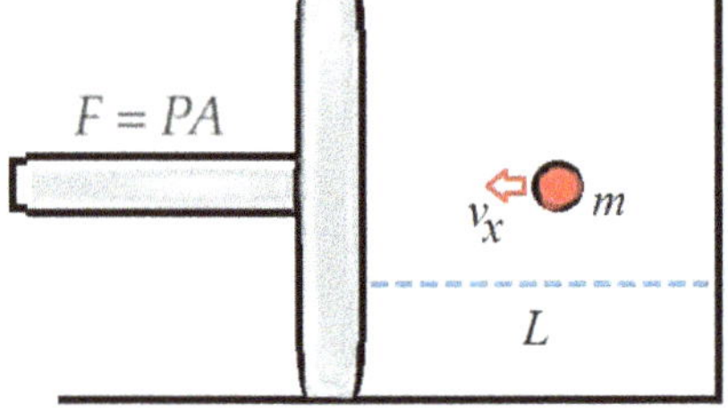

Figure 4.1. One particle of an ideal gas in a piston-and-cylinder system.

since we are assuming elastic collisions the particle does not loose or gain kinetic energy which implies its speed will not change.

To find the relationship between temperature (T) and energy (U), fundamentally we want to know two things,

1. How is temperature (T) related to kinetic energy (KE) of the gas particle, and
2. What is the pressure exerted on the piston by the particle?

To answer the first question, we use the ideal gas law (equation (4.6)) and to answer the second question, we start with the fact that the pressure exerted on the piston by one particle is negligible. However, once in a while a particle can exert quite a large force as it crashes into the piston and bounces off. What we really need to find is the '*average*' pressure (p_{av}) exerted on the piston over a given period of time, in terms of the x-component of force exerted by the particle on the piston,

$$p_{\text{av}} = \frac{F_{\text{x on piston}}}{A} = -\frac{F_{\text{x on particle}}}{A}$$

From Newton's third law of motion ($F_{\text{x on piston}} = -F_{\text{x on particle}}$) and from Newton's second law ($\Sigma F = m\vec{a}$) and knowing that acceleration is the rate of change of velocity, we can write the '*average*' pressure (p_{av}) exerted on the piston over a period of time as,

$$p_{\text{av}} = -\frac{m}{A}\left(\frac{\Delta v_x}{\Delta t}\right) \tag{4.11}$$

Please note that the collision of the particle with perpendicular walls will not affect the motion of the particle in the y-direction.

The next important question is, what is the relevant '*time interval*'. To simplify the problem, let us take the time interval to be the time it takes for the particle to undergo one round trip starting from the piston. Then, we can write,

$$\begin{aligned} \text{Speed } (v_x) &= \frac{\text{Distance}}{\text{time}} = \frac{2L}{\Delta t} \\ \Rightarrow \Delta t &= \frac{2L}{v_x} \end{aligned} \tag{4.12}$$

During this time interval, particle undergoes one collision with the opposite wall, hence change in velocity along the x-axis is,

$$\Delta v_x = v_{xf} - v_{xi} = -v_x - v_x = -2v_x \tag{4.13}$$

Putting values from equations (4.12) and (4.13) into equation (4.11), we get:

$$p_{\text{av}} = \frac{mv_x^2}{V} \tag{4.14}$$

where $AL = V$ is the volume of the cylinder. Equation (4.14) gives us the mathematical relationship between the average pressure and velocity. From equation (4.14) and knowing the kinetic energy is given as ($KE = 1/2\ mv^2$), we can also write:

$$p_{av} = \frac{2KE_x}{V} \tag{4.15}$$

where KE_x represents the kinetic energy of the particle in the x-direction. Equation (4.15) gives the pressure exerted on the piston by one particle moving in the x-direction.

For N particles: We can generalize the above model to include N non-interacting particles, each of mass (m) contained in the cylinder. Assume each particle periodically collides with the piston. Then the average pressure from equation (4.14), can be re-written as,

$$p_{av} V = mv_{1x}^2 + mv_{2x}^2 + mv_{3x}^2 + \cdots \tag{4.16}$$

and for N particles we can write,

$$p_{av} V = N \times (mv_x^2) \tag{4.17}$$

If we assume that the number of particles is large enough and collisions are frequent enough, that the pressure is essentially continuous, we can then say that the pressure (p) is given as,

$$pV = N \times (mv_x^2) \tag{4.18}$$

Equation (4.18) gives the pressure exerted on the piston by N particles.

The ideal gas law states ($pV = NkT$) and treating it as an experimental fact we can rewrite equation (4.18) as,

$$NkT = N \times (mv_x^2) \tag{4.19}$$

which can be written in terms of average kinetic energy in the x-direction as,

$$\frac{1}{2}NkT = KE_x \tag{4.20}$$

We find that the values of kinetic energies in all other directions (y- and z-directions) give the same value as equation (4.20), which implies that the total translational kinetic energy ($KE_{\text{translation}}$) in 3-D is given as:

$$KE_{\text{translation}} = K_x + K_y + K_z$$
$$= \frac{1}{2}NKT + \frac{1}{2}NKT + \frac{1}{2}NKT$$
$$KE_{\text{translation}} = \frac{3}{2}NkT$$

or,

$$KE_{\text{translation}} = \frac{3}{2}NT = \frac{3}{2}nRT \tag{4.21}$$

This is what we were looking for. Equation (4.21) gives us the relationship between temperature (T) and kinetic energy (KE) for an ideal gas. It is interesting to note, we started from a simple one-particle gas then we talked about N particles inside a cylinder, and came up with a conclusion that the average translational KE of particles is given by a simple constant times the temperature. This result gives us a nice relation between the energy and the temperature of a system[2]. It is worth noting here that the average speed of particles in a gas is,

$$\frac{1}{2}mv^2 = \frac{3}{2}kT$$

$$v_{\text{av}} = \sqrt{\frac{3kT}{m}}$$

This equations says that temperature of a gas is a direct measure of the average translational kinetic energy of its particles.

4.4 State properties of an ideal gas

In this section we will discuss some state properties of ideal gases. We know that we can specify the complete state of a system if we specify any two properties of the system. In other words, if we can write energy as a function of variables, $U = U(T, V)$, or $U = U(p, V)$ or $U = U(p, T)$ we can completely specify the state of any system. In order to calculate (ΔU) for an ideal gas, we must first restrict ourselves to consider the following processes only:

1. Processes that do not involve chemical reactions.
2. Processes that do not involve changes in phase of the system.

4.4.1 Internal energy (ΔU) in ideal gases

Consider the expansion of an ideal gas from some initial state with volume (V_1) and temperature (T_1) to some final state with volume (V_2) and temperature (T_2). Let's assume internal energy (U) is a function of T and V (i.e., $U = U(T, V)$). For ideal gases from section 4.2.1 we know that the particles of ideal gases do not interact with each other. This means the internal energy (U) does not depend on the distance between the particles themselves. Therefore internal energy (U) cannot be a function of V, which means that ΔU must only be a function of temperature (T), i.e., $U = U(T)$. Therefore we can find the internal energy under various conditions using the First law of thermodynamics ($\Delta U = Q + W$) as,

1. **ΔU for a constant volume processes:** Under constant volume processes ($\Delta V = 0$), and in temperature ranges where heat capacity at constant volume (C_V) is constant, the internal energy (ΔU) of an ideal gas is given as,

[2] Lingering issues: Going back to equation (4.21), our main result. In order to derive this result we made a number of assumptions. Are those assumptions only true for real gases? Strictly speaking, this derivation will break down when particles exert forces on each other, or when collisions are inelastic. Problems are also seen when the gas becomes dense. At these limits ideal gas law will break down and not hold.

$$\Delta U = Q_V = C_V \Delta T \tag{4.22}$$

where Q_V is the heat exchanged at constant volume.

2. **ΔU for a constant pressure processes:** Under constant pressure processes ($\Delta p = 0$) in the a temperature range over which heat capacity at constant pressure (C_p) is constant, the internal energy of an ideal gas is given as,

$$\Delta U = Q_p + p\Delta V \tag{4.23}$$

where Q_p is the heat exchanged at constant pressure and $p\Delta V$ is the compression–expansion or boundary work.

3. **ΔU for an isothermal processes:** Under isothermal processes ($\Delta T = 0$), $Q = -W = -p\Delta V$, and the internal energy of an ideal gas is given as

$$\Delta U = 0 \tag{4.24}$$

4. **ΔU for an adiabatic processes:** Under adiabatic processes, $Q = 0$ and the internal energy of an ideal gas is given as

$$\Delta U = W = C_p \Delta T \tag{4.25}$$

4.4.2 Heat capacity for an ideal gas

Consider the internal energy as a function of temperature and pressure, $U = U(T, V)$, and use the chain rule to evaluate how internal energy (ΔU) changes with respect to T and V as,

$$\Delta U(T, V) = \left(\frac{\partial U}{\partial T}\right)_V dT + \left(\frac{\partial U}{\partial V}\right)_T dV \tag{4.26}$$

From this equation and using equipartition[3] of energy we can calculate the molar heat capacity of an ideal gas. The measured heat capacity of a substance depends on the manner by which heat is added to the system. For a gas it can be done in two ways: one, when the volume is held constant; or two, when the pressure is held constant during the process.

1. **Heat capacity at constant volume (C_V):** Consider a piston and cylinder system that contains 'n' moles of a gas. The First law for a quasi-static process, with ($dW = -pdV$) is given as,

$$\Delta U = Q + W = Q - pdV \tag{4.27}$$

If we fix the piston at a certain distance in the cylinder such that the volume cannot change, then $dV = 0$ and the work done equals zero. The second term

[3] Equipartition theorem says that the energy of a molecule is shared equally on average among all the independent ways in which the molecules can absorb energy.

in equation (4.27) disappears ($\Delta U = Q$), so from equation (4.26) we can re-write heat (Q) as,

$$Q = \left(\frac{\partial U}{\partial T}\right)_V dT$$

which can be further rearranged to be,

$$\left(\frac{\partial Q}{\partial T}\right)_V = \left(\frac{\partial U}{\partial T}\right)_V \tag{4.28}$$

This equation (4.28) implies that any increase in energy is due to the transfer of energy in the form of heat only. We can therefore define the heat capacity for constant volume processes as,

$$C_V \equiv \left(\frac{\partial U}{\partial T}\right)_V \tag{4.29}$$

Ideal gases can be monoatomic, diatomic or polyatomic. In a monoatomic gas, atoms are not bound to each other. Thermodynamic properties of monatomic gases are simple when compared to polyatomic gases because monatomic gases do not have rotational or vibrational energy. Examples of monatomic gases include noble gases, such as: helium, neon, argon, krypton, radon, etc. Diatomic molecules have two atoms that may be the same element, e.g., hydrogen (H_2) or oxygen (O_2) or different elements, e.g., carbon monoxide (CO) or nitric oxide (NO). Molecules of polyatomic gases have internal variables associated with internal energy states, for example, the rotational state, the vibrational state, the electronic state and the nuclei state.

- *Heat capacity (C_V) for mono-atomic ideal gas:*In a mono-atomic ideal gas the heat capacity at a constant volume, is constant over a wide range of temperature and is very nearly equal to,

$$C_V \equiv \frac{\partial}{\partial T}\left(\frac{3}{2}NkT\right) = \frac{3}{2}Nk = \frac{3}{2}nR \tag{4.30}$$

  and the *heat capacity per mole for a monoatomic ideal gas* is $(C_V/n) = 1.5R = 12.5$ J mol^{-1} K^{-1}.
- *Heat capacity (C_V) for a diatomic ideal gas:*For a diatomic ideal gas the heat capacity at constant volumes is also nearly constant at ordinary temperatures,

$$C_V \equiv \frac{\partial}{\partial T}\left(\frac{5}{2}NkT\right) = \frac{5}{2}Nk = \frac{5}{2}nR \tag{4.31}$$

  and the *heat capacity per mole for a diatomic ideal gas* $(C_V/n) = 2.5R =$ 20.8 J mol^{-1} K^{-1}.

- *Heat capacity (C_V) for a polyatomic ideal gas:*In a polyatomic gas, such as CO_2, NH_3, the heat capacity at a constant volume, is also nearly constant at ordinary temperatures,

$$C_V \equiv \frac{\partial}{\partial T}(3NkT) = 3Nk = 3nR \tag{4.32}$$

Then *heat capacity per mole for a polyatomic ideal gas* will be $= 3R = 24.9\ \text{J mol}^{-1}\ \text{K}^{-1}$.

From these equations we note that the heat capacity is independent of the temperature and these results agree with experimental observations of low-density gases at reasonably high temperatures.

4.4.3 Heat capacity at constant pressure (C_P)

Figure 4.2 shows two isotherms[4] that differ in temperature by ΔT. From the figure, we can see points (1) and (2) are at the same volume, whereas points (2) and (3) have the same pressure value. Hence path (1–2) gives the constant volume process and path (2–3) gives the constant pressure process. Compression–expansion work done along the path (1–2) is zero and work done along path (2–3), for one mole of an ideal gas, is given as,

$$W_{12} = 0$$
$$W_{23} = -p\Delta V = -R\Delta T$$

We know from above that the internal energy of an ideal gas depends only on temperature ($U = U(T)$). Since both isotherms correspond to the same temperature change we can say that internal energy of both paths is the same. We can mathematically express this as,

$$\Delta U_{13} = \Delta U_{24}$$

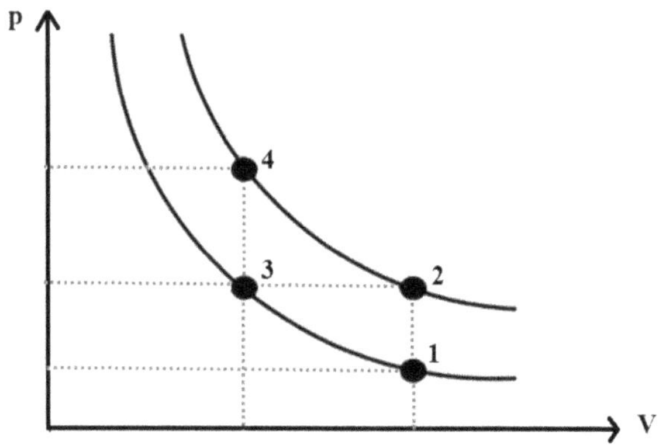

Figure 4.2. Isotherms

[4] Isotherms are a line on pV-graph connecting points that have the same temperature at any given time.

Additionally, we know that heat transferred in a constant pressure process in terms of molar heat capacity at constant pressure C_P is,

$$Q = C_P \Delta T \tag{4.33}$$

Using the expression 4.33, and $W = -R\Delta T$, we can rewrite the First law of thermodynamics as,

$$\begin{aligned} \Delta U_{13} &= Q + W \\ C_V \Delta T &= C_P \Delta T - R\Delta T \\ C_V &= C_P - R \end{aligned}$$

which can also be written as,

$$C_P = C_V + R \tag{4.34}$$

- *Heat capacity (C_P) for monoatomic ideal gas:*For a mono-atomic ideal gas the heat capacity at a constant pressure is given by,

$$C_p \equiv \frac{3}{2}R + R = \frac{5}{2}R \tag{4.35}$$

  and C_P per mole for monoatomic ideal gas is numerically equal to 20.8 J mol^{-1} K^{-1}.
- *Heat capacity (C_P) for diatomic ideal gas:*For a diatomic ideal gas the heat capacity at a constant pressure is almost constant at normal room temperatures, however, it increases steadily at higher temperatures.

$$C_p \equiv \frac{5}{2}R + R = \frac{7}{2}R \tag{4.36}$$

  and C_P per mole for diatomic ideal gas is numerically equal to 29.1 J mol^{-1} K^{-1}.
- *Heat capacity (C_P) for polyatomic ideal gas:*For polyatomic gases the heat capacities for both constant volume (C_V) and constant pressure (C_P) processes vary with temperature and with each gas used as the system. However, we can still approximate it as,

$$C_p \equiv 3R + R = 4R \tag{4.37}$$

  C_P per mole for polyatomic ideal gas is numerically equal to 33.3 J mol^{-1} K^{-1}.

4.4.4 Relationship between C_V and C_P

For an ideal gas, $(\frac{\partial U}{\partial T})$ is identical for both constant pressure and constant volume. The relationship between C_V and C_P is given as,

$$C_p = C_V + nR \tag{4.38}$$

where n is the number of moles in the gas, mathematically given as ($n = \frac{N}{N_A}$) and N_A is the Avogadro's number given as 6.022×10^{23} molecules per mole.

4.4.5 Specific heat ratio

Another parameter of interest for ideal gases is called the '*specific heat ratio*', denoted as γ, and mathematically given as,

$$\gamma = \frac{C_p}{C_V} \tag{4.39}$$

The specific heat ratio is also called the 'ratio of specific heats' or '***adiabatic exponent***'. Specific heat ratios (γ) and heat capacities per mole for various types of gases are given in table 4.1.

4.4.6 Enthalpy change (ΔH) in ideal gases

In practice, we cannot really measure the enthalpy of systems, we can only measure changes in enthalpy (ΔH). When enthalpy is positive (i.e., ΔH is greater than zero), the system absorbs heat and a chemical reaction where $\Delta H > 0$ is called an '*endothermic reaction*'. When enthalpy is negative (i.e., ΔH is less than zero), the system releases heat and a chemical reaction where $\Delta H < 0$ is called an '*exothermic reaction*'.

For ideal gases, we start with asking a question, how many variables are required to define the change in enthalpy (ΔH) for an ideal gas? To answer this question, lets start by redefining enthalpy ($H \equiv U + pV$) from chapter 3, equation (3.27) as a change in enthalpy (ΔH) as,

$$\Delta H = \Delta U(T) + \Delta(pV) \tag{4.40}$$

For an ideal gas we know the ideal gas law from equation (4.7) as ($pV = nRT$), and put this value in equation (4.40) to get,

$$\Delta H = \Delta U(T) + nR\Delta T \tag{4.41}$$

From this equation (4.41) we clearly see that H is a function of temperature (T) alone. Hence the answer to the question on how many variables are required to define the change in enthalpy for an ideal gas is that we only need one variable to define enthalpy (ΔH) for an ideal gas, which is the temperature, [$\Delta H = \Delta H(T)$].

Table 4.1. Properties of ideal gases per mole.

	γ	C_V J mol^{-1} K^{-1}	C_P J mol^{-1} K^{-1}
Monoatomic gas	$\gamma = 1.67$	12.5	20.8
Diatomic gas	$\gamma = 1.40$	20.8	29.1
Polyatomic gas	$\gamma = 1.33$	24.9	33.3

We can now find the ΔH for the various processes,

1. **ΔH for a constant volume process:** For constant volume processes ($\Delta V = 0$) over the range of temperatures over which C_V is constant, the change in enthalpy is given as,

$$\Delta H = Q_V + V\Delta P \tag{4.42}$$

   where Q_V is the heat exchanged at constant volume.

2. **ΔH for a constant pressure process:** For constant pressure processes ($\Delta p = 0$) over the range of temperatures over which C_P is constant, the change in enthalpy is given as,

$$\Delta H = C_p\Delta T \tag{4.43}$$

3. **ΔH for a isothermal process:** For isothermal processes ($\Delta T = 0$), the change in enthalpy is given as,

$$\Delta H = 0 \tag{4.44}$$

Worked example 4.3. Two tanks that are connected by a valve that can be opened or closed make up a system as shown in figure 4.3. One tank contains 5 kg of nitrous oxide (N_2O) at 70 °C and 1 atm. The other tank has 15 kg of N_2O at 20 °C and 1.2 atm (1.21×10^5 Pa). When the valve is opened gases mix with each other while the total amount of N_2O in the system does not change. Suppose the final equilibrium temperature is 40 °C and c_v of N_2O is 690 J kg^{-1} K^{-1}. Using the ideal gas model, determine,

1. The final equilibrium pressure.
2. The heat transfer for the process.

Solution:

1. The final equilibrium pressure.

   For tank 1 and 2, we can write the general gas equation as:

$$pV = \frac{m}{M}RT$$

   where m is the mass in [kg] and M is the molar mass in [kg kmol^{-1}]. Since we only have one type of gas we can ignore molar mass for these calculations. Then,

Figure 4.3. Example: Tanks A and B with valve assembly.

$$V_1 = \frac{m_{1i}RT_{1i}}{P_1} = 0.141 \text{ m}^3$$
$$V_2 = \frac{m_{2i}RT_{2i}}{P_2} = 0.301 \text{ m}^3$$

The final pressure can be found from ideal gas equation of state:

$$p_f = \frac{m_f RT_f}{V_1 + V_2}$$

Solving the above equation, with $m_{1i} = 5\text{kg}$, $m_{2i} = 15\text{kg}$, $m_f = 20\text{kg}$, $T_{1i} = 343$ K, $T_{2i} = 293$ K and $T_f = 313$ K, we see that the final pressure, $p_f = 1.17 \times 10^5$ Pa = 1.17 atm.

2. Heat transfer for the process.
   From First law we can write,

$$\Delta U = Q$$

since $W = 0$.

$$m_1 c_v(T_f - T_{i1}) + m_2 c_v(T_f - T_{i2}) = Q$$

Putting values and solving we see that $Q = 103.5$ kJ. The plus sign indicates that the heat transfer is into the system.

Concept questions 4.1

1. Why does the temperature in a rising cloud fall as it goes to higher altitudes?
2. What happens to the air temperature in a valley when cold air blowing from the mountain tops descends into the valley?
3. Why do we blow on a cup of tea with our lips puckered to cool it down quickly and blow on our hands with our mouth open to warm up our hands.

Exercises 4.1

1. 72 liters of an ideal gas is held at a pressure 3 atm and a temperature of 220 K; how many moles of gas are present in the container?
2. The best vacuum that has been achieved in labs corresponds to about 10^{-18} atm. Calculate the number of molecules per cm^3 present in the vacuum at room temperature (21 °C).
3. A balloon has 3 L of air at sea level with a pressure of 100 atm. Suppose the balloon rises in the atmosphere to a certain height and expands to volume of 10 L.
   What is the pressure in the balloon at the new volume?
   What is the temperature at the new height?

4.5 Various processes ideal gas can undergo

In this section we will discuss the various processes a system that is made up of an ideal gas can undergo.

4.5.1 Work done 'on' or 'by' an ideal gas

Recall from chapter 2, 'work' is defined as any other way of transferring energy in or out of a system other than spontaneous heat flow. One way to do work on a system is to compress it and reduce its volume, this is called 'compressional work'. Conversely, when a system expands, work is done on the environment and the work done is called 'expansion work'.

In particular, we are interested in the reversible compression–expansion work done by an ideal gas which proceeds slowly enough that the gas has time to respond to the change and the gas remains approximately in equilibrium. This is known as a *quasi-static compression.* Suppose we have an ideal gas enclosed in the usual cylinder-and-piston system. Work done *on* the ideal gas in this cylinder-and-piston system is mathematically given as,

$$W = -\int_{V_i}^{V_f} pdV \tag{4.45}$$

where the final volume of state is V_f and the initial volume of the system is V_i. Pressure (p) is a scalar quantity[5]. As the gas expands, $V_f > V_i$ and dV is positive, temperature of the gas decreases, hence the internal energy decreases, which implies the work done on the gas is negative. During a compression, $V_f < V_i$ hence dV is negative, temperature of the gas goes up, hence internal energy goes up and work done on the gas is considered positive. The magnitude of work done will be given by the area under the curve in a pV-curve as shown in figure 4.4.

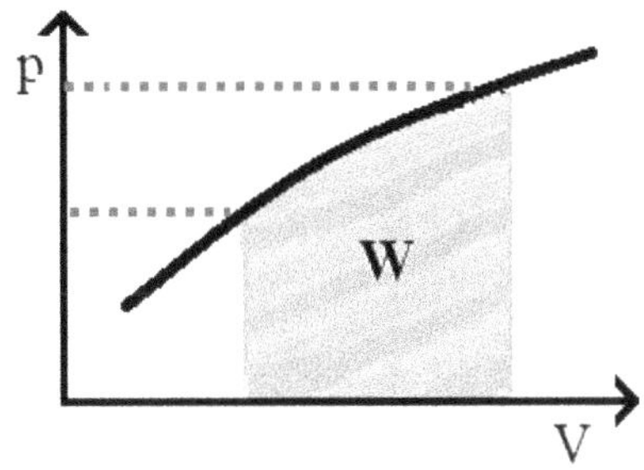

Figure 4.4. Magnitude of work done on the system is equal to the area under the curve.

[5] Please note, scalars can have negative values. Pressure can be negative as it occurs in liquids at least. That is how trees get water to their tops.

4.5.2 Ideal gas undergoing a cyclic process

Cyclic processes are those where the system returns to its initial state after a series of processes or operations. Since the process starts and ends at the same point the change in internal energy for the process is zero ($\Delta U = 0$). Then from the First law of thermodynamics (equation (2.29)) we can write,

$$0 = Q_{\text{total}} + W_{\text{total}} \tag{4.46}$$

where Q_{total} and W_{total} represent the total heat and work done in the complete cycle. From this equation (4.46) we see that when work (W) is positive, heat (Q) will be negative and vice versa. Please note that for processes done in a counterclockwise direction, work (W) will be positive and heat (Q) will be negative, whereas for cycles performed in clockwise direction (W) will be negative and (Q) will be positive. This is shown in figure 4.5.

4.5.3 Ideal gas undergoing a free-expansion process

In order to understand a free-expansion process, consider a thermally isolated box with a hatch in the middle that can be opened and closed. In a free-expansion process, initially the gas is supposed to be only on one side of the box with the hatch closed. Once the hatch is opened the gas moves freely and expands into the initially evacuated half of the box, as shown in figure 4.6.

For this process, $W = 0$, because no weights are lifted and $Q = 0$, since the box is thermally isolated from the surroundings. Hence from First law of thermodynamics we get,

$$\Delta U = 0 \tag{4.47}$$

This means the internal energy of an ideal gas under free-expansion remains constant and since internal energy depends on temperature, the temperature will also remain constant.

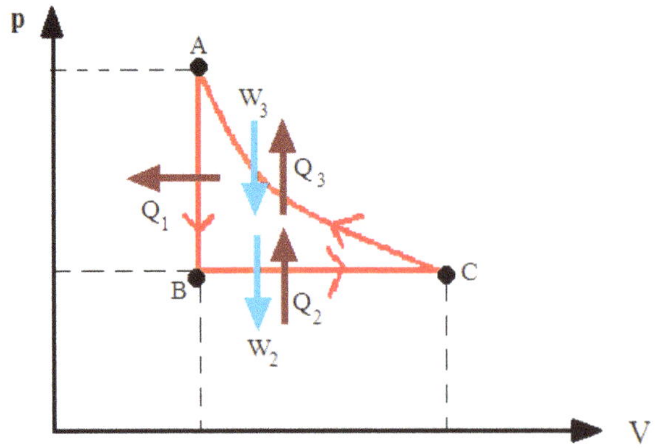

Figure 4.5. pV graph for a cyclic process. Path AB is a constant volume process, path BC is a constant pressure process and path AC represents an isothermal process.

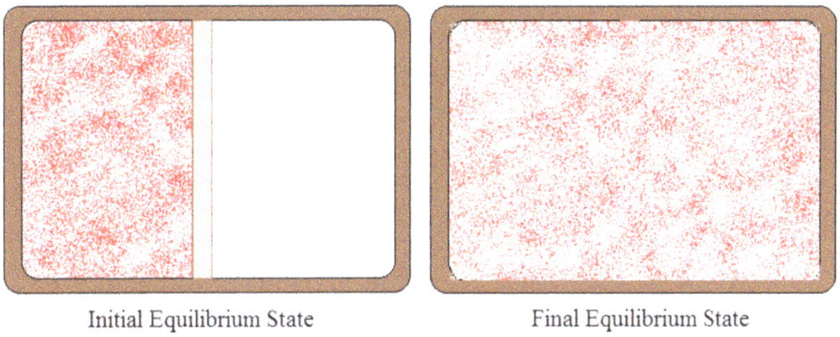

Figure 4.6. Free expansion.

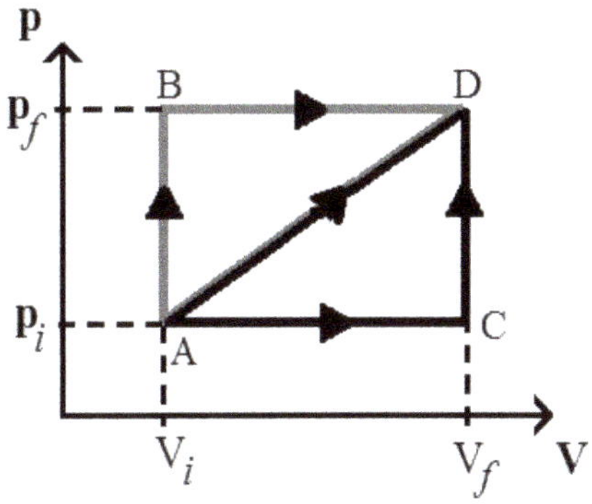

Figure 4.7. pV graph. Work done from point A to point D.

4.5.4 Ideal gas undergoing a constant volume process

When the volume of a gas remains constant, $\Delta V = 0$; which implies work = 0, as is seen in the vertical segments AB and CD in figure 4.7. From the First law of thermodynamics (equation (2.29)) we get:

$$\Delta U = Q \tag{4.48}$$

This means all the heat that enters the ideal gas will be stored as internal energy.

4.5.5 Ideal gas undergoing a constant pressure process

When the pressure of a gas remains constant during the transformation, equation (4.45) gives us:

$$W = -\int_{V_\mathrm{i}}^{V_\mathrm{f}} pdV = -p\int_{V_\mathrm{i}}^{V_\mathrm{f}} dV = -p(V_\mathrm{f} - V_\mathrm{i}) \tag{4.49}$$

Examples of this are segments AC and BD in figure 4.7. Please note that work done *on* the gas will be negative in both of these segments because volume increases in both processes. From the equation of the First law of thermodynamics (equation (2.29)) we get:

$$\Delta U = Q + W \tag{4.50}$$

4.5.6 Ideal gas undergoing an isothermal process

In an isothermal process, temperature (ΔT), hence the internal energy (ΔU) of the system remains constant. In general, heat must flow in or out of the system to keep the temperature of the system constant. This is achieved by keeping the system in thermal contact with a '*heat reservoir*' which allows the system to remain in thermal equilibrium with the environment. Typically, isothermal processes require the process to proceed slowly enough so the necessary heat exchanges can take place. In an isothermal process, $\Delta T = 0$, which implies $\Delta U = 0$ and from the equation of the First law of thermodynamics (equation (2.29)) we can write equation (4.51). This is represented by the curve on a pV diagram as shown in figure 4.8.

$$0 = Q + W \tag{4.51}$$

For compressional work, the work is done on the gas ($W > 0$), which implies an equivalent amount of heat (Q) is released by the gas to the environment to keep $\Delta T = 0$ and the work done on the gas will not be stored as internal energy of the system. In an isothermal process, pressure is a function of volume ($p = p(V)$) and the ideal gas law can be written as:

$$p = p(V) = \frac{NkT}{V}$$

and the work done on the gas ($W = -\int pdV$) will be given as

$$\begin{aligned} W &= -\ NkT \int_{V_1}^{V_2} \frac{dV}{V} \\ &= -\ NkT\ \ln V|_{V_1}^{V_2} \end{aligned}$$

Hence the work done on the gas will be given as

$$W = -NkT\ \ln\frac{V_2}{V_1} \tag{4.52}$$

4.5.7 Ideal gas undergoing an adiabatic process

In an adiabatic process (compression or expansion), no heat flows in or out of the gas, so the (compression) work is responsible for all of the changes in internal energy. That means $dQ = 0$ which implies from First law of thermodynamics

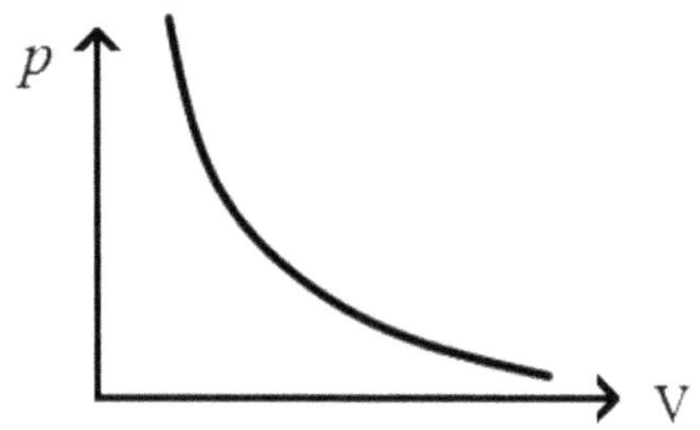

Figure 4.8. Isotherm, a curve on a pV diagram for an isothermal process.

$dU = dW = -pdV$. The pV graph is shown in figure 4.9. Please note that the isotherms are shown in figure for comparison.

Physically, an adiabatic process can occur under the following conditions:

1. When the system is insulated and unable to exchange heat with its environment, or
2. When the process occurs quickly enough for no heat exchange to take place (but slowly enough to still be quasi-static).

Adiabatic changes are usually accompanied by changes in temperature. An example of an adiabatic process is the vertical flow of air in the Earth's atmosphere. Air will expand as it rises and hence will cool, whereas the opposite will happen on its way down.

For an adiabatic process, the system has to be insulated from the surroundings such that no heat enters or leaves that system ($Q = 0$). Then the First law of thermodynamics (equation (2.29)) becomes:

$$\Delta U = W \tag{4.53}$$

and for ideal gas, $\Delta U = nC_V\Delta T$. So for small changes equation (4.53) becomes,

$$\begin{aligned} dW &= nC_V dT \\ -pdV &= nC_V dT \end{aligned} \tag{4.54}$$

1. **Relation between p and V for an adiabatic process**

   To find the relation between p and V for an adiabatic process we assume that the process is carried out in such a way that the pressure is always well-defined. For an ideal gas the ideal gas law can be written in differential form,

$$\begin{aligned} d(pV) &= d(nRT) \\ pdV + Vdp &= nRdT \\ Vdp &= -pdV + nRdT \\ Vdp &= dW + nRdT \\ Vdp &= nC_V dT + nRdT \\ Vdp &= n(C_V + R)dT \end{aligned}$$

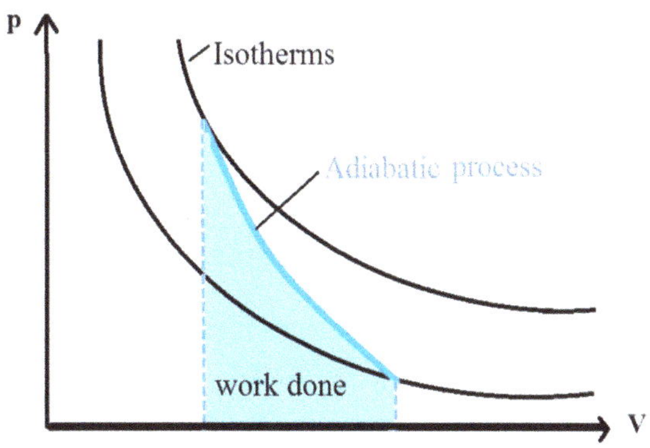

Figure 4.9. pV graph for an adiabatic process. The work done during the process is the area under the curve between the starting and ending volumes.

From equation (4.34), $C_V + R = C_p$, hence the above equation becomes,

$$Vdp = nC_p dT \tag{4.55}$$

If we take the ratio of equations (4.55) and (4.54), we get

$$\frac{Vdp}{-pdV} = \frac{nC_p dT}{nC_V dT}$$

Recognizing that $\frac{C_p}{C_V} = \gamma$, as the specific heat ratio from section 4.4.5 we can rewrite the above equation as,

$$\frac{dp}{p} = \gamma \frac{-dV}{V}$$

Integrating and simplifying we get,

$$\ln\left(\frac{p_f}{p_i}\right) = -\gamma \ln\left(\frac{V_2}{V_1}\right)$$

$$\frac{p_f}{p_i} = \left(\frac{V_2}{V_1}\right)^{\gamma}$$

where V_2 represents the volume of state 2 and V_1 represents the volume of state 1. This implies that for adiabatic compression we can say,

$$pV^{\gamma} = \text{constant} \tag{4.56}$$

where γ is called the '*adiabatic exponent*'. This expression gives the relationship between the pressure and the volume of an ideal gas undergoing an adiabatic process. Please note equation (4.56) defines a family of curves on the pV diagram in such a way that every adiabatic process can be represented as a segment of one of these curves.

2. **Relation between p and T for adiabatic process**

   A common observation is that the temperature of a gas increases as you compress it, whereas the temperature of a gas lowers as you expand it. This implies if we compress a gas via an adiabatic compression then $V_i > V_f$ hence the temperature $T_f > T_i$. To find the relation between p and T, start from equation (4.56), for the adiabatic compression,

$$PV^{\gamma} = \text{constant}$$

From the ideal gas law, $[PV = nRT]$ we can say ($V = \text{const}\frac{T}{p}$) if we ignore the constants n and R, and the above equation becomes,

$$P\left(\frac{T}{P}\right)^{\gamma} = \text{constant}$$

$$P^{1-\gamma}T^{\gamma} = \text{constant}$$

$$P_2^{1-\gamma}T_2^{\gamma} = P_1^{1-\gamma}T_1^{\gamma}$$

$$T_2^{\gamma} = T_1^{\gamma}\left(\frac{P_1}{P_2}\right)^{1-\gamma}$$

where P_2 represents the pressure of state 2 and P_1 represents the pressure of state 1.

Rearrange and simplify to get:

$$\frac{T_2}{T_1} = \left(\frac{P_1}{P_2}\right)^{\frac{1-\gamma}{\gamma}} \tag{4.57}$$

Worked example 4.4. An insulated cylinder is fitted with a piston and it contains 0.11 mol of oxygen at room temperature. The gas is compressed from 4 m^3 to 1 m^3 while its pressure increases from 10 to 40 Pa. Compare the work done along the three paths shown in figure 4.10.

Solution: We will consider the various paths individually as follows:

1. **Path 1, [A–C–B]:** In this path, in segment A–C, pressure is constant, hence $\Delta p = 0$, and in segment C–B, volume is constant, hence $\Delta V = 0$;

$$W_{AC} = -P_i(V_f - V_i) = -(10\text{ Pa})(1 - 4)\text{ m}^3 = -30\text{ J}$$
$$W_{CB} = 0$$
$$W_{total} = W_{AC} + W_{CB} = -30\text{ J}$$

2. **Path 2, [A–B]:** From the figure 4.10 we see this is an isotherm. Hence,

$$W_{AB} = -P_i V_i \ln\frac{V_f}{V_i} = -(10\text{ Pa})(4\text{ m}^3)\ln\left(\frac{1}{4}\right) = -55\text{ J}$$

3. **Path 3, [A–D–B]:** In this path, in segment A–D, volume is constant, hence $\Delta V = 0$, and in segment D–B, pressure is constant, hence $\Delta P = 0$;

$$W_{AD} = 0$$
$$W_{DB} = -P_f(V_f - V_i) = -(40\text{ Pa})(1 - 4)\text{ m}^3 = -120\text{ J}$$
$$W_{total} = W_{AD} + W_{DB} = -120\text{ J}$$

It is important to note that in all three paths work is negative and the magnitudes increase according to the area under each path on the pV-graph.

Worked example 4.5. Suppose a system is made up of 2.5 moles of an ideal gas for which $c_V = 20.8$ J mol^{-1} K^{-1} is taken through the cycle, as shown in figure 4.11

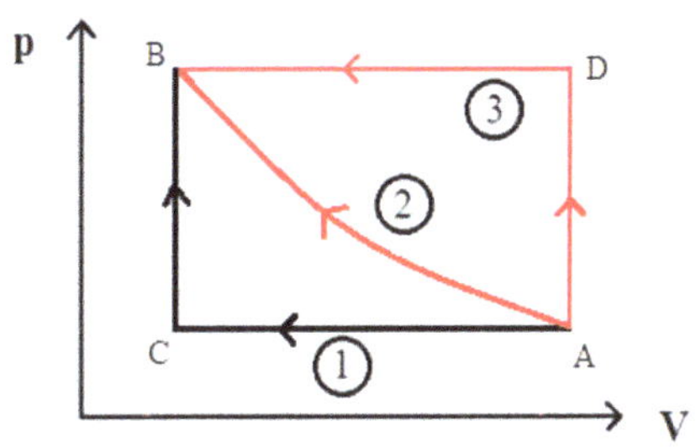

Figure 4.10. pV graph. Work done along the three paths 1 [A–C–B], path 2 [A–B], and path 3 [A–D–B].

following the path of the arrows. The path (point 3 to point 1) corresponds to a curve given by ideal gas equation, where $T = T_1 = T_3$. Calculate q, W, ΔU and ΔH for each segment and the total cycle.

Solution: Before we begin we must find quantities that we do not have based on the provided data. We get the following values from figure 4.11,

Volume	Pressure
$V_1 = 1\ \text{L} = 0.001\ \text{m}$	$P_1 = 16.6 \times 10^5\ \text{N m}^{-2}$
$V_2 = 25\ \text{L} = 0.025\ \text{m}$	$P_2 = 16.6 \times 10^5\ \text{N m}^{-2}$
$V_3 = 25\ \text{L} = 0.025\ \text{m}$	$P_3 = 0.6 \times 10^5\ \text{N m}^{-2}$

Using the ideal gas law ($pV = nRT$), we can find $T_1 = 77.9$ K, $T_2 = 1996$ K and $P_3 = 63410.6\ \text{N m}^{-2}$. We will consider the various paths individually as follows:

1. **Step [1–2]:** During this step the pressure is constant, hence $\Delta p = 0$.
   Internal energy: $\Delta U_{12} = nc_V\Delta T = nc_V(T_2 - T_1) = 96$ kJ.
   Work done: $W_{12} = -p_{12}(V_2 - V_1) = -38.4$ kJ.
   Heat: $q_{12} = \Delta U - W = 134.4$ kJ
   Enthalpy: $\Delta H_{12} = \Delta U + nR(T_2 - T_1) = 134.4$ kJ
2. **Step [2–3]:** During this step the volume is constant, hence $\Delta V = 0$.
   Work done: $W_{23} = 0$.
   Internal energy: $\Delta U_{23} = nc_V\Delta T = nc_V(T_3 - T_2) = -96$ kJ.
   Enthalpy: $\Delta H_{23} = \Delta U + nR(T_3 - T_2) = -134.4$ kJ
   Heat: $q_{23} = \Delta U - W = -96$ kJ
3. **Step [3–1]:** During this step the temperature is constant, hence $\Delta T = 0$.
   Internal energy: $\Delta U_{31} = 0$
   Enthalpy: $\Delta H_{31} = 0$
   Work done: $W_{31} = nRT\ \ln\frac{V_2}{V_1} = 5.15$ kJ.
   Heat: $q_{31} = \Delta U - W = -5.15$ kJ

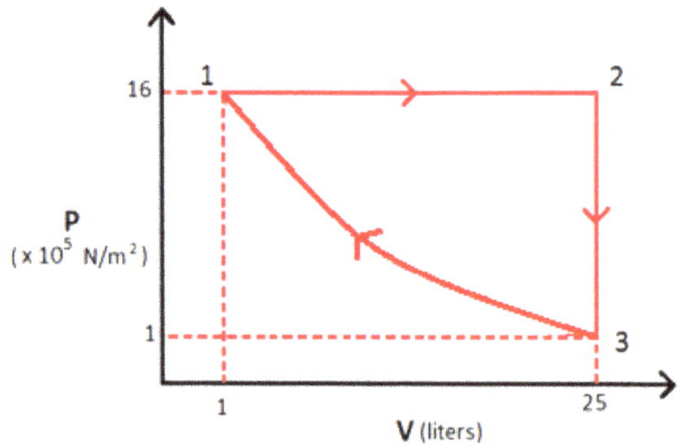

Figure 4.11. Example: Total work performed during the cycle.

4.6 Summary

Things to keep in mind when solving problems:

1. For an ideal gas, in general,

$$\text{work} = -\int P_{\text{ext}} dV \tag{4.58}$$

2. For reversible processes:

$$\text{work} = -\int \frac{nRT}{V} dV \tag{4.59}$$

3. Constant volume process, $W = 0$.
4. Constant pressure process,

$$\text{work} = -p_{\text{ext}}(V_{\text{f}} - V_{\text{i}}) \tag{4.60}$$

5. Reversible and isothermal processes: $\Delta U = \Delta H = 0$, because ΔU and ΔH depend on T only.

$$\text{work} = -nRT \int \frac{1}{V} dV \tag{4.61}$$

$$\text{work} = -nRTln\left(\frac{V_2}{V_1}\right) \tag{4.62}$$

6. Adiabatic process, $Q = 0$.

$$p_{\text{f}} V_{\text{f}}^{\gamma} = p_{\text{i}} V_{\text{i}}^{\gamma} \tag{4.63}$$

$$T_{\text{f}} = T_{\text{i}}\left(\frac{V_1}{V_2}\right)^{\gamma-1} \tag{4.64}$$

7. For ideal gases: $C_p = C_v + nR$, so if we know C_v or C_p we can find the other quantities.

Process			First law	ΔU	Work	Heat (Q)
Adiabatic	$Q = 0$	$pV^{\gamma} = \text{const}$	$\Delta U = W$	$nC_V\Delta T$	$W = \frac{1}{\gamma-1}(p_2V_2 - p_1V_1)$	0
Isochoric	const V	$P_1T_2 = P_2T_1$	$\Delta U = Q$	$nC_V\Delta T$		$nC_V\Delta T$
Isobaric	const P	$V_1T_2 = V_2T_1$	$\Delta U = Q + W$	$nC_V\Delta T$	$W = -P\Delta V = -nR\Delta T$	$nC_p\Delta T$
Isothermal	const T	$P_1V_1 = P_2V_2$	$Q = -W$	$nC_V\Delta T$	$W = -nRT \ln \frac{V_2}{V_1}$	
Cyclic			$Q = -W$	$nC_V\Delta T$		

Concept questions 4.2

1. When work is dissipated into internal energy, the change in disorderly motion of molecules decreases/increases/remains the same? Explain in your own words.
2. Why do adiabatic processes proceed fast while isothermal processes are slow?
3. Which type of ideal gas will have the smallest value of $(C_P - C_V)$? Explain in your own words.
   (a) Monoatomic
   (b) Diatomic
   (c) Polyatomic
   (d) the value will be the same for all.
4. Explain why it is not a good idea to lick a metal stop sign at the side of the road when it is snowing outside. How do you release yourself without hurting yourself?

Exercises 4.2

1. Starting from the First law of thermodynamics, prove that the relation between temperature (T) and volume (V) for an ideal gas with constant heat capacities undergoing a reversible, adiabatic compression or expansion is:

$$\frac{T_2}{T_1} = \left(\frac{V_1}{V_2}\right)^{\gamma-1} \tag{4.65}$$

2. Develop an analogous expression to relate temperature (T) and pressure (P).

$$\frac{T_2}{T_1} = \left(\frac{P_2}{P_1}\right)^{\frac{\gamma-1}{\gamma}}$$

3. Find the temperature after compression for the following cases:
   (a) A monoatomic ideal gas initially at temperature 20° is suddenly compressed to one tenth its original volume.
   (b) A diatomic ideal gas initially at temperature 20° is suddenly compressed to one tenth its original volume.

Further reading

- Cengel Y and Boles M 2006 *Thermodynamics: An Engineering Approach* 5th edn (New York: McGraw-Hill Higher Education)
- Resnick R, Halliday D and Krane K 2002 *Physics* vol 1 5th edn (New York: Wiley)
- Schroeder D 2000 *Thermal Physics* (Reading, MA: Addison-Wesley Longman)
- Sears F and Salinger G 1975 *Thermodynamics, Kinetic Theory and Statistical Thermodynamics* 3rd edn (Reading, MA: Addison-Wesley Publishing Company)

IOP Publishing

Thermodynamics and Statistical Mechanics
An introduction for physicists and engineers
Samya Zain

Chapter 5

Second and Third laws of thermodynamics

5.1 The Second law of thermodynamics

In this chapter we will talk about the Second and Third laws of thermodynamics. You may have studied these from a traditional point of view, which is okay. But now let's try a little different approach to get to the Second law of thermodynamics.

You may be asking, we already have the Zeroth and the First law of thermodynamics, then why do we need a Second or a Third law. The answer to this question is very important and needs more consideration. You and I know from our real life experience that some processes do not happen in real life. For example, a dropped stone always falls toward the Earth, it never stays in the air from where you dropped it, neither does it float away during its way to the floor, unless you are watching a magic trick!

To explain the Second law of thermodynamics, let me start with a story of a peddler man.

5.1.1 Story of a peddler man

Imagine you live out on the prairie in the 1800s. There is no electricity, no running water and no wifi. Life is hard and you only have limited funds and you need them to buy food. Imagine one day a peddler man, like the one shown in figure 5.1 comes to your door and wants to sell you a magic box, like the one in figure 5.2. The peddler man calls it a '*majik box*'. He says that this 'majik box' takes in compressed air from the top and gives you hot air from the pipe on the right side and cold air from the pipe on the left side. So in the winter the '*majik box*' will heat your rooms and in the summer it will cool your rooms and all you have to do is supply the compressed air. You seem to have an unlimited amount of compressed air in your shed, so that will not be a problem.

However, the peddler man has two conditions for sale: one, you cannot ever open the box because if you do it will take the magic away; and two, you cannot touch the

doi:10.1088/978-0-7503-3083-1ch5

Figure 5.1. The peddler man with his '*majik box*'. [Courtesy: M Humayun Rauf, Lahore, 2020.]

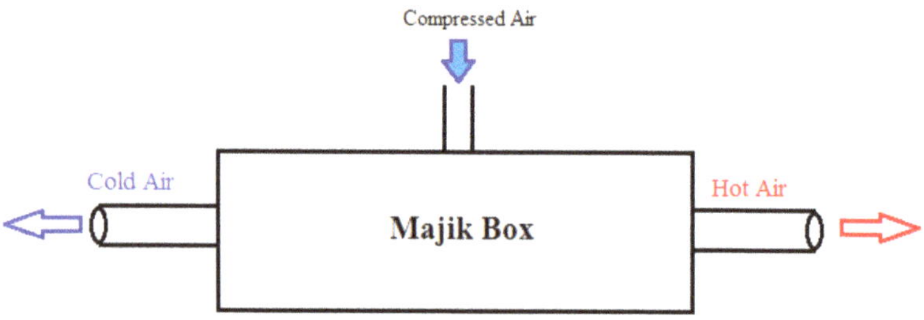

Figure 5.2. Peddler man's majik box.

'*majik box*' until you buy it. It seems too good to be true; the question is, do you trust the peddler man and buy it with the little money you have?

Luckily, you are a physics major and you know there must be a way to figure out if the processes claimed by the peddler man can actually happen. What do you know from your previous knowledge and experience to figure out if the '*majik box*' is real or not? To find a meaningful answer to this question let's try a few common experimental facts:

1. You know that 'all engines have moving parts'. This '*majik box*' has no moving parts. Therefore, the '*majik box*' is not an engine.
2. You also know that 'heat cannot flow from cold to hot objects without some external help'. The '*majik box*' is insulated so no heat flows. Additionally, there is no obvious connection between this statement and the accomplishments the peddler man claims the '*majik box*' to have.

3. You also know that 'all processes left to their own devices will always go toward the equilibrium state'. The question is, which one of the possible states is the actual equilibrium state of the '*majik box*'? The hot air stream, the cold air stream or the initial regular compressed air? This is a question we are not sure about.

What we do know is that all of the statements above are absolutely true. But how are they relevant to our current dilemma, is the '*majik box*' real or not? Which of these statements will you use to make an educated guess on whether or not to spend your hard earned money on the '*majik box*'. Additionally, you might be wondering why am I talking about these as my launching point for my definition of the Second law of thermodynamics? Well, it is obvious that the above statements are qualitative statements and do not appear to be useful for now. So for the moment, let's look for a quantitative solution to our problem, which is to figure out the validity of the '*majik box*'. The quantitative solution means that we should be looking for a '*property*' of state. The property we are looking for has to be something we should not be able to measure directly, since the peddler man will not let us open the box to look inside of it. However, we guess that this property must be a function of other variables. Let's try to find this property.

To find such a property, we can carry out a series of controlled experiments and make as many measurements as we can without touching the '*majik box*'. We can then examine the data to see if we can reach some logical conclusions. The simplest setup, as always, would be a piston-and-cylinder system with an enclosed ideal gas. We can start with a fixed amount of gas in the piston-and-cylinder system at some initial temperature (T_1) and initial pressure (P_1) and end at a final state with temperature (T_2) and pressure (P_2). However, we change the experiment each time such that the path taken from initial state to final state is different. This can be easily done by changing the amounts and relative rates of heat and work added or extracted from the system.

We find, as always, that the only processes that provide precise and detailed measurements are the ones that are carried out slowly (isothermally) and reversibly or very nearly reversibly. After much experimentation, many observations and data collection we get the following results:

1. Change in volume (ΔV) is the same for all possible paths, since initial (V_1) and final volume (V_2) are the same for all processes. Hence volume (V) is a property of the system and is a function of T and p. To put it more clearly we can say that an equation of state exists that connects p, V and T for an ideal gas. We already knew that, but this does give us confidence that we are on the right path.
2. We also note that even though heat (Q) and work (W) may be different for each process, the sum ($Q + W$) is the same for all processes. This means ($Q + W$) must also be a measure of a property that changes between initial and final state. We know that this property is the change in internal energy from initial to final state, $\Delta U = U_2 - U_1$. Eureka, we have just '*rediscovered the First law*

of thermodynamics'! Again we may not have learnt anything new but we gain more confidence that our method produces results that we know to be true.

3. So next we plot (T versus p), (T versus V), (p versus V), (T versus Q), (T versus W), (p versus Q), (p versus W), etc and examine the plots for clues about the existence of 'the property of state' we are looking for. We know that whenever we find a quantity that is same for all processes between two fixed states, the quantity is independent of the path and must depend only on the initial and final state of the system. Think back to the case of work done in a conservative field, which is also independent of the path taken.

   Without going into much detail, I will tell you that you will find the property of state when you find the particular plot that gives curves which have the same area under the curve. However, all the plots we have so far do not seem to give a particular property that gives the same area under the curve and hence we conclude that we did not find a 'state property' from the many plots given above.

   Never one to give up, and with your hard-earned money on the line, you next plot ($1/p$ versus W), and note that all of these plots have the same area under the curve. Finally, we have found what we were looking for. The same area under the curves means that $\left(\int_0^W \frac{1}{p} dW\right)$ is the same for all processes, yay! You decide to give this property a name, and call it '*Xtropy*' (ΔX) and write it as,

$$\int_0^W \frac{1}{p} dW = \Delta X \tag{5.1}$$

   However, you note once you have written this equation on paper that you have seen this before; ΔX is nothing other than ΔV, which reduces equation (5.1) to,

$$\int_0^W \frac{1}{p} dW = \Delta V \tag{5.2}$$

   or

$$W = p\Delta V \tag{5.3}$$

   Again this is nothing new, '*Xtropy*' (ΔX) is the boundary work done on a gas. It is something we knew all along.

4. So after all this you still have not discovered anything new. However, these experiments do give you an idea. Since T is related to Q in the same way as p is related to W, you guess, it might be interesting to plot ($1/T$ versus Q) for the various experiments. When you do that, you see experimentally that the area under the curves you get is the same. So you conclude that $\left(\int_0^Q (1/T) dQ\right)$ is the same for all processes and must be a measure of a property of state of the system. You next decide to call this state property

'*Ntropy*' or as it is more commonly known '*Entropy*' (S) of a system, defined as:

$$\int_0^Q \frac{1}{T} dQ = \Delta S$$

or

$$Q = T\Delta S \tag{5.4}$$

So finally you have found something new, '*entropy*' is really new and we know very little about it except: one, it is an equation that tells us how to calculate '*changes*'; two, it is a property of the material in our piston-and-cylinder system; and three, it is a function of temperature (T) and heat (Q) of our system. Basically, we know as much about entropy as we do about energy, which if you think about it, was as much as we knew about the sugar cubes and lumps from chapter 2! That is, very little!

The difference is that energy has several forms, kinetic, potential, thermal, etc which may be the reason that energy seems less abstract, whereas entropy is just entropy. Hence we now have a new property of state, that may or may not be helpful in determining if we want to buy the '*majik box*' from the peddler man. Before we answer this question let's concentrate on our new-found property of state and define it properly.

5.1.2 Visualizing entropy

Since entropy is a really new and strange property of state it is at times helpful to visualize entropy as a fluid, having the following properties:

1. Entropy can change form and move around. Imagine whenever energy enters or leaves an object in the form of heat, it is required by '*some law*' to carry some entropy with it. Mathematically, change in entropy (ΔS) is given by the amount (Q/T)
2. Entropy is only half conserved. Entropy can never be destroyed, however it is '*created*' whenever heat flows between objects at different temperatures. The only time that entropy is not created is when the objects are in thermal equilibrium and there is no temperature difference between them. In this limit, however, there is no tendency of heat to flow in the first place.

It is important to remember that primarily the net increase in entropy is the driving force behind the flow of heat and the allowed direction of processes in thermodynamics. Fundamentally, though, entropy is not a fluid at all and this model is '*simply wrong*', just like the '*heat-fluid*' theory was! But it is there to help us visualize, like the equator line on a world map.

5.2 Officially defining entropy

There are two related definitions of entropy, one in terms of thermodynamics and the other in terms of statistical mechanics.

1. **Macroscopic approach:** The macroscopic approach, also called the classical approach, involves the transfer of heat and the temperature at which the transfer happens. The classical thermodynamics definition of entropy developed first in the course of history.
2. **Microscopic approach:** The microscopic approach, also called the statistical approach, involves counting the ways in which atoms or molecules that make up the system can be arranged.

In this chapter we focus on the classical thermodynamic definition and leave the statistical definition of entropy till chapters 11 and 12.

5.2.1 Entropy principle

We know that all naturally occurring processes go in one direction and do not naturally go the other way. For example, a hot cup of tea sitting on the kitchen counter will get cold but will never get hot on its own; similarly ice cubes taken out of a freezer and put on the same counter will melt but never refreeze on their own, even when both processes are occurring at the same time. These processes are called '*irreversible*' or 'one-way' processes.

It is worth noting that the First law of thermodynamics places no restriction on the direction of a process, but satisfying the First law does not ensure that the process actually occurs. This is a limitation of the First law, it does not tell us whether a process can or cannot take place. We remove this inadequacy to the First law by introducing the Second law of thermodynamics. In a nutshell, a process cannot occur unless it satisfies both the First and Second laws of thermodynamics.

Entropy principle states:

> '*If an irreversible process occurs in a closed system, the entropy of the system always increases*'.

In section 5.1.1, we said we know as much about entropy as we know about energy. This is not quite true; we know energy follows the law of conservation of energy which says no matter what changes occur within a closed system, the energy of that system remains constant. Entropy, on the other hand, does not obey any conservation laws, there is no law of conservation of entropy. What we do know is that for any irreversible process the entropy of the system always increases and it is not the energy of the system that controls the direction of irreversible processes, it is the '***entropy***' (S) of the system that controls the direction of irreversible processes.

5.2.2 Change in entropy of a system

In the First law of thermodynamics, we focused on the change in energy (ΔU) instead of talking about the total energy of a system (U). Similarly, here we will focus on the changes in entropy (ΔS) of a system rather than talking about total entropy (S) of the system.

For irreversible processes, $\Delta S > 0$ always, which implies only forward processes are allowed. If backward processes were allowed, say reformation of an egg after it has been scrambled, then $\Delta S < 0$ and this would violate the entropy principle.

In a reversible process[1], say a system consisting of a metal plate and a heat source (at temperature, T), when you increase the temperature of the heat source by a small amount (dT) the Zeroth law of thermodynamics says that an amount of heat (Q) will be added from the heat source to the metal plate. If we decrease the temperature of the heat source by an amount (dT) an equal amount of heat (Q) will need to be transferred back to the heat source from the metal plate. Both of these steps restore the system to its original state. The process is therefore called a '*reversible*' process. The change in entropy of such a reversible system is defined as:

$$\Delta S \equiv \int_{\mathrm{i}}^{f} \frac{Q}{T} \tag{5.5}$$

where (Q) is the small amount of heat energy transferred in or out of the closed system and T is the temperature measured in kelvin (K). From this definition, the SI units of the change in entropy (ΔS) is joule per kelvin.

We know that temperature is measured in kelvin and is always positive, hence the algebraic sign of (ΔS) depends on the sign of heat (Q). When heat is added (reversibly) to an object the entropy of the object increases ($\Delta S > 0$); conversely if heat is removed from an object the entropy of the object decreases ($\Delta S < 0$).

1. **Reversible, isothermal process:** In a reversible isothermal process, the temperature of an object remains constant as heat is added to the object. The change in entropy of the system (ΔS) will be given as,

$$\Delta S = \frac{Q}{T} \tag{5.6}$$

An example of such a process is a phase change. When temperature (T) of the system is changing it is usually more convenient to write the change in entropy of the system (ΔS) in terms of heat capacity at constant volume as,

$$\Delta S = \frac{C_V dT}{T} \tag{5.7}$$

2. **Reversible, isochoric process:** In a reversible isochoric process, volume is kept constant and the system does no other form of work apart from compression–expansion work. In this case the change in entropy (ΔS) is given as,

$$\Delta S = \frac{Q}{T} = \frac{dU}{T} \tag{5.8}$$

Since heat and temperature are usually pretty easy to measure, this equation allows us to compute the change in entropy for multiple processes.

[1] A reversible process is a process in which by making small changes in the system and its environment we can return the system to its original state.

3. **Reversible, adiabatic process:** In a reversible adiabatic process, no heat enters or leaves the system ($Q = 0$), so the change in entropy (ΔS) is zero.

$$\Delta S_{\text{total}} = 0 \tag{5.9}$$

These processes occur at constant entropy and are called '*isentropic processes*'.

There are a number of ways to increase the entropy of a system. We can increase the entropy of a system by adding particles or adding energy or by increasing the volume of container. We can also add entropy by decomposing the molecules into their component forms.

Worked example 5.1. An insulating cylinder that has 2 kg of water is placed on a hot plate. Initially both the water and hot plate are at room temperature (20 °C). Then the temperature of the hot plate is raised slowly to 100 °C and the water begins to boil. What is the entropy change of water during this process?

Solution: This is a reversible process in which the water and hot plate are in thermal equilibrium at all times. This means that as we lower the temperature of the hotplate the temperature of water also goes down. If we choose the water as our system we can write the change in entropy as,

$$\Delta S = \int_{\mathrm{T_i}}^{T_\mathrm{f}} \frac{Q}{T}$$

We know $Q = mcdT$, where m is the mass of water, c is the specific heat of water and $c = \frac{C}{m}$.

$$\Delta S = mc \int_{\mathrm{T_i}}^{T_\mathrm{f}} \frac{1}{T} dT = mc \ln \frac{T_\mathrm{f}}{T_\mathrm{i}}$$

Putting numerical values, $m = 2$ kg; $T_\mathrm{f} = 100 + 273 = 373$ K; $T_\mathrm{i} = 20 + 273 = 293$ K; $c = 4186$ J kg^{-1} K^{-1}; in the equation we find that ΔS is numerically equal to 1820 J K^{-1}.

5.2.3 Macroscopic view of entropy of the system

Rudolf Clausius defined entropy to be '*what*' increases by a factor (Q/T) whenever heat (Q) enters a system at temperature (T). Please keep in mind that while this definition does not tell us what entropy '*is*', it is still sufficient for many purposes, especially when we are not concerned with the microscopic makeup of the system. As we saw in the last section, change in entropy is formally defined as,

$$\Delta S = \frac{Q}{T} \tag{5.10}$$

To understand this better, lets consider what happens when Object A (at temperature $T_1 = 50$ °C $= 323$ K) and Object B (at temperature $T_2 = 20$ °C $= 293$ K) are put in thermal contact with each other. Experimental observation tells us that heat flows from object A (at higher temperature) to object B (at the lower temperature). Suppose during a certain time interval (Δt), 2000 J of heat flows from object A to B. Then the change in entropy for each object is:

$$\begin{aligned} &\text{For Object A: } \Delta S_A = -\frac{2000 \text{ J}}{323 \text{ K}} = -6.2 \text{ J K}^{-1} \\ &\text{For Object B: } \Delta S_B = +\frac{2000 \text{ J}}{293 \text{ K}} = +6.8 \text{ J K}^{-1} \end{aligned} \tag{5.11}$$

This equation shows that object A loses entropy since the heat flows out of it, whereas object B gains entropy because heat flows in. But in section 5.2.2 we said that entropy of a system always increases. What is happening here? Object A loses entropy, can that be true? To understand this it is important to remember that it is the total change in entropy of the system that counts and in this case the total change in entropy is [$-6.2 + 6.8 = +0.6$ J K^{-1}]. That is, the total entropy of the system increases even when the entropy for object A goes down, as it should be. It is important to remember that although entropy may decrease in a part of a closed system, there will '*always*' be an equal (or larger) entropy increase in another part of that system so the entropy of the system as a whole never decreases.

5.2.4 Second law of thermodynamics

Equation (5.11), says that change in entropy of the closed system consisting of objects A and B is the sum of the two entropies for objects A and B, which is zero. Object A loses entropy as Object B gains entropy. Based on this we can re-define entropy change to include both reversible and irreversible processes. This extended statement of the change in entropy is called the *Second law of thermodynamics* and is written as:

> '*When changes occur in a closed system its entropy ether increases (for irreversible process) or remains constant (for reversible processes), it never decreases.*'

Mathematically, the Second law of thermodynamics is written as:

$$\Delta S \geqslant 0; \text{ No exception has ever been found} \tag{5.12}$$

It is important to note that this generalization is quantitative and when first proposed by Clausius in 1865 CE was simply a guess! He did not offer any proof and left it to time to prove him right or to prove him wrong.

The use of the Second law of thermodynamics is not just limited to finding the direction of process. The Second law also asserts that energy has '*quality*' as well as '*quantity*'. The First law is concerned with the quantity of energy and the transformation of energy from one form to another regardless of its quality. For engineers and scientists, it is very important to preserve the '*quality*' of energy and the Second law provides a way to determine the quality of energy and measure the energy loss during a process. In engineering systems, the Second law is used to find the theoretical limit to the performance of heat engines and refrigerators, as shown in chapters 6–9. In chemical thermodynamics, the Second law is used to predict the degree of completion of chemical reactions, discussed in chapter 10.

5.2.5 'Quantity' versus 'quality' of energy

The law of conservation of energy says that the '*quantity*' of energy is always conserved. What is not conserved is the '*quality*' of energy. We can think of the '*quality*' of energy as the energy that can be converted into '*useful work*'. One unit of high-quality energy can be more valuable than four units of lower-quality energy. For example, smaller amounts of thermal energy at a high temperature (high-quality energy) are more useful to power plants than a large amount of thermal energy at low temperatures (lower-quality energy).

Let me explain with a simple example, consider two students; A and B in our thermodynamics class. Student A has 100 friends who always show up for fun times but they always seem to be too busy to help student A with homework or a trip to a doctor. On the other hand student B has only 5 friends who are never too busy to help her with homework or drive her to the doctor's office. Who has more friends? The First law of thermodynamics, which considers only the '*quantity*' will say that student A has more friends. However, the Second law of thermodynamics which considers '*quality*' as well, will say student B has more friends. This means quantity alone is not a complete view and we have to worry about quality of energy as well, that is, we need both the First and Second laws to evaluate thermodynamical systems.

5.2.6 Entropy balance relation

Entropy is a property of state and unlike energy, entropy does not exist in various forms. Hence to find the change in entropy of a system during a process we have to evaluate the entropy of the system at the beginning and at the end of the process and take their difference. The Second law of thermodynamics states that the entropy can be created but it cannot be destroyed. Therefore, the change in entropy of a system during a process is greater at the end than it was before the process started. The increase of entropy for any system is expressed as,

$$\Delta S_{\text{system}} = S_{\text{in}} - S_{\text{out}} + S_{\text{gen}} \tag{5.13}$$

where S_{in} is the total entropy entering the system, S_{out} is total entropy leaving the system and S_{gen} is the entropy generated during the process. This equation (5.13) is called the '*entropy balance relation*' and is applicable to any system undergoing any process. The entropy balance equation is stated as:

> '*the entropy change of a system during a process equals the net entropy transferred through the system boundary and the entropy generated within the system*'

Entropy is a property of state, and the value of a property does not change unless the state of the system changes. Therefore, the change in entropy (ΔS_{system}) of a system is zero if the state of the system does not change during a process. For example, the entropy change is zero during steady operation of devices such as turbines, heat pumps, and heat exchangers, etc. Entropy can be transferred to (S_{in}) or from a system (S_{out}) across its boundary by two mechanisms, heat transfer and/or mass flow, which are explained below:

- **Heat transfer:** Heat transfer to a system increases the entropy of the system and heat transferred out of a system decreases the entropy. In fact, heat rejection is the only way the entropy of a fixed mass can be decreased. The ratio of the heat transfer (Q) at a location to the absolute temperature ($|T|$) at that location is called the '*entropy flow*' or '*entropy transfer*', mathematically given as,

$$S_{\text{heat}} = \frac{Q}{T} \tag{5.14}$$

  The direction of entropy transfer is the same as the direction of heat transfer since temperature (T) is always a positive quantity. When the temperature (T) is not constant, entropy transfer during a process is found by integrating over Q as,

$$S_{\text{heat}} = \int \frac{dQ}{T} \tag{5.15}$$

- **Mass flow:** Both entropy and energy are carried in or out of a system by streams of matter (mass). The rate of entropy and the rate of energy transport into or out of a system are proportional to the mass flow rate $\left(\dot{m} = \frac{dm}{dt}\right)$. When a mass ($m$) enters or leaves a system, entropy in the amount of (ms) enters or leaves with it, where (s) is the specific entropy (defined as entropy per unit mass).

$$S_{\text{mass}} = ms \tag{5.16}$$

5.2.7 Special cases

Lets consider some specific cases for the change in the entropy of a system:

1. **Isolated system:** For an isolated system we know there is no exchange of matter or exchange of energy with the surroundings. This means any changes that happen in this system cannot cause changes to the surroundings. In this case,

$$\Delta S_{\text{system}} \geqslant 0 \quad \text{and} \quad \Delta U_{\text{system}} = 0 \tag{5.17}$$

2. **System in equilibrium:** For a system in equilibrium, the properties of the system do not change. Hence we can say,

$$\Delta S_{\text{system}} = 0 \quad \text{and} \quad \Delta U_{\text{system}} = 0 \tag{5.18}$$

Worked example 5.2. A frictionless piston-cylinder device contains a saturated liquid–vapor mixture of water at 100 °C. During a constant-pressure process, 600 kJ of heat is transferred to the surrounding air at 25 °C. As a result, part of the water vapor contained in the cylinder condenses. Assuming the process is internally reversible and the water pressure and temperature remains constant everywhere, including the boundaries. Determine,

(a) The entropy change of the water, and
(b) The total entropy generation during this heat transfer process.

Solution: Suppose the water in the cylinder forms a closed system since no mass crosses the system boundary during the process. Saturated liquid–vapor mixture of water loses heat to its surroundings, and some of the vapor condenses. Also the entropy of the system decreases during the process because of heat loss. $T_{\text{system}} = 100\ °\text{C} + 273 = 373\ \text{K}$.

(a) The entropy change of the water will be given as:

$$\Delta S_{\text{sytem}} = \frac{Q}{T} = \frac{-600\ \text{kJ}}{373\ \text{K}} = -1.61\ \text{kJ K}^{-1}$$

(b) To find the total entropy generated during this process, we consider the extended system, which includes the water, the piston-cylinder device, and the region immediately outside the system that experiences a temperature change so that the entire boundary of the extended system is at the surrounding temperature of 25 °C = (25 + 273) K. The entropy balance for this extended system (system + immediate surroundings) yields

$$\Delta S_{\text{system}} = S_{\text{in}} - S_{\text{out}} + S_{\text{gen}}$$

$$\Delta S_{\text{system}} = -\frac{Q_{\text{out}}}{T} + S_{\text{gen}}$$

$$-1.6\ \text{kJ K}^{-1} = -\frac{600}{(25 + 273)\ \text{K}} + S_{\text{gen}}$$

$$S_{\text{gen}} = 0.40\ \text{kJ K}^{-1}$$

The entropy generation in this case is entirely due to irreversible heat transfer through a finite temperature difference.

5.3 Entropy as a property of state for an ideal gas

In order for something to be a useful quantity we have to make sure that it is a true property of a given equilibrium state of a system. Here we prove that entropy is a property of state for an ideal gas.

5.3.1 Change in entropy for an ideal gas at constant pressure

The First law of thermodynamics says,

$$dU = Q - pdV \tag{5.19}$$

To find the change in entropy of an ideal gas as a function of temperature and pressure (written as $S = S(T, p)$) we begin with the ideal gas law in differential form:

$$d(pV) = d(nRT)$$

For one mole of ideal gas, we can write:

$$pdV + Vdp = RdT$$

Dividing throughout by (pV):

$$\frac{dV}{V} + \frac{dp}{p} = \frac{R}{pV}dT$$

Putting the value of dV into equation (5.22) we get,

$$dS = \frac{C_V}{T}dT - R\left(\frac{dT}{T} - \frac{dp}{p}\right) \tag{5.20}$$

Rearrange, knowing $(C_V - R = C_P)$

$$dS = \frac{C_P}{T}dT - R\frac{dp}{p}$$

and integrate to get:

$$\Delta S = C_P \ln \frac{T_2}{T_1} - R \ln \frac{P_2}{P_1} \tag{5.21}$$

where P_2 and P_1 are the pressures at the final and initial states, respectively, and T_2 and T_1 are the temperatures of the final and initial states, respectively.

5.3.2 Change in entropy for an ideal gas at constant volume

For an ideal gas, $dU = nC_V dT$, using constant volume process, and $PV = nRT$, so equation (5.19) can be written as,

$$nC_V dT = Q - \frac{nRT}{V}dV$$

Dividing throughout by (T), and knowing $\left(\frac{Q}{T} = \Delta S\right)$ from the Second law, we can rewrite the above equation as,

$$\Delta S = \frac{nC_V}{T}dT + \frac{nR}{V}dV \tag{5.22}$$

Integrating from the initial state, with temperature (T_i) and volume (V_i), to the final state with temperature (T_f) and volume (V_f), we can write the change in entropy of a system at constant volume as,

$$\Delta S = nC_V \ln\frac{T_f}{T_i} - nR \ln\frac{V_f}{V_i} \tag{5.23}$$

5.3.3 Change in entropy for an ideal gas under isothermal processes

We know from chapter 4, for ideal gases under isothermal processes the temperature of the initial and the final states is the same. Under this condition, the entropy change of isothermal expansion process, from equation (5.23), is,

$$\Delta S = nR\ \ln\frac{V_f}{V_i} \tag{5.24}$$

5.3.4 Peddler man again

After all these new introductions and information we are now ready to go back and talk about the peddler man and his '*majik box*' and see if we can find out if the '*majik box*' actually works and help you make a life-changing dcision about its purchase. We can do that by calculating an actual numerical value for the change in entropy. Suppose the '*majik box*' encloses an ideal gas, just to make our life easier. For ideal gases we know two things:

1. The equation of state for one mole is, ($pV = RT$), and
2. The change in internal energy is, ($dU = C_V dT$).

Starting from First law of thermodynamics in section 5.3, we developed equation (5.21) for the calculation of changes in entropy of an ideal gas with heat capacity at constant pressure (C_P), as,

$$\Delta S = C_P\ \ln\left(\frac{T_2}{T_1}\right) - R\ \ln\left(\frac{P_2}{P_1}\right) \tag{5.25}$$

where T_1 and P_1 are temperature and pressure of the initial state and T_2 and P_2 are temperature and pressure of the final state.

We now use this equation (5.25), to calculate the change in entropy for the processes claimed by the peddler man. To check the claims made by the peddler man we have to answer two basic questions: one, is the mass during the process conserved (i.e., $\Delta m = 0$)? and two, is the Second law violated or not (i.e., $\Delta S \geqslant 0$)? To answer these questions, let's start with some assertions; peddler man says that his device can take in compressed air at 4 atm and 70 °F to produce two equal streams, a hot stream at 1 atm and 140 °F and a cold stream at 1 atm and 0 °F, as we see in figure 5.3. Let's consider the total process in two parts,

1. **Part I: Hot stream:** one mole of air goes from state 0 (at 4 atm and 70 °F) to state 1 (at 1 atm and 140 °F). Taking $C_p = 29$ KJ kmol^{-1} K^{-1} and $R = 8.314$ J mol^{-1} K^{-1}, we get the change in entropy from equation (5.25) as,

$$\Delta S_1 = +15.1 \text{ J K}^{-1} \tag{5.26}$$

2. **Part II: Cold stream:** One mole of air goes from state 0 (at 4 atm and 70 °F) to state 2 (at 1 atm and 0 °F), produces a change of entropy, given numerically as,

$$\Delta S_2 = +7.4 \text{ J K}^{-1} \tag{5.27}$$

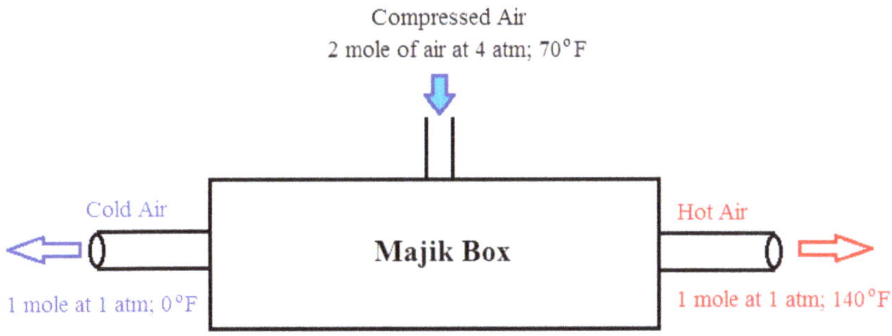

Figure 5.3. Peddler man's majik box.

This means for the total system the change in entropy is,

$$\Delta S_{\text{total}} = +22.5\ \text{J K}^{-1} > 0 \tag{5.28}$$

Equation (5.28) shows that this process can absolutely theoretically work. This is great news, so we have just proved that the '*majik box*' can work, at least theoretically. So thermodynamics helped us answer the question; the device peddler man is selling does indeed theoretically work!

However, suppose the peddler man had said that 2 moles of air at 1.1 atm and 70 °F will produce 1.6 moles of air at 1 atm and 0 °F and 0.4 moles of air at 1 atm and 350 °F. In this case, we would calculate,

$$\Delta S_{\text{total}} < 0 \tag{5.29}$$

which cannot happen since ΔS_{total} can never be negative.

Building the 'majik box'

Let's see how we could make such a '*majik box*'. In order to make the '*majik box*' the peddler man needed to have a turbine, a heat pump, a refrigerator and a compressed air source. Compressed air would be run through a turbine, where the turbine would produce work at the expense of the internal energy of the air. The temperature of air drops as it expands in the turbine to atmospheric pressure. This air stream is further split into two parts. Work from the turbine is then used to operate the refrigerators and the heat pump inside the '*majik box*' that cool and heat the air, respectively, by extracting heat from one stream and discharging heat to the other. Hence we get a device that is mechanically and thermodynamically possible and is schematically shown in figure 5.4.

But is this device real?

Yes, believe it or not, this device is real and it is called '*Hilsch–Ranque Vortex tube*'. It was invented in 1930s by a French physicist named Georges Ranque. It remained obscure till World War II, when a German scientist, Rudolf Hilsch, built a number of them and distributed them to German laboratories. Each of these devices consisted of a straight tube divided into two parts with an opening. Compressed air at some initial temperature was fed into the opening tangentially to the tube. The

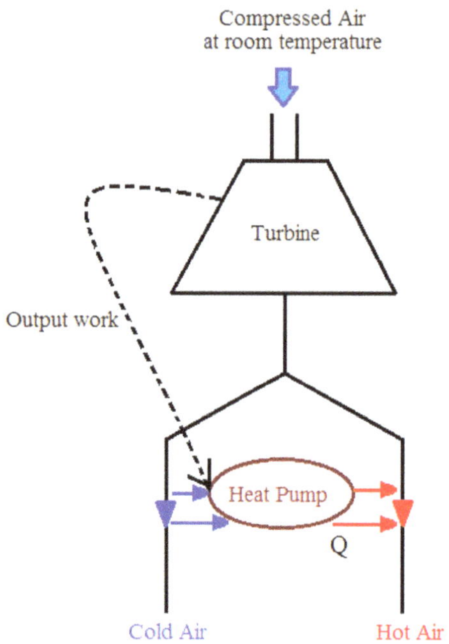

Figure 5.4. Peddler man's majik box—mechanical point of view.

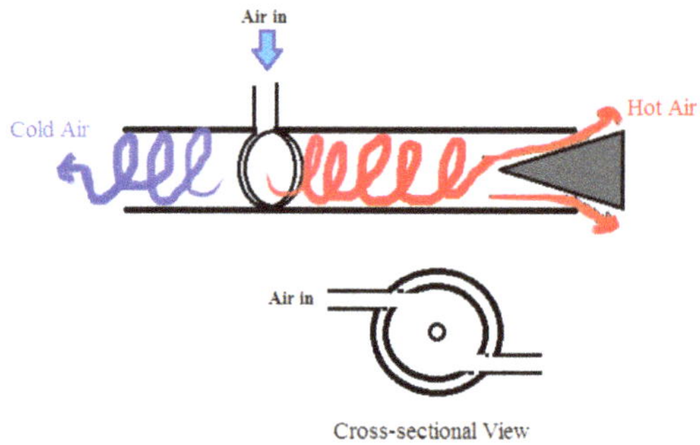

Figure 5.5. Vortex tube.

compressed air expands as it enters the tube and speeds up. This introduces a rotary motion in air and therefore it was called the '*vortex tube*', as shown in figure 5.5.

Expanding air gains kinetic energy at the expense of internal energy and lowers the temperature of the air. This temperature change is pretty uniform over the cross-section of the tube. The velocity and hence kinetic energy varies from point to point in the tube. The air at the center of tube has the least velocity and hence the least kinetic energy, whereas the velocity close to the walls of the tube is the highest and the kinetic energy is maximum.

The air near the center of the opening will leave the tube as a cold air stream from the '*cold end*', whereas air in the outer parts of the tube with high kinetic energy loses its kinetic energy as it swirls down the tube away from the opening and converts kinetic energy to internal energy as a result of turbulence, internal friction and mixing which raises the air temperature coming out of the '*hot end*', as seen in figure 5.5. Thus air temperature rises and this comes out at the hot side. The air emerging from the '*hot end*' can reach temperatures of 200 °C (392 °F), and the air emerging from the '*cold end*' can reach 50 °C (58 °F).

Think about it, this device from a practical point of view is very attractive. It is cheap to build and since it has no moving parts it is very easy to maintain. Vortex tubes are currently used mostly for cooling cutting tools during machining, as in CNC machines, since most machine shops already have compressed air available.

Concept questions 5.1

1. The entropy of an isolated system always *increases/decreases* and becomes a *minimum/maximum* at the state of equilibrium.
2. According to entropy principle, the entropy of an isolated system can never *increase/decrease* and remains constant only when the process is *irreversible/reversible*.
3. Are there any natural processes that are reversible?
4. Why are most natural processes irreversible?

Exercises 5.1

1. An ideal gas undergoes a reversible isothermal expansion at 100 °C. How much heat must be absorbed if the entropy of the gas increases by 50 J K^{-1}?
2. A 2 kg block of copper at room temperature is increased reversibly in temperature to 120 °C. Given the specific heat of copper is 0.385 J g^{-1} K^{-1}. Find,
   (a) Heat absorbed.
   (b) Change in entropy

5.4 Thermodynamic identity

There is an equation that connects internal energy, the theoretical definition of temperature and pressure. To derive it we consider a process in which we change both the energy (ΔU) and the volume (ΔV) of a system by small amounts and ask the question, by how much will the entropy of the system change?

To answer this question, we assume that we can divide the process into two steps: step one, where the volume is kept constant ($dV = 0$); and step two, where the energy is kept constant ($dU = 0$). Then the total entropy change will be the sum of the changes in step 1 and step 2:

$$\Delta S = \Delta S_1 + \Delta S_2 \tag{5.30}$$

Multiply and divide the first term with ΔU and second term by ΔV to give us:

$$\Delta S = \left(\frac{\Delta S}{\Delta U}\right)_1 \Delta U + \left(\frac{\Delta S}{\Delta V}\right)_1 \Delta V \tag{5.31}$$

Since all changes are small we can rewrite partial derivatives instead to give us:

$$dS = \left(\frac{\partial S}{\partial U}\right)_1 dU + \left(\frac{\partial S}{\partial V}\right)_1 dV \tag{5.32}$$

To find a thermodynamic identity which is true for infinitesimal changes, we run the experiment first at constant volume and then at constant temperature, as:

1. Step 1: At constant volume (V) the First law of thermodynamics ($dU = Q$) gives:

$$dS = \frac{Q}{T} = \frac{dU}{T} \Longrightarrow \frac{dS}{dU} = \frac{1}{T} \tag{5.33}$$

2. Step 2: At constant temperature implies that $\Delta U = 0$, the First law of thermodynamics gives, $W = -Q$,

$$dS = \frac{Q}{T} = \frac{-W}{T} = \frac{pdV}{T} \Longrightarrow \frac{\partial S}{\partial V} = \frac{p}{T} \tag{5.34}$$

   rearranging we get,

Using equations (5.33) and (5.34), we can write equation (5.31) as,

$$\begin{aligned} dS &= \left(\frac{\partial S}{\partial U}\right)_1 dU + \left(\frac{\partial S}{\partial V}\right)_1 dV \\ &= \left(\frac{1}{T}\right) dU + \left(\frac{p}{T}\right) dV \end{aligned} \tag{5.35}$$

Hence, equation (5.35) can be rearranged for infinitesimal changes ($\Delta S \rightarrow dS$) to give:

$$dU = TdS - pdV \tag{5.36}$$

This equation is called the '*thermodynamic identity*' and is always true for any infinitesimal changes in any system, given T and p are well-defined and other relevant variables (say number of particles in a system) do not change.

1. **Isochoric process:** During isochoric processes ($\Delta V = 0$), equation (5.36) reduces to:

$$TdS = dU$$

   which is a definition of temperature as,

$$T = \frac{dU}{dS} \tag{5.37}$$

2. **Isothermal process:** During isothermal processes ($\Delta U = 0$), equation (5.36) reduces to:

$$TdS = pdV$$

which is a definition of pressure as,

$$p = T\frac{dS}{dV} \tag{5.38}$$

5.4.1 Mechanical equilibrium

We know that when systems at different temperatures are brought into thermal contact, one system spontaneously gives up heat and other spontaneously gains heat until they are both at thermal equilibrium. So we can say that spontaneous exchange of energy between systems is governed by temperatures. Similarly we can say that the exchange of volume between two systems is governed by their pressures. Hence there must be a close relationship between pressure and entropy, similar to the relation $\left(\frac{\partial S}{\partial U} = \frac{1}{T}\right)$.

Let's consider two systems, system A and system B that are free to exchange both energy and volume between them, by utilizing a moveable piston, while the total energy and volume of the system remains the same. The total entropy is a function of internal energy (U) and total volume (V). The systems reaches equilibrium when S_{total} gets to its maximum value. At equilibrium,

$$\left(\frac{\partial S_{\text{total}}}{\partial U_A}\right) = 0; \quad \text{condition 1}$$

$$\left(\frac{\partial S_{\text{total}}}{\partial V_A}\right) = 0; \quad \text{condition 2}$$

Condition 1 is equivalent to saying that the two systems are at the same temperature, while condition 2 implies,

$$\frac{\partial S_{\text{total}}}{\partial V_A} = 0$$

$$\left(\frac{\partial S_{\text{A}}}{\partial V_A} + \frac{\partial S_{\text{B}}}{\partial V_A}\right) = 0$$

$$\left(\frac{\partial S_{\text{A}}}{\partial V_A} - \frac{\partial S_{\text{B}}}{\partial V_B}\right) = 0$$

The last step uses the fact that the total volume is fixed, hence $dV_A = -dV_B$, therefore, we can conclude at equilibrium:

$$\frac{\partial S_{\text{A}}}{\partial V_A} = \frac{\partial S_{\text{B}}}{\partial V_B}; \quad \text{at equilibrium} \tag{5.39}$$

From experience, we know that when two systems are in mechanical equilibrium their pressures are equal. Therefore, pressure must be a function of $(\partial S/\partial V)$. To figure this out, we have to look at the units. Entropy has units of (J K^{-1}), whereas $(\partial S/\partial V)$ has units (J K^{-1} m^{-3}). We can rearrange (J K^{-1} m^{-3} = N m^{-2} K^{-1}) or (Pa K^{-1}), which are the units for the ratio pressure/temperature. Hence we can say:

$$\left(\frac{\partial S}{\partial V}\right)_{U,N} = \frac{p}{T} \tag{5.40}$$

This is not quite the definition of pressure, but it has the same qualities as pressure and hence it is probably the same thing as force per unit area!

5.5 Third law of thermodynamics

For any process we can consider a sequence of tiny steps that take us from some initial state to some final state. We can compute the change in entropy (ΔS) for each step and then add them up all together to get the total change in entropy as:

$$\Delta S = S_{\mathrm{f}} - S_{\mathrm{i}} = \int_{\mathrm{T_i}}^{T_{\mathrm{f}}} \frac{C_V}{T} dT \tag{5.41}$$

Often heat capacity, C_V, is fairly constant over the temperature range of interest and can be taken out of the integral. In other cases, for example. at low temperatures, C_V changes quite a bit and must be taken into account when integrating. If you are lucky enough to know C_V all the way down to absolute zero, you can calculate the '*total entropy*' of the system by simply taking the lower limit of the integral in equation (5.41) as absolute zero ($T = 0$ K), this is called '*entropy at absolute zero*' ($S(0)$). At absolute zero temperature a system should settle into its unique lowest-energy state, and the total change in entropy of the system can be written as,

$$\Delta S = S_{\mathrm{f}} - S(0) = \int_{0}^{T_{\mathrm{f}}} \frac{C_V}{T} dT \tag{5.42}$$

In principle $S(0)$ is zero and equation (5.42) is called the '**Third law of thermodynamics**'.

5.5.1 Residual entropy

In practice there are many reasons why $S(0)$ may not be zero. This is called *residual entropy*. Residual entropy is the difference in entropy between a non-equilibrium state and state of a substance close to absolute zero.

1. The first example of residual entropy was given by Linus Pauling in 1935 CE to describe the water-ice system. In an ice crystal, water molecules can orient themselves in many different ways. Each orientation has a unique energy but there is only one orientation that has the lowest possible energy; this orientation is called the 'ground state'. In practice, orientations of systems are random or nearly random. One would have to wait a really, really long time for the crystal to rearrange itself into the true ground state.

2. The second type of residual entropy comes from the alignments of nuclear spins. At absolute zero, the spins align either parallel or anti-parallel to their neighbors and entropy disappears. In practice, residual entropy usually does not happen until the temperature reaches very close to 1 K.
3. Another form for residual entropy arises from the mixing of various isotopes of elements. Most elements have more than one stable isotope and these isotopes are mixed together randomly and each has a unique entropy of mixing. At absolute zero we have a unique lowest-energy state called the '*ground state*', where all isotopes are unmixed. In practice, however, we find that the atoms for most elements are always stuck at their random sites in the lattice, with the possible exception of helium, which remains a liquid at absolute zero, allowing two isotopes (^{3}He and ^{4}He) to arrange themselves in an orderly way.

5.5.2 Another statement of Third law

Entropies for many substances are computed by measuring the heat capacities and using equation (5.41) and can be found in most chemistry textbooks. By convention, most tabulated entropies include residual entropies due to molecular orientation, but do not include residual entropies due to mixing or nuclear-spin orientations.

One thing we have not talked about is what happens to the right-hand side of equation (5.41), when temperature (T) goes to zero, since T is in the denominator of this equation. In this case the integrand will diverge and hence either S_f will be infinity or $S(0)$ will be negative infinity. However, we know that entropy must always be finite and positive. The only way for this to happen is if C_V goes to zero at $T = 0$,

$$\text{As } T \to 0; \quad C_V \to 0 \tag{5.43}$$

This result is also called the '**Third law of thermodynamics**'.

5.6 Finally ...

After all this, let's go back to the First and Second laws of thermodynamics. The First law says that '*quantity*' of energy is always preserved during a process, but according to the Second law the '*quality*' of energy generally decreases. For example, after the transfer of 100 kJ of energy as heat from a hot medium to a cold one will still have 100 kJ of energy, but this energy will be at the lower temperature and thus at a lower quality. This decrease in '*quality*' of energy is always accompanied by an increase in entropy and the Second law becomes,

$$\Delta S_{\text{total}} \geqslant 0 \tag{5.44}$$

The subscript 'total' in this equation means that we have taken into account all the changes in both the system and its surroundings, so we can write:

$$(\Delta S_{\text{system}} + \Delta S_{\text{surrounding}}) \geqslant 0 \tag{5.45}$$

The Second law applies to both system and surroundings taken together and does not impose any general restraint on the system alone. It can also be applied to the First law as:

$$(\Delta U_{\text{system}} + \Delta U_{\text{surrounding}}) = 0 \tag{5.46}$$

Concept questions 5.2

1. Heat flows from Sun to the Earth. Show that the entropy of the Sun–Earth system increases during this process.
2. Why does the following statement not contradict the Second law of thermodynamics: 'Cosmic rays continually decrease the entropy of the Earth on which they fall.'
3. A dropped stone always falls to the ground. Why do air molecules not fall to the ground?

Exercises 5.2

1. One mole of ideal gas at p_0, V_0, and T_0 is heated at constant volume to $2T_0$, then allowed to expand at constant temperature to $2V_0$ and finally it is allowed to cool down at constant pressure to T_0. What is the net entropy change for this ideal gas process?
2. An ideal gas undergoes an isothermal expansion at room temperature increasing its volume from 1 L to 4 L. The entropy change of the gas is 30 J K^{-1}. How many moles of gas are present in the sample of the gas?
3. At very low temperatures the molar specific heat of solids is approximately proportional to T^3. Calculate the entropy change of 8 moles of aluminum when its temperature is raised from 5 K to 10 K. Assume the constant of proportionality for aluminum is numerically equal to 0.000 0314 J mol^{-1} K^{-4}.

Further reading

- Cengel Y and Boles M 2006 *Thermodynamics: An Engineering Approach* 5th edn (New York: McGraw-Hill Higher Education)
- Fermi E 2012 *Thermodynamics* (New York: Dover Publications)
- Resnick R, Halliday D and Krane K 2002 *Physics* vol 1 5th edn (New York: Wiley)
- Schroeder D 2000 *Thermal Physics* (Reading, MA: Addison-Wesley Longman)
- Van Ness H C 1969 *Understanding Thermodynamics* (New York: McGraw-Hill Book Company)

IOP Publishing

Chapter 6

Application of the laws of thermodynamics

The study of heat pumps and heat engines is essential in the analysis of classical thermodynamics. The first real steam engine was invented by Newcomen in 1712 CE, which was a very inefficient device. Fifty years later James Watt introduced a number of innovations that made the steam engine a more practical device. James Watt, around the 1760s, mass produced his steam engines, which were used widely. In the 1820s Sadi Carnot worked extensively with steam engines and knew the issues that reduce efficiencies in practical engines. Carnot wondered if there was a limit to the improvement of any real engine. Hence, he came up with the idea of an ideal engine and in 1824 CE published a paper which is considered to be the first theoretical consideration of heat engines. We will discuss the Carnot cycle and Carnot engine in section 6.4.

6.1 Introduction

In chapter 4, section 4.5, we found boundary work from a compressed gas by means of an adiabatic expansion in a piston–cylinder system and found the maximum work done for a reversible process is given by the formula: $W = -\int pdV$.

However, we know that nearly all processes in real life are irreversible. Scrambling an egg is easy; however, making a scramble back into a full egg cannot be done. An example of reversible versus irreversible processes is shown in figure 6.1. Additionally everything eventually 'runs down', for example batteries, oscillators, etc. Okay, so if all real processes are irreversible than why even talk about reversible processes? Well, the answer is, the study of reversible processes provides us with best case scenarios and allows us to do the calculations. Hence, we use a reversible process as a benchmark and measure the performances of real engines against it.

There are specific cases where we do not need the reversibility condition; these include processes where the energy balance can be written in a way that it contains only one unknown term. In these situations, we do not have to worry about the system being reversible or irreversible since conservation of energy holds for both

doi:10.1088/978-0-7503-3083-1ch6

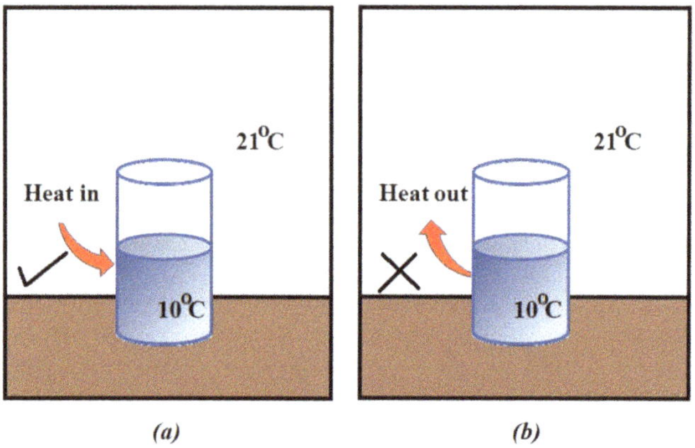

Figure 6.1. (a) Irreversible versus (b) impossible heat transfer.

types of processes. In processes where the energy balance contains more than one unknown term the assumption of reversibility is usually essential to do any calculations at all. These conditions basically split thermodynamics into two parts,

1. **Theoretical thermodynamics:** Theoretical thermodynamics mainly deals with properties of materials and their experimental measurements. First, we derive the equation for a reversible process. Once it is derived, we find that most equations just contain properties of the system and do not depend on the kind of process considered. This shows us that thermodynamics is backwards! We derive an equation for a special case then conclude it must be general!
2. **Applied thermodynamics:** Applied thermodynamics deals with the solution of practical problems through properties provided by experimental measurements.

   Most processes in real life are irreversible. What do we do now? Well, we ignore irreversibilities! It might be strange to think, but this is okay and here is how we do it:

   (a) Assume all problems are reversible.
   (b) Find the solution to the problem at hand, but know that this method will not give completely accurate answers.
   (c) Once you have an answer, ask how wrong the answer is.
   (d) Make the necessary corrections to get a realistic but approximately correct answer.

We will see this method in work while dealing with multiple cycles in the following sections.

6.2 Thermodynamic cycles

A thermodynamic cycle is a series of thermodynamic processes done on a system that return the system to its initial state. From the First law of thermodynamics we know the net heat input in any cycle equals the net output work. Heat and work are not zero over a cycle, but rather they depend on the process followed.

An ideal cycle is defined for processes that are '*quasistatic*' and locally reversible. This means there are no dissipative forces, like friction present in the cycle. Most classical thermodynamics problems are solved for the ideal case where we ignore all drag forces such as friction. Once we have the solution for the ideal friction-free situation we can generalize for real-life situations.

6.2.1 Types of thermodynamic cycles

There are two main types of thermodynamic cycles and I will call them work-to-heat-transfer cycles and heat-transfer-to-power cycles.

1. **Work-to-heat-transfer cycles:**

   Work-to-heat-transfer cycles use mechanical work input to transfer heat from low temperature (sinks) to high temperature (sources)[1]. These are called '*temperature cycles*' for short. Examples of work-to-heat-transfer cycles include heat pumps, where a heat pump is called a 'heater' when we want to warm a heat source, for example when warming the inside of a home on a cold day, or a 'refrigerator' when we want to cool the heat source, as in the normal operation of a refrigerator or a freezer. In both cases the operating principle is the same; heat is moved from a cold place to a warm place which does not happen naturally. We will talk more about examples of work-to-heat-transfer cycles in chapter 7, section 7.1.
2. **Heat-transfer-to-power cycles:**

   Thermodynamical cycles that convert heat inputs into mechanical work outputs are called 'heat-transfer-to-power' cycles. These are also called '*power cycles*' and are the basis for the operation of heat engines, which supply most of the world's electric power and run our cars.

   Heat cycles can be divided according to the type of heat engine they model. The most common cycles are:

   (a) The Otto cycle models internal combustion engines, section 8.2.
   (b) The Diesel cycle models Diesel engines, section 8.3.
   (c) The Brayton cycle models jet engines, section 8.4.
   (d) The Stirling cycle models Sterling engines, section 9.1.
   (e) The Rankine cycle models steam turbines, section 9.3.

If the cyclic process moves clockwise around the loop on a pV diagram, it represents a power cycle that runs heat engines and boundary work is positive. If the cyclic process moves counterclockwise, it represents a heat pump and boundary work is negative.

[1] It is important to keep in mind that if you are transferring heat from something, it is the 'source' no matter if it is low temperature or high.

6.3 Heat-transfer-to-power cycles

We know that heat spontaneously flows from a hot object to a cold object. A device that exploits this temperature gradient between a hot '*source*' and a cold '*sink*' and uses it to perform mechanical work is called a '*heat engine*'.

Ideally we want all the heat transferred to be converted to useful work. Unfortunately, in reality, only a small part of the heat energy absorbed is converted into useful work by a heat engine. This is because when heat is absorbed, entropy is brought with it, and this entropy needs to be removed from the system before the cycle can begin again. This removal is achieved by releasing some of the heat into the surroundings, which implies that the efficiency of a heat engine is the difference between heat absorbed and waste heat expelled.

6.3.1 Reservoir

In thermodynamics, a '*thermal energy reservoir*', or a '*reservoir*' is anything that is large enough that its temperature does not change noticeably when heat enters or leaves it. A reservoir in engineering systems is a hypothetical body with a relatively large thermal energy capacity[2] that can supply or absorb finite amounts of heat without undergoing any change in temperature. A reservoir that supplies energy in the form of heat is called a '*source*', and a reservoir that absorbs energy in the form of heat is called a '*sink*'.

- **Hot reservoir** (T_{H}): Sources are called hot reservoirs, written usually as (T_{H}). In an engine, heat (Q_{H}) is absorbed by the engine from a hot reservoir at temperature (T_{H}).
- **Cold reservoir** (T_{C}): Sinks are called cold reservoirs, where waste heat (Q_{C}) is dumped. The temperature of cold reservoirs is usually written as T_{C}. In practice, surrounding environments usually form sinks for the engines, such as the air around a car engine, or lakes and rivers around power plants.

6.3.2 A basic heat engine

A basic heat engine consists of a gas or a liquid, called a '*working fluid*', confined by a piston-and-cylinder system. Gases expand when they are heated and as the gas expands in the cylinder it moves the piston outwards. As the gas cools it contracts and moves the piston inwards. A practical heat engine goes through cycles where heat is either transferred to or is taken out of it while undergoing a cycle. The piston in our system moves back and forth as the gas is heated and/or cooled.

In a full cycle of a heat engine, three things happen:

1. Heat (Q_{H}) is added at a relatively high temperature (T_{H}) from a source, for example, solar energy, oil furnace, nuclear reactor, etc. Heat energy (Q_{H}) taken from the high temperature source is written in terms of entropy (S_{H}) from the Second law as,

$$Q_{\mathrm{H}} = T_{\mathrm{H}}\Delta S_{\mathrm{H}}$$

[2] Thermal energy capacity is defined as mass of the object times the specific heat.

2. Some of the energy from this input heat (Q_H) is used to perform mechanical (or useful) work (W), usually in the form of rotating a shaft.
3. The rest of the heat (Q_C) is removed at a relatively cold temperature (T_C) to the immediate surroundings, for example, the atmosphere near a car engine, or a lake or a river near a power plant, etc. In terms of entropy (S_C), heat energy Q_C delivered to the heat sink is given as,

$$Q_C = T_C \Delta S_C$$

An energy-flow diagram for a heat engine is given in figure 6.2.

From figure 6.2 and conservation of energy in a complete cycle, we can write the heat absorbed (Q_H) by the engine from the hot reservoir as,

$$Q_H = Q_C + W. \tag{6.1}$$

A heat-engine cycle cannot be completed without rejecting some heat (Q_C) to a low-temperature sink (T_C). Here, W is the useful work extracted from the engine, mathematically defined as, $W \equiv -\oint pdV$. The negative sign is used since in work done by the engine, pressure and change in volume are inversely proportional to each other.

6.3.3 Laws of thermodynamics revisited for engines

Before we talk about the efficiencies of engines, let's quickly revisit the laws of thermodynamics. Historically, thermodynamics grew out of our efforts to construct more and more efficient heat engines. Thermodynamics is summarized by its four laws we have discussed in the previous chapters. These laws are established based on many experimental observations and basically describe what is '*not possible*' when you wish to convert heat into useful work. I will summarize these laws for heat engines:

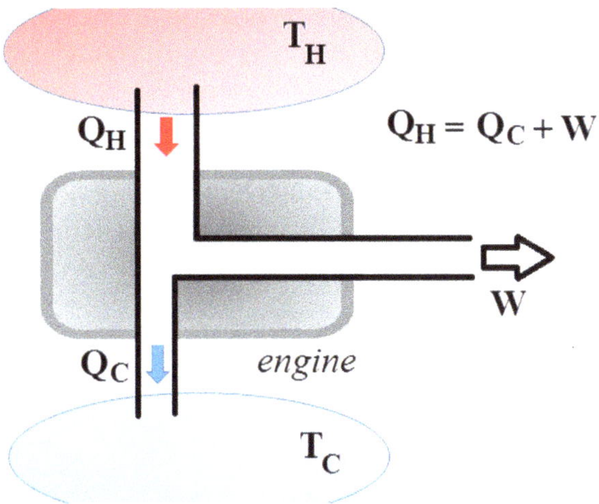

Figure 6.2. Energy-flow diagram for a heat engine.

- **Zeroth law:** You cannot ignore the temperature of your heat engine.
- **First law:** You cannot win, that is, you cannot extract more useful work than your heat intake.
- **Second law:** You cannot even break even, that is, you will always '*lose*' unless you reach absolute zero temperature.
- **Third law:** You can never reach absolute zero temperature.

6.3.4 Efficiency (η) of a heat engine

The efficiency of a heat engine relates how much useful work is obtained for a given amount of heat energy input. In general, the efficiency of a given heat transfer process is defined as the ratio of '*what you get out*' to '*what you put in*'. The efficiency (η) of the engine is defined as,

$$\eta = \frac{\text{Benefit}}{\text{Cost}} = \frac{\text{Useful work}}{\text{Heat input}} = \frac{W}{Q_H} = \frac{Q_H - Q_C}{Q_H} = 1 - \frac{Q_C}{Q_H} \tag{6.2}$$

However, the theoretical maximum efficiency of any heat engine depends on the temperatures it operates between. Mathematically, in reversible processes, the change in entropy of the cold reservoir and the hot reservoir are related as, ($\Delta S_C = -\Delta S_H$), which keeps the overall change of entropy equal to zero and we can rewrite equation (6.2) in terms of measurable quantities, T_C and T_H, which are just the absolute temperatures, in kelvin, of the hot source and the temperature of the cold sink respectively.

$$\eta_{max} = 1 - \frac{T_C \Delta S_C}{-T_H \Delta S_H} = 1 - \frac{T_C}{T_H} \tag{6.3}$$

The First law of thermodynamics says that the efficiency of a heat engine (η) cannot be greater than 1, since you cannot get more useful work out of the engine than the heat you put in. The Second law of thermodynamics makes this statement even worse, we cannot even get $\eta = 1$, unless $T_C = 0$ K and T_H is infinity. We know that both of these, $T_C = 0$ K and $T_H =$ infinity, are not possible in practice, so what we are saying is that we cannot even break even! For example, if you have $T_C = 300$ K and $T_H = 500$ K, then from equation (6.3) we find max $\eta = 40\%$, which implies 60% of input heat is wasted!

Worked example 6.1. A reversible heat engine operates between two heat reservoirs, one at $T_1 = 1000$ K and the other at $T_2 = 300$ K. If 100 J of heat is transferred to the heat engine from the reservoir at T_1, how much work is done by the engine?

Solution: Use equation (6.2) and given values,

$$\frac{W}{Q_H} = 1 - \frac{T_C}{T_H}$$

$$W = Q_1\left(1 - \frac{T_2}{T_1}\right)$$

$$W = 100\left(1 - \frac{300}{1000}\right)$$

Solving it we get that the work done by the engine is numerically equal to 70 J.

Worked example 6.2. The engine given in the above example 6.3.1 would produce more work if the cold reservoir temperature was decreased to 200 K by means of refrigeration of some kind. An engineer proposes that we could use a pump to remove heat from the cold reservoir (at 300 K) and put it into a 'cold box' held at 200 K. Calculate the net work (W) of the process and compare the result with the answer to the above example.

Solution:

$$\text{Network}(W) = |W_{\text{engine}}| - |W_{\text{heat pump}}|$$
$$= |Q_1|\left(1 - \frac{T'}{T_1}\right) - |Q'|\left(1 - \frac{T_2}{T'}\right)$$
$$= |Q_1|\left(1 - \frac{T'}{T_1}\right) - |Q_1\frac{T'}{T_1}|\left(1 - \frac{T_2}{T'}\right)$$
$$= Q_1\left(1 - \frac{T_2}{T_1}\right)$$

which is the same as we found in example 6.3.1. The engine operates reversibly between the same two temperatures T_2 and T_1. We get that the work done by the engine is numerically equal to 80 J.

Worked example 6.3. Consider the simple heat engine that consists of a piston–cylinder device with a load that can be raised and lowered. The working fluid (a gas) contained within the cylinder is initially held at 30 °C. Suppose 100 kJ of heat is transferred to the gas working fluid from a source at 100 °C, which causes the piston to expand till its maximum extension, taking its load with it. At this point, the load is removed, and the gas temperature is measured to be 90 °C. The work done on the load during this expansion process is equal to the increase in its potential energy, say 20 kJ. Under these conditions, the amount of heat supplied to the gas is greater than the work done because part of the heat supplied is used to raise the temperature of the gas. Is it possible to transfer the 80 kJ of excess heat at 90 °C back to the reservoir at 100 °C for later use?

Solution: Let's start with the assumption that it is possible to transfer the 80 kJ of excess heat at 90 °C back to the reservoir at 100 °C for later use. In this case we will have a heat engine that can have a thermal efficiency of 100% under ideal conditions. This is not possible, because heat is naturally always transferred from a high temperature region to a low-temperature region, and never the other way around. Therefore, we cannot cool a gas from 90 °C to 30 °C by transferring heat to a reservoir at 100 °C. Instead, we have to bring the system into contact with a low-temperature reservoir, say at 20 °C, so that the gas can return to its initial state by rejecting its 80 kJ of excess energy as heat to this reservoir. This energy cannot be recycled, and has to be rejected to the sink and it is aptly called waste energy.

Worked example 6.4. Heat is transferred to a heat engine from a furnace at a rate of 80 kW (kilo-watt) per hour[3]. If the rate of waste heat rejection to a nearby river is 50 kWh, determine the net power output and the thermal efficiency for this heat engine. Assume, heat losses through the pipes and other components are negligible.

Solution: In this engine, the furnace serves as the reservoir and the river is the sink. We are given,

$$\dot{Q}_{\mathrm{H}} = 80\ \mathrm{kW\ and}\ \dot{Q}_{\mathrm{C}} = 50\ \mathrm{kW}$$

The net power output of this heat engine is,

$$\dot{W}_{\mathrm{out}} = (\dot{Q}_{\mathrm{H}} - \dot{Q}_{\mathrm{C}}) = (80 - 50)\ \mathrm{kW} = 30\ \mathrm{kW}$$

Then the thermal efficiency is,

$$\eta = \frac{\dot{W}_{\mathrm{out}}}{\dot{Q}_{\mathrm{H}}} = \frac{30\ \mathrm{kW}}{80\ \mathrm{kW}} = 0.375$$

That means the heat engine converts 37.5% of the heat it receives to work.

6.4 Carnot cycle

The simplest and the most ideal of heat engines is called the '*Carnot engine*'. Nicolas Leonard Sadi Carnot proposed the Carnot cycle in 1824 CE as a theoretical ideal thermodynamic cycle. Carnot believed that heat, like electricity, was a fluid that flowed from hot things to cold things and he called one complete cycle for a Carnot engine a '*Carnot cycle*'. The Carnot cycle is the most efficient cycle for converting thermal energy into useful work, as seen in Diesel engines or steam engines. Conversely, the Carnot cycle is the most efficient cycle by which a given amount of work creates a temperature difference, for example, during the refrigeration process.

Carnot considered his system to be made up of a regular piston-and-cylinder device, containing an ideal gas, in thermal contact with two reservoirs, one at a high temperature (T_{H}), called *heat source* and the other at low temperature (T_{C}) called the *heat sink*.

We know that any ideal reversible heat engine must operate in such a way that it takes in heat at temperature (T_{H}) only and discards heat at temperature (T_{C}) only. This means any steps during which heat is transferred must be isothermal (at the same T) and any steps where heat is not transferred must be adiabatic ($Q = 0$). We also want this engine to operate in a cycle, which means the system must return to its initial state. These requirements put constraints on the types of steps the Carnot cycle may contain. Carnot concluded that his cycle consists of four steps, two

[3] The kilowatt-hour is commonly used as a billing unit for energy delivered by electric companies to consumers. A 100 watt light bulb uses 100 watts of power. To convert the power in watts to kilowatt-hours, multiply 100 watts by 1 h, then divide by 1000 to find the energy usage in kWh. If electricity costs $0.50 per kWh, then a 100 watt light bulb will cost 5 cents to run for an hour.

isothermal steps, one of which is at T_H and the other which is at T_C, and two adiabatic steps that separate the two isothermal steps.

6.4.1 Carnot's idea

In Carnot's times '*heat-fluid theory*' was widely accepted and heat was considered a fluid that flows from the object at higher temperature to the other object at lower temperature. Carnot thought that a steam engine and a water wheel employed the same principles with a few differences, mainly that steam engines use steam and water wheels use water for their operations.

In a water wheel, water falls through a gravitational potential difference (gh) and the potential energy is transformed into mechanical work by the rotating wheel. Carnot said in his heat engine the heat fluid falls through a temperature difference ($T_H - T_C$) and the heat exchanged is transformed into mechanical work by the heat engine.

Carnot also believed there was an absolute zero of temperature and he therefore reasoned, if you cooled the '*heat-fluid*' down to absolute zero, it would give up all its heat energy. So, the maximum possible amount of energy you can extract by cooling '*heat-fluid*' from temperature (T_H) to temperature (T_C) is just,

$$\text{Max amount of energy extracted} = \frac{T_H - T_C}{T_H}$$

This is the maximum efficiency of a perfect reversible engine. Of course, today we know that the heat-fluid view is not right, but this result is!

Carnot also concluded that heat engines are not perfect engines. They are limited in their energy-conversion efficiency by the operating between temperatures T_H and T_C and that the maximum efficiency of the Carnot engine is obtained when the working fluid is assumed to follow the following ideal processes discussed in section 6.4.2 below.

6.4.2 Processes in a Carnot cycle

The discussion of a Carnot cycle must start with a few assumptions, which include: one, all processes are internally reversible; two, heat exchanges have no losses; and three, working fluid in the system behaves as an ideal gas and circulates continuously in a closed loop.

The Carnot cycle consists of the following processes shown in figure 6.3 and whose details are as follows:

1. **At Point (1):** The piston is at the inner limit of its stroke. The system is in thermal equilibrium with the heat reservoir (at T_H).
2. **From Points (1) to (2):** The ideal gas expands reversibly and isothermally (T_H) and does work against the piston. Heat (Q_H) is added to the gas to maintain the temperature at T_H, otherwise the temperature would have dropped since internal energy would have dropped. Heat added to the ideal gas is mathematically given as,

$$Q_{12} = W_{12} = -\int pdV = Q_H = nRT_H \ln \frac{V_2}{V_1}$$

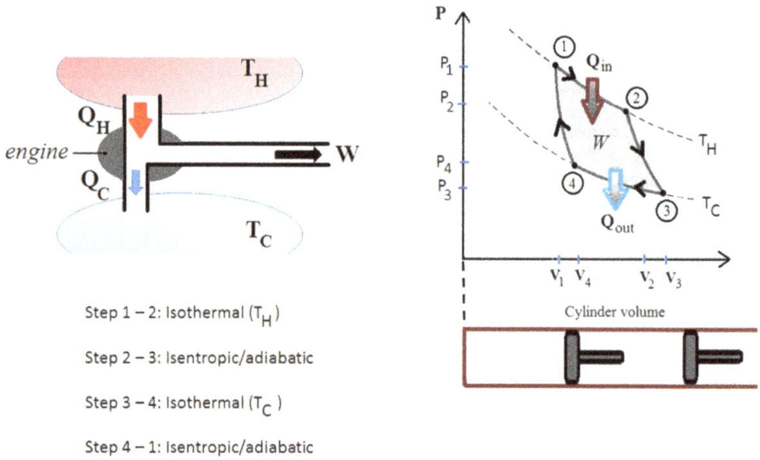

Figure 6.3. Steps in a Carnot cycle.

3. **From Points (2) to (3):** The gas expands reversibly and adiabatically. The temperature of the system at T_{H} drops to T_{C} and we get mechanical work output as,

$$W_{23} = nC_V(T_3 - T_2)$$

where C_V is the molar heat capacity at constant volume.

4. **At Point (3):** The piston at point (3) is at the outer limit at its extension and it starts its journey back into the cylinder. It can go back either through points $3 \rightarrow 2 \rightarrow 1$, however, this will give us no useful work, or it can go through points $3 \rightarrow 4 \rightarrow 1$, in which case we get useful work. So that is what the piston does, it makes its return journey following the path: points $3 \rightarrow 4 \rightarrow 1$.
5. **From Points (3) to (4):** From points (3) to (4), the piston follows a process that is reversible and isothermal and it occurs at temperature (T_{C}). It is a compression process and hence it requires work input. This work input raises the temperature of the gas unless heat is transferred to the cold reservoir at T_{C}. This means heat (Q_{C}) must be extracted to make the process isothermal, which is given mathematically as,

$$Q_{34} = Q_{\mathrm{C}} = -nRT_{\mathrm{C}} \ln \frac{V_4}{V_3}$$

6. **From Point (4) to (1):** The process from points (4) to (1) occurs under compression. It is reversible and adiabatic, which means $T_{\mathrm{C}} \rightarrow T_{\mathrm{H}}$ and input work is given as,

$$W_{\mathrm{in}} = -nC_V(T_1 - T_4)$$

6.4.3 Efficiency of Carnot cycle

The Carnot cycle is shown in the shaded region of figure 6.3. This area represents useful work ($W = -\int pdV$), taken around the entire cycle. This is the total work the engine produces in one cycle. From the First law of thermodynamics we get,

$$\begin{aligned} dU &= Q - W \\ 0 &= (Q_H - Q_C) - W \\ W &= Q_H - Q_C \end{aligned} \tag{6.4}$$

Then thermal efficiency will be,

$$\eta_{Carnot} = \frac{W}{Q_H} = \frac{Q_H - Q_C}{Q_H} = 1 - \frac{Q_C}{Q_H} \tag{6.5}$$

Please note that Q_C is the heat leaving the engine; this will always have a negative value, which ensures that η_{Carnot} is always less than 1. One thing to keep in mind is that the only physical parameters we specified for the Carnot engine cycle were the temperatures of the source at (T_H) and sink at (T_C). Clearly, the ratio (Q_C/Q_H) depends only on T_H and T_C. The nature of this dependency turns out to be very simple and is given by:

$$\frac{Q_C}{Q_H} = \frac{T_C}{T_H} \tag{6.6}$$

Hence, equation (6.5) may also, for all practical purposes, be rewritten as:

$$\eta_{Carnot} = 1 - \frac{T_C}{T_H} \tag{6.7}$$

6.4.4 Carnot's conclusions

Carnot reached some conclusions by a few rational deductions, namely,

1. The efficiency of an engine that operates under real conditions, like under the effect of drag forces, is always less than one.
2. Among all engines exchanging heat at different temperatures, the one with highest efficiency is the one that exchanges heat at the two extreme temperatures (the hottest and the coldest)
3. All reversible engines working between the same temperature extremes have the same efficiencies.

Please keep in mind that a Carnot cycle does not make a practical real engine. One of the biggest reasons is that heat transfer only happens during the isothermal steps, which occurs very, very slowly. This means that it will take a really long time to complete this step and an engine made of two isothermal steps will proceed too slowly to be of any practical use. It is important to remember that Professor Schroeder once said, 'So don't bother installing a Carnot engine in your car; while it would increase your gas mileage, you would be passed on the highway by pedestrians'!

Worked example 6.5. Prove directly (by calculating heat taken in and heat expelled) that a Carnot engine using ideal gas as the working substance has an efficiency of $1 - \frac{T_C}{T_H}$.

Solution: The heat added to the system is given by,

$$Q_H = \int_{V_1}^{V_2} pdV = nRT_H \ln \frac{V_2}{V_1}$$

and the heat leaving the system is mathematically given by,

$$Q_C = \int_{V_3}^{V_4} pdV = -nRT_C \ln \frac{V_4}{V_3} = nRT_C \ln \frac{V_3}{V_4}$$

Using these equations we can calculate thermal efficiency as:

$$\eta = 1 - \frac{Q_C}{Q_H} = 1 - \frac{T_C \ln \frac{V_3}{V_4}}{T_H \ln \frac{V_2}{V_1}}$$

For an adiabatic process we know, $TV^{\gamma-1}$ = constant. For step (2) to (3) we can write,

$$T_H V_2^{\gamma-1} = T_C V_3^{\gamma-1}$$

For step (4) to (1) we get,

$$T_H V_1^{\gamma-1} = T_C V_4^{\gamma-1}$$

Divide the above equations,

$$\frac{V_2^{\gamma-1}}{V_1^{\gamma-1}} = \frac{V_3^{\gamma-1}}{V_4^{\gamma-1}}$$

This implies,

$$\frac{V_3}{V_4} = \frac{V_2}{V_1}$$

Which means the thermal efficiency has to be,

$$\eta = 1 - \frac{T_C}{T_H}$$

Concept questions 6.1.

1. What is the most efficient heat engine?
2. Can a heat engine be 100% efficient? Explain in terms of the Second law of thermodynamics.
3. What factors reduce the efficiency of a heat engine from its ideal value?
4. Why is the Carnot cycle not encountered in real life?
5. For a Carnot cycle which of the following statements is true:
   (a) it is bounded by two isotherms and two adiabats on a pV graph;
   (b) consists of two isothermal and two constant volume processes;
   (c) is any four sided process on a pV graph;
   (d) only exists for an ideal gas;
   (e) has an efficiency equal to the enclosed area on a pV graph.

Exercises 6.1.

1. A heat engine absorbs 50 kJ of heat and exhausts 40 kJ of heat in each cycle. Calculate,
   (a) efficiency of the cycle.
   (b) work done by engine per cycle.
2. Engine (I) receives three times the heat input and exhausts two times the heat while producing five times the work per cycle as compared to engine (II). Calculate the efficiency of each engine.
3. In a Carnot cycle the isothermal expansion takes place at 140 °C and the isothermal compression occurs at 25 °C. During the expansion, 2000 J of heat are transferred to the working fluid. Determine,
   (a) the work done by the gas during the isothermal expansion.
   (b) the heat rejected from the gas during the isothermal compression
   (c) work done on the gas during the isothermal compression.

6.5 Fuels and combustion processes

Materials that can be burned to release thermal energy are called '*fuels*'. Most familiar fuels are liquid hydrocarbon fuels that are obtained from the Earth-drilled crude oil[4] by a process called '*distillation*'. Hydrocarbons, as the name suggests, consist primarily of compounds made of hydrogen and carbon atoms and are denoted as C_nH_m, where n and m are integral values. For example, gasoline is called octane, with chemical formula (C_8H_{18}), and diesel is called dodecane, with chemical formula ($C_{12}H_{26}$)[5]. It is important to remember that both the naturally occurring crude oil extracted from the Earth and petroleum products resulting from refining the crude oil are called '*petroleum products*'. The word petroleum comes from the Latin words *petra* which means rock and *oleum* which means oil, to literally translate as 'rock oil'.

Additionally, over a million vehicles in the world run on compressed natural gas (CNG) stored in the gas phase at pressures between 150 and 250 atm or on liquefied natural gas (LNG) stored in the liquid phase at −162 °C. In countries where natural gas reservoirs are readily available, CNG is cheaper for the local population as compared to the cost of imported and refined gasoline. The use of natural gas is also much friendlier to the environment than burning gasoline and/or diesel fuel. Another common household fuel is propane, also called liquefied petroleum gas (LPG), which is used in many households across the US for heating houses in the winter and heating water in water heaters.

6.5.1 Distillation process

The distillation process, also called '*fractional distillation*', is centuries old, and is used in many different applications from alcohol distillation to oil extraction. In this

[4] Crude oil, also called fossil fuel, is formed when large quantities of dead organisms are buried under rock under intense heat and pressure.

[5] 'oct' is Latin for 8 and 'dodeca' is Latin for 12.

section we concentrate on the distillation process as it relates to fuel requirements in machines.

Crude oil, as discussed above, is a mixture of different chemicals called hydrocarbons. Crude oil is first heated and then put into a distillation column, also known as a still, which turns the liquid crude oil into a mixture of gases in the column. Hot gases rise up in the column and cool as they rise until they condense and turn back into liquid again. These condensed liquids are recovered at different temperatures that correspond to different heights, as shown in figure 6.4.

Lighter products, such as butane and other liquid petroleum gases (LPG), and gasoline blending components are recovered at the lowest temperatures, <85 °F. Mid-range products that include jet fuel and kerosene are recovered between 350 °F and 450 °F, and distillates, such as heating oil and diesel fuel, are recovered at the mid-range temperatures (450–650 °F). The heaviest products such as heavy gas oil are recovered between 650 °F and 1050 °F and residual fuel oil is recovered at temperatures over 1000 °F.

6.5.2 Combustion process

A chemical reaction during which a fuel is oxidized and a large quantity of energy is released is called '*combustion*'. Simply put combustion is a process of burning fuel which requires an oxidizer to make the fuel burn. As a result of the combustion process energy is released, along with some byproducts.

During a combustion process, the components that exist before the reaction are called '*reactants*' and the components that exist after the reaction are called '*products*'. For example, in the reaction, $2H_2 + O_2 \rightarrow H_2O$, the H_2 and O_2 are '*reactants*', since they exist before combustion and water (H_2O) is the '*product*' since it exists after combustion. It is worth remembering that we needed to add (× 2) in the above reaction to balance the equation because the principle of conservation of mass

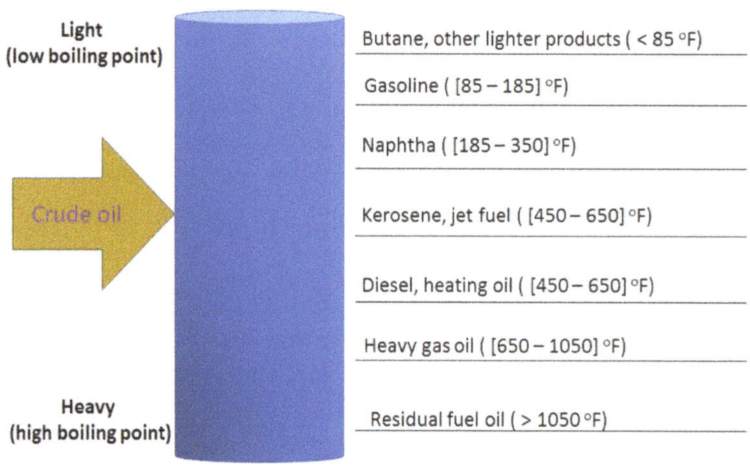

Figure 6.4. Crude oil distillation.

(or the mass balance) states, the total mass of each element must be conserved during a chemical reaction.

The most common oxidizer used in combustion processes is regular air, which is made up by volume of approximately 21% oxygen, 78% nitrogen, 0.9% argon, and traces of carbon dioxide, helium, neon, and hydrogen each. Pure oxygen (O_2) as an oxidizer is used where air cannot be used, such as during cutting and welding. In the analysis of combustion processes, gases that exist in trace amounts can be neglected. Therefore, air can be approximated as 21% oxygen and 79% (nitrogen + argon) by number of moles. Therefore, each mole of oxygen entering a combustion chamber is accompanied by (0.79/0.21) = 3.76 moles of nitrogen. That is,

$$1 \text{ kmol } O_2 + 3.76 \text{ kmol } N_2 = 4.76 \text{ kmol air} \tag{6.8}$$

It is worth remembering here that the number of atoms must be preserved during a chemical reaction, but the total mole numbers need not be. Having nitrogen in the combustion chamber is very helpful for the combustion process because nitrogen absorbs a large amount of the chemical energy released during combustion. Other than that, nitrogen behaves as an inert gas and does not react with other elements[6].

The fuel in combustion chamber must be brought above its respective ignition temperature to start the combustion process, since bringing a fuel in contact with oxygen is not enough to start a combustion process. The minimum ignition temperature varies with the substance used, for example, it is approximately 260 °C for gasoline, 580 °C for hydrogen and 630 °C for methane in air.

A quantity used in the analysis of combustion processes to quantify the amounts of fuel and air needed is called the **air–fuel ratio**, denoted as AF[7]. AF is usually expressed as,

$$\text{AF} = \frac{\text{mass of air}}{\text{mass of fuel}} = \frac{m_{\text{air}}}{m_{\text{fuel}}} \tag{6.9}$$

or in terms of molar mass (M):

$$\text{AF} = \frac{\text{mole numbers of air}}{\text{mole numbers of fuel}} = \frac{nM_{\text{air}}}{\Sigma(n_i M_i)_{\text{fuel}}} \tag{6.10}$$

where n is number of moles.

A combustion process is considered '*complete*' when all the combustible components of the fuel are completely burned during the combustion process. The minimum amount of air needed for the complete combustion of a fuel is called the '**stoichiometric**' or the '**theoretical air**'. The theoretical air is also referred to as the '*chemically correct amount of air*'. The ideal combustion process during which the fuel is burned completely with 100 percent theoretical air is called the '**stoichiometric**'

[6] Please note, at very high temperatures, such as those reached in internal combustion engines, a small fraction of nitrogen reacts with oxygen and forms hazardous gases like nitric oxide (NO).

[7] The reciprocal of air–fuel ratio is called the fuel–air ratio and is denoted as FA.

or '**theoretical combustion**' of that fuel. For example, the theoretical combustion of methane is,

$$CH_4 + 2[O_2 + 3.76N_2] \rightarrow CO_2 + 2H_2O + 7.5N_2 \tag{6.11}$$

Because the products of the combustion contain no unburned methane (CH_4), or carbon (C), hydrogen (H_2), carbon monoxide (CO), OH, or free O_2 this combustion process is considered a complete combustion.

The combustion process is considered '*incomplete*' if the combustion products contain unburned fuel or components such as C, H_2, CO, or OH at the end. Incomplete combustion may occur due to many reasons, such as,

1. Insufficient supply of air or oxygen in the combustion chamber.
2. Insufficient mixing of fuel and the oxygen during contact time duration.
3. Dissociation, a process where the heat is absorbed rather than ejected in the combustion chamber. Dissociation can be considered as the disintegration of combustion produce at high temperatures.

Worked example 6.6. Octane (C_8H_{18}) is burned with 250 percent theoretical air, which enters the combustion chamber at 25 °C, as seen in figure 6.5. Assuming complete combustion and all gases are ideal and the combustion takes place at a total pressure of 1 atm, determine the air–fuel (AF) ratio and the fuel–air (FA) ratio of the products (=CO_2, H_2O, O_2, and N_2).

Given, the molar mass of C = 12 kg kmol^{-1}, H_2 = 2 kg kmol^{-1}, and air = 29 kg kmol^{-1}.

Solution: The combustion equation in this case,

$$\begin{aligned} C_8H_{18} + 2.5a_{th}[O_2 + 3.76N_2] &\rightarrow 8CO_2 \\ &+ 9H_2O + 1.5a_{th}O_2 + (2.5 \times 3.76)a_{th}N_2 \end{aligned} \tag{6.12}$$

where a_{th} is the stoichiometric coefficient for air, is determined by balancing the equation for O_2 as,

$$\text{Balance } O_2\text{: } 2.5a_{th} = 8 + \frac{9}{2} + 1.5a_{th}$$
$$\Longrightarrow a_{th} = 12.5$$

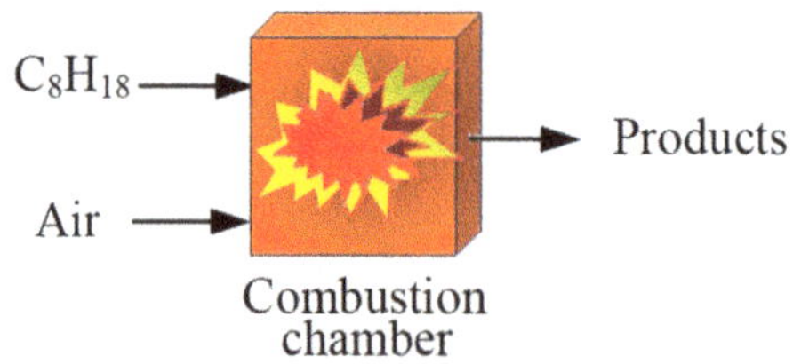

Figure 6.5. Combustion chamber schematic.

Substituting value in equation (6.12).

$$C_8H_{18} + (2.5 \times 12.5)[O_2 + 3.76N_2] \rightarrow 8CO_2 + 9H_2O + 1.5a_{th}O_2 + (2.5 \times 3.76)(12.5)N_2$$

$$C_8H_{18} + (31.25)[O_2 + 3.76N_2] \rightarrow 8CO_2 + 9H_2O + (1.5 \times 12.5)O_2 + (117.5)N_2$$

This means, given, the molar mass of C = 12 kg kmol^{-1}, H_2 = 2 kg kmol^{-1}, and air = 29 kg kmol^{-1}, we can write,

$$AF = \frac{m_{air}}{m_{fuel}} = \frac{(31.25 \times 4.76\ \text{kmol})(29\ \text{kg kmol}^{-1})_{air}}{(8\ \text{kmol} \times 12\ \text{kg kmol}^{-1})_C + (9\ \text{kmol} \times 2\ \text{kg kmol}^{-1})H_2}$$
$$= 37.8\ \frac{\text{kg of air}}{\text{kg of fuel}}$$

The fuel–air ratio (FA) is the inverse of the air–fuel ratio (AF),

$$FA = \frac{1}{AF} = \frac{1}{37.8} = 0.026\ \text{kg fuel/kg air}.$$

6.5.3 Actual combustion processes

In actual combustion processes, more air is used than the required '*theoretical air*' amount. This is usually done to either control the temperature of the combustion chamber or to increase the chances of a complete combustion. The amount of air in excess of the '*theoretical air*' amount is called '***excess air***'. The amount of excess air in a process is usually expressed as a '*percent excess air*' or '*percent theoretical air*', for example, 200 percent excess air is equivalent to 300 percent theoretical air, or 50 percent excess air is equivalent to 150 percent theoretical air. The '*theoretical air*' itself can be expressed as either 0 percent excess air or equivalently 100 percent theoretical air.

Amounts of air less than the stoichiometric amount are called '***deficiency of air***' and are often expressed as '*percent deficiency of air*', for example, 90 percent theoretical air is equivalent to 10 percent deficiency of air.

Worked example 6.7. Comparison between '***excess air***' and '***deficiency of air***'.

300 percent theoretical air is equivalent 200 percent excess air.
150 percent theoretical air is equivalent to 50 percent excess air.
100 percent theoretical air is equivalent to 0 percent excess air.
90 percent theoretical air is equivalent to 10 percent deficiency of air.
60 percent theoretical air is equivalent to 40 percent deficiency of air.

The amount of air used in combustion processes is also expressed in terms of the '***equivalence ratio***', which is the ratio of the actual fuel–air ratio to the stoichiometric fuel–air ratio.

$$\text{Equivalence ratio} = \frac{\text{actual fuel air ratio}}{\text{stoichiometric fuel air ratio}}$$

When dealing with actual combustion processes in practical engines things get complicated. Actual combustion processes are never complete even when excess air is present. It is therefore impossible to predict the composition of the products based only on the mass balance equation alone. The only option, therefore, is to measure the amount of each component in the product directly.

Suppose we want to measure the amount of CO_2 in our sample. We do this by using a device called the '*Orsat gas analyzer*' where we measure the amount of each component in our sample individually. To measure the amount of each component the combustion gases are cooled first to standard atmospheric conditions, then the sample is interacted with a chemical which absorbs the component we are interested in, in this case the CO_2. The volume of the remaining gas is then measured again at standard atmospheric conditions. The reduction in volume of the gas gives the volume fraction of CO_2 in the sample, which is equivalent to the mole fraction, given as,

$$(\text{Mole fraction})_{CO_2} = y_{CO_2} = \frac{\Delta V_{CO_2}}{V_{\text{before}}} \tag{6.13}$$

Worked example 6.8. Suppose a sample of combustion gas has one liter (L) of CO_2 before reacting with a chemical that absorbs carbon dioxide in the '*Orsat gas analyzer*' at 1 atm and 25 °C. After the reaction we measure 0.9 L of CO_2 at 1 atm and 25 °C. Find the mole fraction of the CO_2 in the combustion gas.

Solution: The mole fraction of the CO_2 in the combustion gas is calculated by taking the ratio of the CO_2 before and after the reaction, as,

$$y_{CO_2} = \frac{\Delta V_{CO_2}}{V_{\text{before}}}$$

$$y_{CO_2} = \frac{1 - 0.9}{1} = \frac{0.1}{1} = 0.1$$

The mole fraction of the CO_2 in the combustion gas is 0.1.

The volume fraction of the other gases in the combustion gas can be calculated by repeating this procedure for each individual gas.

Worked example 6.9. In a combustion chamber, ethane (C_2H_6) is burned steadily at a rate of 8 kg h^{-1} with air that enters the combustion chamber at a rate of 176 kg h^{-1}. Determine the percentage of excess air used during this process.

Assume, combustion is complete and the combustion products contain only (CO_2, H_2O and N_2).

Solution: First we have to write the theoretical combustion equation, which in this case will be,

$$C_2H_6 + a_{th}[O_2 + 3.76N_2] \rightarrow 2CO_2 + 3H_2O + 3.76a_{th}N_2 \quad (6.14)$$

where a_{th} is the stoichiometric coefficient for air, determined by the O_2 balance in equation (6.14) as,

$$\text{Balance } O_2\text{:} \quad a_{th} = 2 + 1.5$$
$$\Longrightarrow a_{th} = 3.5$$

Substituting values in equation (6.14).

$$C_2H_6 + 3.5[O_2 + 3.76N_2] \rightarrow 2CO_2 + 3H_2O + (3.76 \times 3.5)N_2$$
$$C_2H_6 + 3.5[O_2 + 3.76N_2] \rightarrow 2CO_2 + 3H_2O + (13.125)N_2$$

This means, given, the molar mass of C = 12 kg kmol^{-1}, H_2 = 2 kg kmol^{-1}, and air = 29 kg kmol^{-1}, we can write,

$$AF_{th} = \frac{m_{air}}{m_{fuel}} = \frac{(3.5 \times 4.76 \text{ kmol})(29 \text{ kg kmol}^{-1})_{air}}{(2 \text{ kmol} \times 12 \text{ kg kmol}^{-1})_C + (3 \text{ kmol} \times 2 \text{ kg kmol}^{-1})H_2}$$
$$= 16.1 \frac{\text{kg of air}}{\text{kg of fuel}}$$

The actual air–fuel ratio used is the ratio of the mass flow rate of air to the mass flow rate of ethane,

$$AF_{real} = \frac{\dot{m}_{air}}{\dot{m}_{fuel}} = \frac{176 \text{ kgh}^{-1}}{8 \text{ kgh}^{-1}} = 22\frac{\text{kg of air}}{\text{kg of fuel}}$$

From these the percentage of theoretical air used is,

$$\text{Percent theoretical air} = \frac{AF_{real}}{AF_{th}} = \frac{22}{16.1} = 137\%$$

This implies the excess air used during this process is 37%.

6.5.4 Enthalpy consideration in a combustion process

During a chemical reaction, some chemical bonds are broken while new ones are formed. The chemical energy associated with these bonds is different for the reactants and the products, which must be accounted for in the energy balance equation.

While analyzing a reacting system, the heat transfer (Q) during the process is equal to the difference between the enthalpy of the products ($H_{product}$) and the enthalpy of the reactants ($H_{Reactant}$), as,

$$Q = H_{product} - H_{Reactant} \quad (6.15)$$

When the reactants and the products are at the same common state, chosen to be at 1 atm and 25 °C, the change in enthalpy during this process is only due to the changes in the chemical composition of the system. This property is called the '***enthalpy of reaction***' (h_R), and is defined as *the difference between the enthalpy of the products and the enthalpy of the reactants at the same state for a complete reaction.*

For a combustion process, the enthalpy of reaction is called the '***enthalpy of combustion***' (h_C), which is the amount of heat released during a steady-flow combustion process when 1 kg (= 1 kmol) of fuel is burned completely.

$$h_C = H_{\text{product}} - H_{\text{Reactant}} \tag{6.16}$$

The enthalpy of combustion is not a useful quantity to know if the combustion is '*incomplete*'. Therefore, we need a more fundamental property to represent the chemical energy of a state. This property is called the '***enthalpy of formation***' ($\bar{h}_f$), defined as, *the enthalpy of a substance at a specified state due to its chemical composition.* The enthalpy of formation ($\bar{h}_f$) of all stable elements[8] at standard state is 0. Please note that sometimes textbooks use superscript (°) to show properties at the standard condition state of 1 atm and 25 °C, ($\bar{h}_f = \bar{h}°_f$ at the standard condition state).

The change in enthalpy during the formation of CO_2 from its elements carbon (C) and oxygen (O_2) at 25 °C and 1 atm in a steady-flow process is found to be −393 520 kJ kmol^{-1}. In this formation the '***enthalpy of formation***' ($\bar{h}_f$) = 0 since both reactants are elements at the standard state. Therefore, the enthalpy of formation of CO_2 at the standard state is,

$$\bar{h}°_f = -393\ 520 \text{ kJ kmol}^{-1}.$$

In other words, 393 520 kJ of chemical energy is released as heat, Q when carbon (C) and O_2 combine to form 1 kmol of CO_2. A negative enthalpy of formation for a compound indicates an *exothermic reaction.* Conversely, a positive enthalpy of formation for a compound indicates an *endothermic reaction.* To take this into account, a term called the '***heating value of the fuel***' is used, expressed in kJ kg^{-1} of fuel, and is defined as *the amount of heat released when a fuel is burned completely in a steady-flow process and the products are returned to the state of the reactants.* In other words, the heating value of a fuel is equal to the absolute value of the enthalpy of combustion of the fuel, as,

$$\text{Heating value} = |h_C| = \left|(H_{\text{product}} - H_{\text{Reactant}})\right| \tag{6.17}$$

[8] The stable form of an element is the chemically stable form of that element at 1 atm and 25 °C. Examples include O_2, N_2, H_2, and C. For carbon the stable form is graphite not diamond at standard conditions. The stable form of nitrogen at the standard state is diatomic nitrogen (N_2), not the monatomic nitrogen (N), since N_2 exists in diatomic form at 1 atm and 25 °C.

Concept questions 6.2.

1. Define air–fuel ratio. How is it different from fuel–air ratio?
2. What are the causes of incomplete combustion?
3. Is the number of atoms of each element conserved during a chemical reaction?
4. Are the total number of moles conserved during a chemical reaction?

Exercises 6.2.

1. What are the approximate chemical compositions of gasoline, diesel fuel, and natural gas?
2. Methane (CH_4) is burned with stoichiometric amount of air during a combustion process. Assuming complete combustion, determine the air–fuel ratio (AF) and fuel–air (FA) ratio. Given, the molar mass of C = 12 kg $kmol^{-1}$, H_2 = 2 kg $kmol^{-1}$, and air = 29 kg $kmol^{-1}$.
3. Propane (C_3H_8) is burned with 75 percent excess air during a combustion process. Assuming complete combustion, determine the air–fuel ratio.

Further reading

- Cengel Y and Boles M 2006 *Thermodynamics: An Engineering Approach* 5th edn (New York: McGraw-Hill Higher Education)
- Resnick R, Halliday D and Krane K 2002 *Physics* vol 1 5th edn (New York: Wiley)
- Schroeder D 2000 *Thermal Physics* (Reading, MA: Addison-Wesley Longman)

Chapter 7

Work-to-heat transfer cycle

In this chapter we discuss work-to-heat-transfer cycles and introduce the concepts of refrigerators and heat pumps. We will also measure of the performance of refrigerators and heaters. As you go through this chapter, please remember that I will not be discussing how to make a refrigerator or a heater, but will discuss the physics behind both refrigerators and heaters.

7.1 Work-to-heat transfer cycles

In this text, I will call the cycles that use mechanical work to transfer heat from low temperature regions to high temperature regions, '*work-to-heat transfer*' cycles. A device that does this is called a '***heat pump***' and the cycles are sometimes also called '***temperature cycles***'.

There are two different devices that use work-to-heat transfer cycles. When we use a work-to-heat transfer cycle to warm a heat source, for example, when we want to increase the temperature of the inside of a warm home on a cold day, the device is called a '**heater**'. When we use a work-to-heat transfer cycle to cool the heat sink, as done in the normal operation of a refrigerator or a freezer, the device is called a '**refrigerator**'. In both cases the operating principle is the same; heat is moved from a cold place to a warm place, which does not happen spontaneously and input work is needed to make it happen, as shown in figure 7.1.

In this chapter we discuss the operations of a refrigerator and a heater by employing the '***refrigeration cycle***'.

7.1.1 Refrigeration cycle

We know from the laws of thermodynamics that heat naturally flows from an object at a higher temperature to an object at lower temperature. In refrigeration, heat is forcibly transferred from a region of lower temperature to a region of higher

doi:10.1088/978-0-7503-3083-1ch7

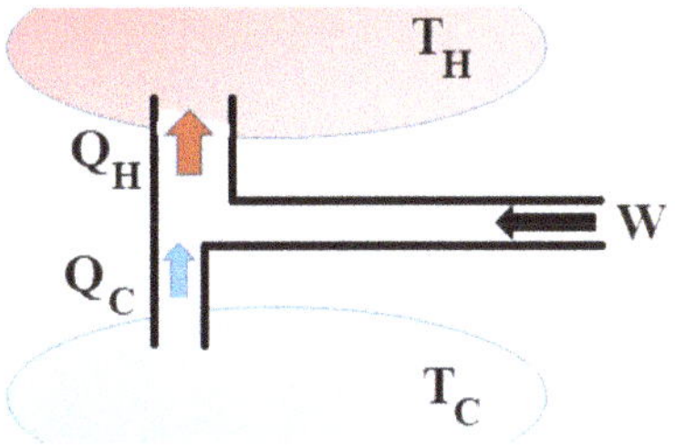

Figure 7.1. Energy-flow diagram for a temperature cycle, T_H and T_C are the temperatures of the source and sink respectively, Q_H and Q_C are heat transferred to the source and sink, respectively, and W is the mechanical work input during this heat transition.

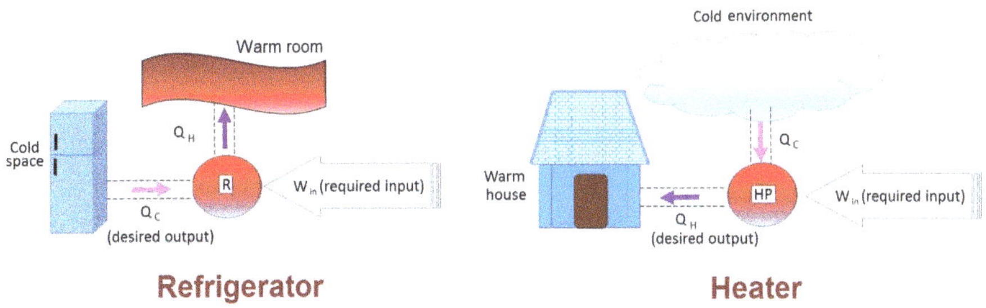

Figure 7.2. Types of heat pumps: Refrigerator versus Heater.

temperature. Devices that produce refrigeration are called '*refrigerators*' and/or '*heaters*', and they operate on the cycles called '*refrigeration cycles*'. Refrigeration is achieved using ***principles of refrigeration*** which state:

1. Liquids absorb heat when they change from a liquid to gas.
2. Gases give off heat when they change from a gas to a liquid.

The most commonly used refrigeration cycle is the '*vapor-compression refrigeration cycle*' in which the working fluid, called the '***refrigerant***', is vaporized and condensed alternately. A schematic of a vapor-compression refrigeration cycle is shown in figure 7.3.

As discussed above, the two main types of devices that use work-to heat transfer cycles and operate using refrigeration cycles are '*refrigerators*' and '*heaters*'.

1. **Refrigerators:** The main purpose of a refrigerator is to remove heat, called the '***cooling load***' ($\dot{Q}_L$), from a low temperature medium (T_C), inside of a fridge, to a high temperature medium, such as a kitchen at room temperature (T_H), as seen in figure 7.2(a). The cooling load is also called the 'refrigeration affect' or 'tons of refrigeration'. Refrigerators are also called '**reverse Carnot cycles**', detailed in section 7.1.3 and the performance of a refrigerator is defined as the coefficient of performance (COP), mathematically given as,

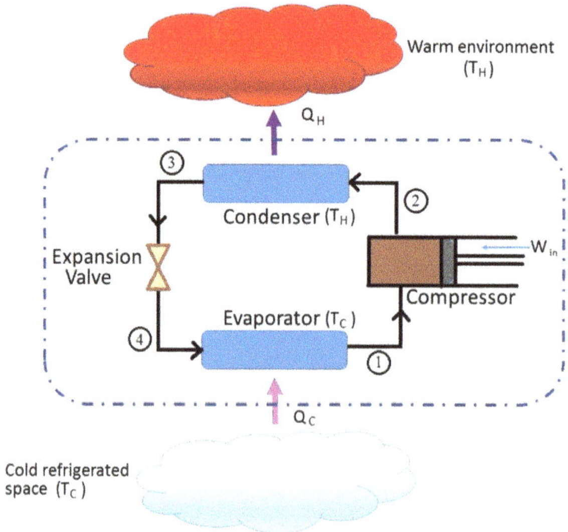

Figure 7.3. Vapor-compression cycle: refrigerator.

$$\text{Coefficient of Performance, } \mathrm{COP_R} = \frac{\text{Desired output}}{\text{required input}} = \frac{\text{Cooling effect}}{\text{work input}} = \frac{Q_\mathrm{C}}{W_\mathrm{in}} \tag{7.1}$$

The cooling load ($\dot{Q}_\mathrm{L}$) is mathematically given as,

$$\dot{Q}_\mathrm{L} = \frac{dQ_\mathrm{L}}{dt} = \frac{dm}{dt}(H_1 - H_4) = \dot{m}(H_1 - H_4) \tag{7.2}$$

where $\dot{m}$ (= dm/dt) is the mass flow and H's are the numerical values of enthalpies at the various steps in the refrigeration cycle. The values of enthalpies for the various refrigerants can be found in most engineering or chemistry textbooks.

2. **Heaters:** The main purpose of a 'heater' is to transfer heat, called the '***heating load***' from a low temperature region to a high temperature region. The performance of a heater is mathematically given as,

$$\text{Coefficient of Performance, } \mathrm{COP_H} = \frac{\text{Desired output}}{\text{required input}} = \frac{\text{Heating effect}}{\text{work input}} = \frac{Q_\mathrm{H}}{W_\mathrm{in}} \tag{7.3}$$

Notice that both Q_H and Q_C represent magnitudes and thus are positive quantities and COP_R and COP_H can be greater than 1. The relationship between COP_R and COP_H for fixed values of Q_H and Q_C under similar operating conditions is,

$$COP_H = COP_R + 1 \tag{7.4}$$

These relations can also be expressed as a rate simply by replacing Q_H by $\dot{Q}_H$ and Q_C by $\dot{Q}_C$.

7.1.2 Refrigerant

Refrigerant is the name given to the working fluid used in the refrigeration cycle. Mechanical refrigeration is accomplished by continuously circulating, evaporating, and condensing a fixed amount of refrigerant in a closed system. Refrigerant transfers heat from an area of low temperature, the refrigerator cabinet, held well below the room temperature, to an area of high temperature, the kitchen, generally held at room temperature or above.

Fluid used as a refrigerant should ideally have a low boiling temperature, since the refrigerant changes back and forth from gas to a liquid during the refrigerator cycle. A low boiling point ensures that less power will be required to change phase between liquid and gas. Most common fluids used as refrigerants include R-12 (CCl_2F_2), R-22 ($CHClF_2$), R-134a (CF_3CH_2F), carbon dioxide, ammonia (NH_4) and CFCs (chlorofluorocarbons), that were developed in the 1930s. The most familiar CFC is 'Freon-12' (CCl_2F_2), and it was very successful as a refrigerant. However, CFCs are bad for the environment since they break down the ozone layer that protects the Earth's atmosphere. Chlorofluorocarbons over the years have steadily been replaced by chlorine-free compounds called '***hydrofluorocarbons***' (HFCs) that contain hydrogen, fluorine, and carbon only and are safer for the Earth.

Critical temperature: The critical temperature ($T_{Critical}$) of a refrigerant is the temperature above which the refrigerant cannot be liquefied regardless of the pressure applied. Ideally the critical temperature of a refrigerant should be as high as possible above the condensing temperature to have the greatest heat transfer at any given temperature. If not, the system will be adversely affected in two ways: one, the system will have excessive power consumption; and two, the system may fail because of pressure and temperature build-up. One solution for both problems is to use hydrocarbons in very hot climates. Hydrocarbons have very high critical temperatures and reject heat 50% faster than chemical refrigerants such as HFCs.

7.1.3 Reverse Carnot cycle

As discussed in section 6.4 a Carnot cycle is a completely reversible cycle that consists of two reversible isothermal processes and two reversible adiabatic processes. A Carnot cycle has the maximum thermal efficiency for any cycle operating under the same temperature limits and serves as a standard against which real thermodynamic cycles can be compared. Since a Carnot cycle is a reversible cycle, we can reverse the processes of a regular Carnot cycle by reversing the directions of heat and work done during the processes. Hence, any cycle that operates in the counterclockwise direction on a pV diagram of a Carnot cycle is called a '***reversed Carnot cycle***', as shown in figure 7.4.

Devices that operate on a reversed Carnot cycle between two constant temperature reservoirs, one at high temperature (T_H) and the other at low temperature (T_C), are called reverse Carnot refrigerators and are as shown in figure 7.5. From the figure, we see heat (Q_C) is absorbed from the cold environment and heat (Q_H) is

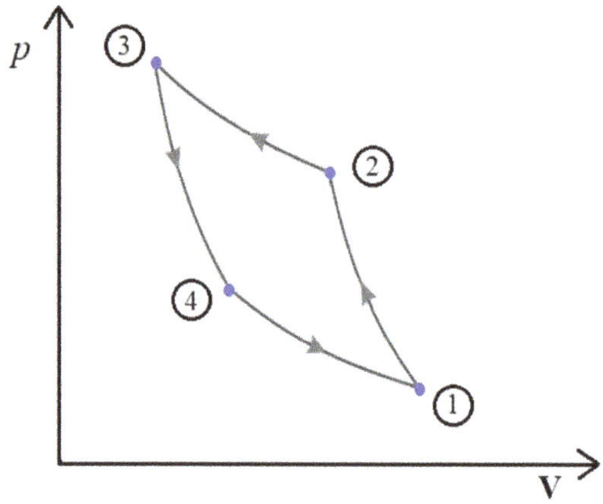

Figure 7.4. pV-diagram for a reversed Carnot refrigerator.

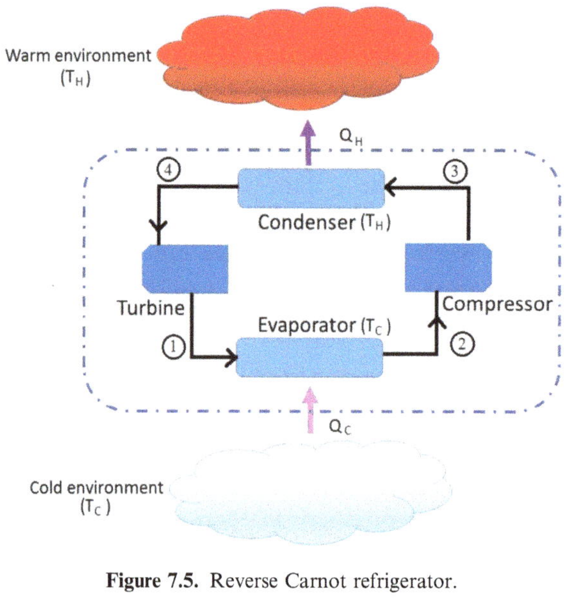

Figure 7.5. Reverse Carnot refrigerator.

rejected to a warm environment. In this case, work input in the amount of W_{rev} is required to make the transfer of heat happen. We know from the First law of thermodynamics the required work is,

$$W_{rev} = -Q_H - Q_C$$

where Q_H is a negative value and Q_C has a positive value.

Similar to the Carnot cycle, the reversed Carnot cycle consists of two isentropic steps and two isothermal steps; however, in this case the process proceeds in a direction $3 \rightarrow 2 \rightarrow 1 \rightarrow 4 \rightarrow 3$, as follows:

1. **Step (3) to (2): Reversible isentropic compression:** In this step the system is insulated and no heat is lost or absorbed. Refrigerant (in gas phase) is compressed slowly until the temperature rises from low temperature (T_C) to high temperature (T_H).
2. **Step (2) to (1): Reversible isothermal compression:** During this step, refrigerant (still in gas phase) is compressed reversibly at constant temperature (T_H) and heat is rejected. From the Second law of thermodynamics, rejected heat is given as,

$$Q_H = T_H \Delta S_{2\rightarrow 1}$$

3. **Step (1) to (4): Reversible isentropic expansion:** Step 1 to 4 is isentropic so no heat is lost or absorbed. Refrigerant (still in gas phase) expands slowly until the temperature drops from T_H to T_C.
4. **Step (4) to (3): Reversible isothermal expansion:** During step 4 to 3, heat is absorbed. Gas is compressed reversibly at the constant temperature T_H.

$$Q_C = T_C \Delta S_{4\rightarrow 3}$$

Please keep in mind that since this is a reversible process, $\Delta S_{total} = 0$, $\Longrightarrow \Delta S_{2-1} = \Delta S_{4-3}$.

The definition of refrigerator or heater depends on our interest. If we are interested in the cooling load, the cycle is called a 'reverse Carnot refrigerator', and if we are interested in the heating load, the cycle is called a 'reverse Carnot heater'.

The coefficient of performance for the reverse Carnot refrigerator is mathematically given as,

$$\mathrm{COP}_{R,\mathrm{Carnot}} = \frac{Q_C}{W_{rev}} = \frac{Q_C}{-Q_H - Q_L}. \tag{7.5}$$

From the Second law of thermodynamics, $Q_H = T_H \Delta S_{2-1}$ and $Q_C = T_C \Delta S_{4-3}$, so we can write the above equation as,

$$\mathrm{COP}_{R,\mathrm{Carnot}} = \frac{Q_C}{-Q_H - Q_C} = \frac{T_C \Delta S_{4-3}}{-(T_H \Delta S_{2-1}) - (T_H \Delta S_{4-3})} = \frac{T_C}{T_H - T_C} \tag{7.6}$$

Since $\Delta S_{2-1} = \Delta S_{4-3}$. This means we can write the coefficient of performance for the reverse Carnot refrigerator in terms of temperatures as,

$$\mathrm{COP}_{\mathrm{R,Carnot}} = \frac{1}{T_{\mathrm{H}}/T_{\mathrm{C}} - 1}. \tag{7.7}$$

The coefficient of performance for the reverse Carnot heater is given as,

$$\mathrm{COP}_{\mathrm{H,Carnot}} = \frac{-Q_{\mathrm{H}}}{W_{\mathrm{rev}}} = \frac{-Q_{\mathrm{H}}}{-Q_{\mathrm{H}} - Q_{\mathrm{C}}}. \tag{7.8}$$

and using the same steps as above, we can write the coefficient of performance for the reverse Carnot heater in terms of temperatures as,

$$\mathrm{COP}_{\mathrm{H,Carnot}} = \frac{T_{\mathrm{H}}}{T_{\mathrm{H}} - T_{\mathrm{C}}} = \frac{1}{1 - T_{\mathrm{C}}/T_{\mathrm{H}}}. \tag{7.9}$$

Please note that both COPs increase as the difference between the two temperatures decreases, that is, as T_{C} rises or as T_{H} falls.

7.2 A refrigerator

A schematic of a real fridge is shown in figure 7.6. A refrigerator is made up of many components but for the study of thermodynamics we consider it to be made of a compressor, a condenser, throttle value and evaporator coils. As you go through this section, please keep in mind that real refrigerators are much more complicated. My purpose in this text is to talk about what refrigerators do, not how to build them.

A refrigerator cycle involves the four processes detailed here (figure 7.7).

- **Step 1: Compressor: adiabatic expansion.** The refrigerant (working fluid) passes through a nozzle and expands into a low-pressure area. The same way that carbon dioxide cools down when it comes out of a fire extinguisher, the refrigerant cools down and turns into a gas. This is essentially an adiabatic

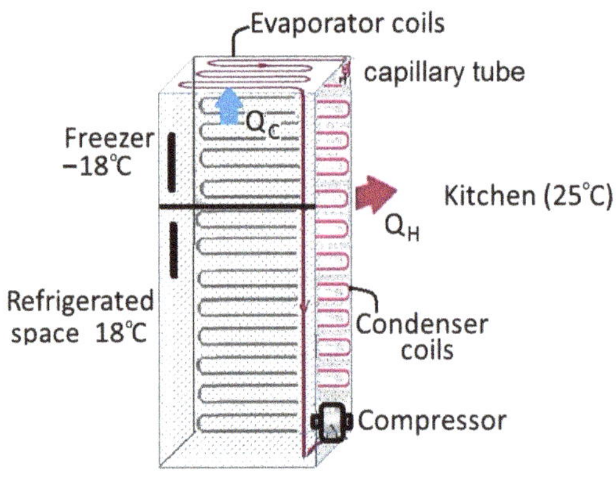

Figure 7.6. Schematic of a real fridge.

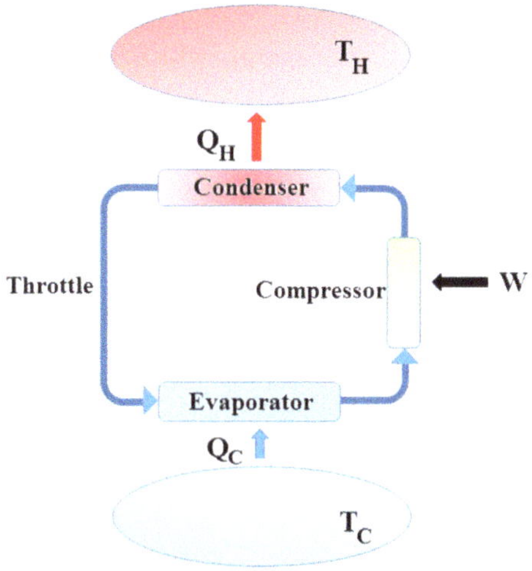

Figure 7.7. Energy-flow diagram for a refrigeration cycle.

expansion and entropy (Sa) during this process is constant. The work done in the compressor during this step in terms of enthalpy is,

$$\begin{aligned}\frac{dW}{dt} &= \frac{dm}{dt}(H_2 - H_1)\\ \dot{W}_{\text{in}} &= \dot{m}(H_2 - H_1)\end{aligned} \tag{7.10}$$

- **Step 2: Condenser/cooler: isobaric expansion.** Next the refrigerant is cooled at constant pressure to a saturated liquid state in the condenser coils. The cool refrigerant in thermal contact with the inner compartment of the fridge heats up as the heat is transferred to it from the fridge. This takes place at constant pressure, so it is an isobaric expansion. The amount of heat transferred is given as,

$$Q_{\text{H}} = \frac{dm}{dt}(H_2 - H_3) = \dot{m}(H_2 - H_3) \tag{7.11}$$

- **Step 3: Throttle Valve:** Refrigerant then passes through the throttle valve, also sometimes called the expansion valve, and is throttled into the evaporator. During this process, $\Delta S > 0$ and work (W) and heat (Q) are both zero. This process is both isochoric ($W = 0$) and adiabatic ($Q = 0$) and change in entropy is positive ($\Delta S > 0$), which heats up the refrigerant and,

$$H_4 = H_3 \tag{7.12}$$

- **Step 4: Evaporator:** The hot refrigerant at constant pressure, in liquid form passes through coils on the outside of the fridge and heat is transferred to the kitchen. This is an isobaric (constant pressure) compression process and heat (Q_C) is rejected to the room, which is mathematically is given as,

$$Q_C = \frac{dm}{dt}(H_1 - H_4) = \dot{m}(H_1 - H_4) \tag{7.13}$$

Evaporation occurs at a low temperature and low pressure while condensation occurs at a high temperature and high pressure. Thus, it is possible to transfer heat from an area of low temperature (refrigerator cavity) to an area of high temperature (kitchen).

7.2.1 Efficiency of a refrigerator

A refrigerator is rated by the '*coefficient of performance*', which is the ratio of the benefit (heat removed from the fridge) to the cost (work required to remove it),

$$\text{Coefficient of Performance (COP)} = \frac{\text{benefit}}{\text{cost}} = \frac{Q_C}{W} = \frac{Q_C}{Q_H - Q_C} \tag{7.14}$$

Please note that COP is the opposite of η for a heat engine which is (W/Q_H). There is no obvious upper limit on this quantity and the First law of thermodynamics allows the value of COP to be greater than 1. The Second law of thermodynamics says that the entropy dumped into the hot reservoir must be at least as much as the entropy absorbed form the cold reservoir.

$$\frac{Q_H}{T_H} \geqslant \frac{Q_C}{T_C} \tag{7.15}$$

Put this value in equation (7.14) to get,

$$\text{Coefficient of Performance (COP)} \leqslant \frac{T_C}{T_H - T_C} \tag{7.16}$$

COP is largest when T_H is almost equal to T_C.

For a typical refrigerator with a freezer, $T_H \approx 25\ °\text{C} = 298$ K and $T_C \approx -18° = 255$ K and COP $\approx$ 5.9. This means for each Joule (J) of electrical energy drawn from the wall, the coolant will take away about 5.9 J of heat from inside the fridge or freezer. In this case, the waste heat dumped into the kitchen is 6.9 J.

Worked example 7.1. We have a working fluid in the ideal compression refrigeration cycle. The refrigerant leaves the evaporator at -20 °C and has a condenser pressure of 900 kPa. The mass flow rate is 3 kg min^{-1}. Assuming the refrigerator is placed in a room with temperature 25 °C, find COP_R and the cooling load. Hint: Suppose that enthalpy at the compressor inlet is given by $H_1 = 235$ kJ kg^{-1} and enthalpy at Compressor exit is given by $H_2 = 275$ kJ kg^{-1} and the enthalpy at throttle exit is $H_4 = 99.6$ kJ kg^{-1}.

Solution:

$$\text{Coefficient of Performance(COP)}_R = \frac{\dot{Q}_L}{\dot{W}_{in}} = \frac{\dot{m}(H_1 - H_4)}{\dot{m}(H_2 - H_1)} = 3.41$$

Cooling load is also called 'tons of refrigeration' and is given as

$$\dot{Q}_C = \dot{m}(H_1 - H_4) = \left(3\frac{\text{kg}}{\text{min}}\right)(235 - 99.6)\frac{\text{kJ}}{\text{kg}} \times \frac{1\ \text{Ton}}{211\ \text{kJ min}^{-1}} = 1.93\ \text{ton}$$

Worked example 7.2. If heat is added to a kitchen fridge at a rate of 300 W, and the fridge has an average internal temperature of 0 °C (= 273 K) in a room with temperature 22 °C (= 295 K), then find the electric power required to run the fridge.

Solution: The optimal COP of the fridge is given as:

$$\text{Coefficient of Performance (COP)}_R = \frac{Q_C}{W} = \frac{T_C}{T_H - T_C} = \frac{273}{295 - 273} = 12.4$$

The electric power required to run the fridge is therefore:

$$\text{power} = \frac{300\ \text{W}}{12.4} = 24\ \text{Watt}$$

In practice, fridges tend to use a lot more power than this. A typical 16 cubic foot fridge-freezer uses around 380 W of power. The difference is probably due to several factors, such as the fridge not having an optimal COP, extra energy being required to drive the compressor, and possibly a higher rate of heat entering the fridge.

7.2.2 Real refrigerators

The steps that a real refrigerator goes through are as follows:

- Beginning at the compressor, the refrigerant (working fluid) is compressed adiabatically which raises its pressure ($P_1 \rightarrow P_2$) and temperature ($T_1 \rightarrow T_2$).
- At the condenser, the refrigerant gives up heat and gradually liquifies in the condenser through a network of pipes in thermal contact with the heat source at (T_H), for example, kitchen or room.

- In through the throttle: the refrigerant next passes through a 'throttling valve' which is a narrow opening or a porous plug. It emerges on the other side at a lower pressure ($P_2 \rightarrow P_1$) and temperature ($T_2 \rightarrow T_1$).
- Finally, the refrigerant absorbs heat and turns back into gas in the evaporator which is just a network of pipes in thermal contact with the sink at T_C.

Coefficient of Performance of a standard refrigerator:
The coefficient of performance of a standard refrigerator is expressed in terms of enthalpies of fluid at various points around the cycle. We know that pressure is constant in the evaporator, and the heat absorbed equals the change in enthalpy as: $Q_C = \Delta H_{1\rightarrow 2} = H_1 - H_4$. Similarly the heat expelled in the condenser is: $Q_H = \Delta H_{2\rightarrow 3} = H_2 - H_3$. Then COP will be given as:

$$\text{Coefficient of Performance} = \frac{Q_C}{W} = \frac{Q_C}{Q_H - Q_C} = \frac{\mathrm{H}_1 - \mathrm{H}_4}{\mathrm{H}_2 - \mathrm{H}_3 - \mathrm{H}_1 + \mathrm{H}_4} \tag{7.17}$$

In general it is known that in the refrigeration cycle, $H_4 = H_3$, so COP reduces to:

$$\text{COP} = \frac{H_1 - H_3}{H_2 - H_1} \tag{7.18}$$

7.2.3 The pV graph for a refrigerator

The pV graph is very useful in calculating the work done. For any kind of heat engine or refrigerator (reverse heat engine), the processes involved form a cycle on the pV graph. The work is the area of the enclosed region on the graph. The diagram for a refrigerator is more complicated because of the two phase changes involved during the real refrigeration cycle, but the pV diagram shown in figure 7.8 is basically what it looks like.

Just like a Carnot cycle, the reversed Carnot cycle is the most efficient refrigeration cycle operating between two particular temperature values. Therefore, we should adapt it as the ideal cycle. However, the reversed Carnot cycle is not a suitable model

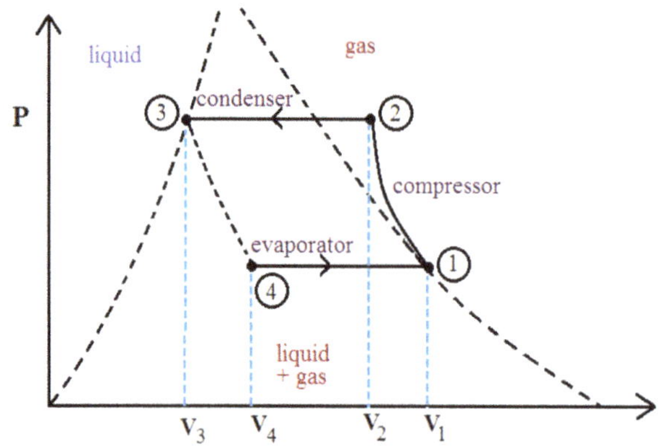

Figure 7.8. pV diagram for a refrigerator.

for a real refrigeration cycle because, even though steps (1 to 2) and (3 to 4), namely the two isothermal heat transfer processes, can be approximated to be ideal by just maintaining a constant pressure to fix the temperature of a two-phase mixture at the saturation value, steps (2 to 3) and (4 to 1) cannot be approximated as ideal in practice. This is because step (2 to 3) requires a compressor to compress liquid–vapor, which means the compressor needs to be able to handle both phases, liquid and gas, whereas, step (4 to 1) is performed using a turbine and involves the expansion of high-moisture-content refrigerant.

It seems like it might be possible to eliminate these issues if we execute the reversed Carnot cycle outside the saturation region. But if we do that it becomes very hard to maintain the isothermal conditions during the heat-absorption and the heat-rejection processes. This means the reversed Carnot cycle is perfect when used as a standard against which actual refrigeration cycles are compared but it cannot be used as a realistic model for real refrigeration cycles.

Worked example 7.3. Can you cool your room by leaving the fridge door open?

Solution: Initially, yes, you can cool your kitchen by leaving your refrigerator door open. But over time, you should not leave a fridge door open in a room to cool the room. Doing so just merges the hot and cold reservoirs into a single environment, which is not a good idea. The fridge will expel heat into the same reservoir from which it extracts it, and the net effect will be zero and eventually no heat will be exchanged. Even worse, the work input to the fridge will be turned into room heat!

7.2.4 Other refrigeration cycles

There are many other refrigeration cycles. I will briefly introduce them here but I would suggest you reference other text for more details. Some common refrigeration cycles include,

1. **Cascade refrigeration systems:** Cascade refrigeration systems can achieve very low temperatures by operating two or more wafer compression systems in series, with the cold part of one fridge acting as the hot part of another. Operating compression systems in series is referred to as '*cascading*' which increases the COP of a refrigeration system.
2. **Gas refrigeration systems:** Heat-transfer-to-power cycles can be used as refrigeration cycles by simply reversing the order of the steps. The reversed Brayton cycle discussed in section 8.4 are called the 'gas refrigeration' cycles. These are used to obtain very low temperatures and are used to cool aircraft. In gas refrigeration systems the output work of the turbine can be used additionally to reduce the input work required by the compressor. The COP of a gas refrigeration cycle is given as:

$$\mathrm{COP} = \frac{\text{Desired output}}{\text{Required input}} = \frac{Q_\mathrm{C}}{W_\mathrm{in}} = \frac{Q_\mathrm{C}}{W_\mathrm{in,compressor} - W_\mathrm{out,turbine}} \tag{7.19}$$

3. **Multistage refrigeration systems:** Multistage refrigeration systems use two compressors, one called the high-pressure compressor and the other called the low-pressure compressor, one after the other to increase the refrigeration effect.
4. **Absorption refrigerators:** The absorption refrigeration cycle is a very economical cycle from the practical standpoint, however, it requires a heat source between 100 °C and 200 °C. The typical refrigerators installed in recreational vehicles (RVs) are absorption refrigerators. Absorption refrigerators use 'generators' between the evaporator and condenser instead of the compressor, which are operated either by electric inputs, solar panels installed on the roofs of RVs or even by natural gas open flames. In the absorption refrigerator, the refrigerant is absorbed by a transport medium and compressed in liquid form. The work input to the pump is usually very small and the COP of an absorption refrigeration system is given as:

$$\mathrm{COP} = \frac{\text{Desired output}}{\text{Required input}} = \frac{Q_\mathrm{C}}{W_{\mathrm{in,pump}} + W_{\mathrm{gener}}} \approx \frac{Q_\mathrm{C}}{W_{\mathrm{gener}}} \tag{7.20}$$

7.3 Air conditioning systems

The working principle of an air conditioner is to collect warm air from a particular space, say homes or offices, at lower temperature and transfer it to a space of higher temperature, say outside summer temperatures. This process is done using the ***principles of refrigeration***, which state that when a liquid evaporates it uses up a great deal of heat, which it draws from anything that happens to be around it, and when the substance changes form gas to liquid state the gas gives off heat to the surroundings.

All air conditioners use the principles of refrigeration and utilize the same cycle consisting of the following processes: compression, condensation, expansion, and evaporation in a closed circuit, as seen in figure 7.9. For the sake of economizing, the same refrigerant is used repeatedly to remove the heat from one area, to cool it, and to expel this heat in another area to warm it.

7.3.1 Parts of an air conditioning system

The main parts of an air conditioning system are shown in figure 7.10 and include a compressor, a condenser, an expansion valve and an evaporator. The operation of each is detailed here:

- **Compressor:** The purpose of the compressor is to circulate the refrigerant in the system under pressure. The compressor concentrates the heat the refrigerator contains.
- **Condenser:** The condenser, which looks like an engine radiator, converts the high-pressure gas from the compressor into a liquid. The condenser does this by utilizing the heat transfer principle that states that heat moves from hot to cold substances naturally. Air passing over the condenser coils carries off the heat and the gas condenses.

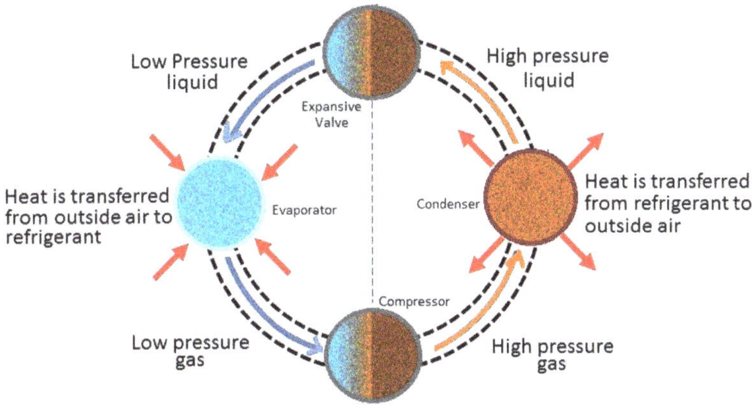

Figure 7.9. Standard air conditioner cycle.

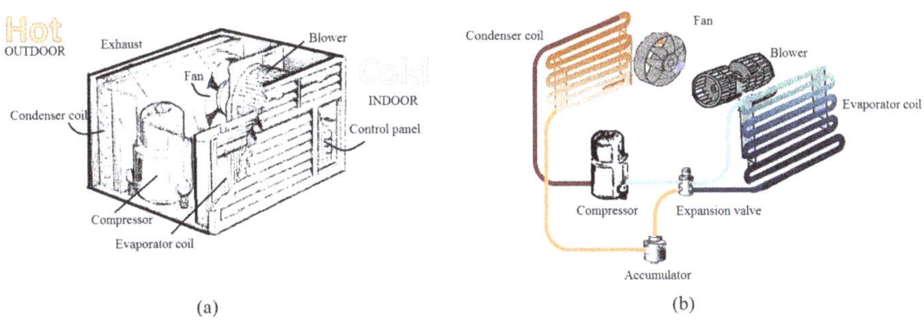

Figure 7.10. Basic air conditioning parts.

- **Expansion valve:** The expansion valve removes the pressure from the liquid refrigerant, which changes the liquid refrigerant from the condenser to vapor. Reducing pressure allows this change of temperature. So the warm liquid entering the expansion valve leaves the valve as a cold gas.
- **Evaporator:** The evaporator works opposite to the way a condenser works. It removes heat from the area to be cooled. By the time the liquid refrigerant reaches the evaporator its pressure has decreased, which allows it to dissipate heat and get much colder. This allows the refrigerant to absorb heat from the warm air and reach its (preferably low) boiling point very quickly. The refrigerant then vaporizes and absorbs the maximum amount of heat. This heat is carried by the refrigerant from the evaporator as a low-pressure gas through a hose pipe to the low side of the compressor, where the whole refrigeration cycle is repeated.
- **Thermostat control:** The opening and closing of electrical contacts in a thermostat are controlled by movement of a temperature-sensitive bellows, similar to the ones discussed in chapter 1. The bellows have a capillary tube filled with refrigerant. The capillary tube is positioned in a way that it may either have the cold air from the evaporator pass over it or it may be connected to the tail pipe of the evaporator. Evaporator temperature affects

the compound in the capillary tube; when the evaporator becomes colder the compound contracts and breaks its connection to the bellows, which breaks the electrical circuit to the compressor and the compressor operation stops. When the evaporator warms, the gas in the capillary tube expands. The expanding gas expands the bellows and connects the electrical circuit to the compressor and the compressor starts up again. This hot-and-cold cycle is repeated as long as the air conditioning is on.

Worked example 7.4. Why can you not put an conditioner in the middle of you dorm room to cool it in the summer time?

Solution: An air conditioner works by using the room as the cold reservoir and the outside of the house as the hot reservoir. The waste heat ejected from the air conditioner is always greater than the heat removed from the cold reservoir. This is why you must put an air conditioner in a window and not the middle of the room so that it can extract heat from the room and expel it to the outside. If you put it in the middle of the room then the AC is expelling the extracted heat back into the same environment, so the net effect will be heating, due to the work input being converted to heat.

7.3.2 A real air conditioning cycle

For a real air conditioning system to operate economically, the same refrigerant is used repeatedly to remove heat from one area and to expel it to another area, in a closed circuit. All air conditioners use the same cycle of compression, condensation, expansion, and evaporation, as discussed here,

- **Step 1—Compressor:** The refrigerant is injected into the compressor as a low-pressure gas, it is compressed and then moves out of the compressor as a high-pressure gas.
- **Step 2—Condenser:** The high-pressure gas flows to the condenser. Here it condenses to a liquid, and gives off its heat to the outside air.
- **Step 3—Expansion valve:** The liquid then moves to the expansion valve, which restricts the flow of the fluid, and lowers its pressure as it comes out of the expansion valve.
- **Step 4—Evaporator:** The low-pressure liquid from the expansion valve gets transferred to the evaporator, where heat from the inside air is absorbed and changes it from a liquid to a gas.
- **Step 0:** As a hot low-pressure gas, the refrigerant moves back to the compressor where the entire cycle starts again.

Note, in figure 7.9 the four-sequence cycle is divided at the center into a high side and a low side. This refers to the pressures of the refrigerant in each side of the system.

Worked example 7.5. When two wires made from different metals are joined at both ends they form a closed circuit. When one of the ends is heated, a current flows continuously in the circuit; this is called the **Seebeck effect**. The circuit thus formed is called a '***thermoelectric circuit***' and a device that operates on this circuit is called a '***thermoelectric device***' or a '***thermoelectric generator***'. The schematic diagram of a thermoelectric generator is shown in figure 7.11. Heat is transferred from a high-temperature source to the hot junction in the amount of Q_H, and it is rejected to a low-temperature sink from the cold junction in the amount of Q_C. Electrons serve as the working fluid in an thermoelectric generator. The net electrical work produced is, $W = Q_H - Q_C$. A thermoelectric refrigerator that resembles a small ice chest is powered by a car battery and has a COP of 0.1. If the refrigerator cools a 0.350 L canned cola from 20 °C to 4 °C in 30 min, determine the average electric power consumed by the thermoelectric refrigerator. Assume that the heat transfer through the walls of the refrigerator is negligible and the properties of canned drink are the same as water at room temperature, density (ρ) = 1 kg L^{-1} and c = 4.18 kJ kg^{-1} °C^{-1}

Solution: The average electric power consumed by the thermoelectric refrigerator in time, $\Delta t = 30$ min × 60 (s min^{-1}) = 1800 s is defined as,

$$\dot{W} = \frac{\dot{Q}_{\text{cooling}}}{\text{COP}_\text{R}} \tag{7.21}$$

where COP_R is given as 0.1 and $\dot{Q}_{\text{cooling}}$ is the cooling rate of the refrigerator and equals the rate of decrease of the energy of the canned drinks as,

$$\dot{Q}_{\text{cooling}} = \frac{Q_{\text{cooling}}}{\Delta t} \tag{7.22}$$

where ΔT = (20-4)°C = 16 °C and,

$$Q_{\text{cooling}} = mc\Delta T = (0.350\ \text{kg})(4.18\ \text{kJ kg}^{-1}\ ^\circ\text{C}^{-1})(16^\circ\text{C}) = 23.4\ \text{kJ}$$

and

$$m = \rho V = (1\ \text{kg L}^{-1})(0.350\ \text{L}) = 0.350\ \text{kg}$$

Putting these values in above equation (7.22),

$$\dot{Q} = \frac{Q_{\text{cooling}}}{\Delta t} = \frac{23.4\ \text{kJ}}{1800s} = 13\ \text{W}$$

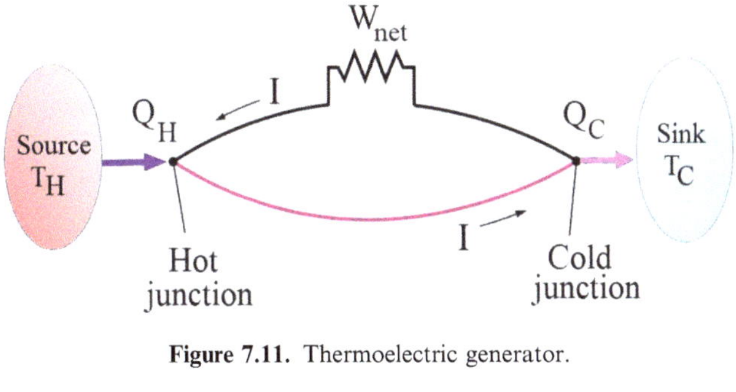

Figure 7.11. Thermoelectric generator.

Putting values in equation (7.21).

$$\dot{W} = \frac{\dot{Q}_{\text{cooling}}}{\text{COP}_{\text{R}}} = \frac{\text{W}}{0.10} = 130\ \text{W}.$$

7.4 Icemaker and ice generator

The first artificial refrigeration was done in 1748 CE by William Cullen at the University of Glasgow. However, it was an American John Gorrie who built the first refrigerator a hundred years later in 1844 CE. The basic components of an ice generator are the compressor, condenser, throttle valve and an evaporator. The refrigeration cycle for an ice machine is given below; it is worth noting that all ice machines basically work in the same way.

1. **Step 1—compressor:** A compressor compresses the low pressure refrigerant vapor to high-pressure vapor then delivers it to the condenser.
2. **Step 2—condenser:** At the condenser high pressure vapor is condensed into high-pressure liquid.
3. **Step 3—throttle valve:** The high-pressure liquid is then pushed through the throttle valve to become a low-pressure liquid.
4. **Step 4—evaporator:** At which point the liquid is conducted to the evaporator where the heat exchange occurs and ice is made.

When a liquid evaporates it uses up heat from its surroundings. During this process, the temperature of nearby water decreases drastically which causes the water to freeze. Basically, evaporation chills the water and eventually freezes it. We know that this effect is increased when we wet the outside of the container that contains the water with ether. However, ether is expensive and highly inflammable. So most ice factories use ammonia (specifically anhydrous ammonia[1]) or sulfurous oxide instead of ether because they both are cheaper and less flammable.

7.4.1 Artificial ice making process

Artificial ice has been made around the world for over 200 years. In making ice on the large scale, either ammonia or sulphureous oxide is used. It is important to note that at room temperatures ammonia exists as a gas or vapor. So as a first step we need to get ammonia into the liquid state, which is done in one of two ways, either by cooling it or by applying pressure to it. The most practical way is in fact by applying a combination of cooling and pressure.

Ammonia gas under pressure is forced through the condenser, a series of coils of pipe, where it is cooled by running water over the pipes. During this process the latent heat of ammonia is removed and carried away by the water. The liquid

[1] Anhydrous ammonia is ammonia perfectly free from water.

ammonia from the condenser is forced into pipes where the liquid ammonia expands and converts back to vapor form which causes the chilling effect.

The two systems generally used to make ice are called the compression system and/or the absorption system, discussed very briefly here. Please refer to other texts for more details.

1. In the **compression system**, the apparatus includes pumps that are driven by steam power, condensers, piping, etc and the cooling agent, anhydrous ammonia. Anhydrous liquid ammonia is filled into cylindrical iron tanks and then passed through a small throttle valve into coils of pipe in the freezing tanks. These tanks are quite large, covering an area about $(20 \times 50)'$ and are 4 feet in depth. The tank contains a solution of salt and water called '***brine***'. About five hundred water cans that need to be frozen go into one tank and float in this brine. Coils of pipes containing ammonia are run between the rows of cans in the tanks. While going through the pipes ammonia evaporates and cools the brine to about -18 °F. This cold brine surrounds the cans containing water and the water in the cans starts to freeze and finally becomes hard blocks of ice in about 60 hours.

   The ammonia gas after passing through the coils of pipe in the freezing tank is drawn through a pipe to a pump where it is compressed and then taken to the condenser to start the cycle again. This way the ammonia can be used over and over again, technically indefinitely.
2. In the **absorption system**, we start with ammonia water. Liquid ammonia is heated in a boiler which gives a mixture that is 90% ammonia gas and 10% steam. This mixed vapor is transferred to a rectifier where most of the steam condenses to water and is returned back to the boiler. The temperature of the rectifier is not low enough to condense ammonia, which gets passed to the condenser. At the condenser, ammonia is condensed by the combined action of cold and pressure. The pressure is supplied by the expansive force of steam of vapor that is constantly driven out of the boiler. Note, no mechanical pressure is needed or used here.

   The liquid ammonia then passes into expansion coils in freezing tanks just like what is done in the compression system. After this, the gas is taken to the absorber, which is similar to a condenser, where it is reabsorbed by water and becomes ammonia water again, and is therefore called the absorption system. Ammonia water is then returned to the boiler by a pump and the cycle begins again.

7.5 Geothermal heat pumps

It is common knowledge that the underground temperature approximately 3 or 6 meters underground remains relatively constant all year-round, at about 10 °C to 16 °C. This means the underground temperature is warmer than the outside air in

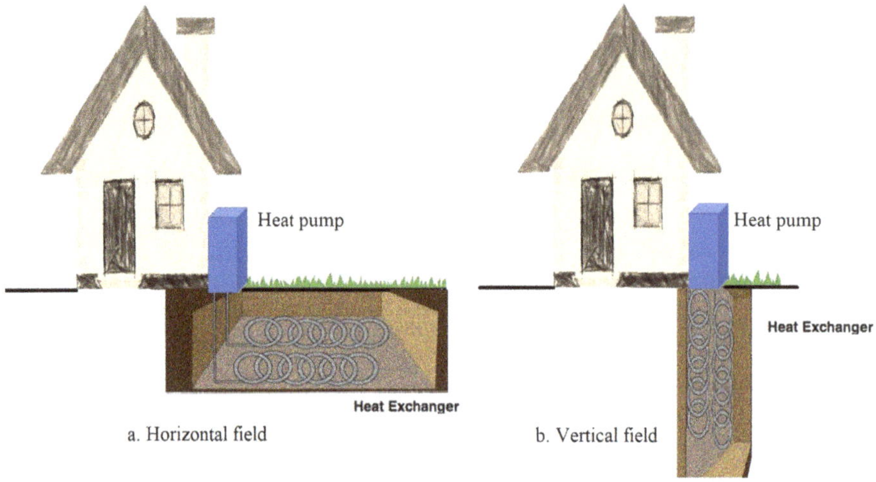

Figure 7.12. A geothermal unit installation, (a) rural, versus (b) urban installation.

the winter when the temperature falls to about 0 °C and is cooler than the outside air in the summer when the outside temperature rises to say 35 °C.

A geothermal heat pump takes advantage of this fact by transferring heat to provide, a source for heating and a sink for cooling our homes. This is usually done by either horizontal ground loops, 100–300 ft of trenches or vertical loops laid vertically to a depth of 100–400 feet. Horizontal ground loops are easier to install but require a large piece of land so are used mostly in rural settings, whereas vertical loops that do not require a lot of space are commonly employed in urban areas. Both are shown in figure 7.12. Geothermal units are also called '*ground-source heat pumps*' and are powered by electricity.

A typical geothermal unit is made up of three basic components:

1. A ground loop system filled with a water solution;
2. A heat exchanger;
3. Ductwork into the building

During the winter months, the water solution in the loop system absorbs heat from underground and becomes warm itself. This warmed water solution gets transferred to a water-to-air heat pump whose purpose is to concentrate the thermal energy and then transfer it to the air in the heating ductwork. The warmed air is then circulated throughout the house to keep the house warm. In the summertime, this process is reversed and the excess heat is pumped away from the home into the ground as seen in figure 7.13. It is important to note that this excess heat can also be used to heat a hot water tank in houses, which in the summer cooling months is used to remove excess heat and give us hot water at no cost.

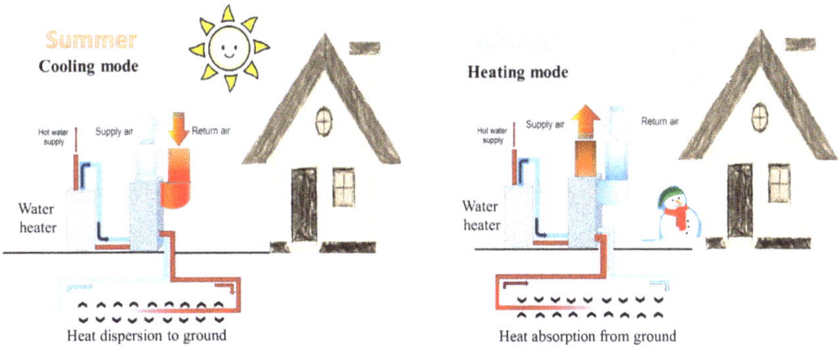

Figure 7.13. Operation of a geothermal heat pump.

Contentcept questions 7.1

1. What is the most widely used refrigeration cycle? What is the advantage of using this particular cycle?
2. The following is a type of expansion device: capillary tube/throttle valve/both/ neither.
3. For a same pressure the saturation temperature of ammonia is higher/lower/the same as the saturation temperature of water.
4. The mechanical work required to run a vapor absorption system is more/the same as/less than the mechanical work required to run a vapor compression system.
5. What is the disadvantage of using ammonia as a refrigerant?

Exercises 7.1

1. The peddler man from chapter 5 wants you to buy a magic heat pump that draws heat from a pond at 1 °C and gives heat to your house at 25 kW at 30°C while only using 2 kW of input electric power. Should you buy this magic heat pump?
2. Suppose we have a fridge (with maximum efficiency of 70%) running in a kitchen with room temperature (25 °C). We want the freezer compartment of the fridge to be maintained at −18 °C and know that the refrigerator motor has an output power of 220 watts. Calculate the work the motor does in an hour at maximum efficiency and the heat that is extracted during the same time.
3. A window air conditioner takes air from a room at 72 °F and transfers it to outside at 100 °F. For each unit of electrical energy required to run the refrigerator how many Joules of heat are transferred from the room?

Further reading

- Cengel Y and Boles M 2006 *Thermodynamics: An Engineering Approach* 5th edn (New York: McGraw-Hill Higher Education)

- Fermi E 2012 *Thermodynamics* (New York: Dover Publications)
- Resnick R, Halliday D and Krane K 2002 *Physics* vol 1 5th edn (New York: Wiley)
- Schroeder D 2000 *Thermal Physics* (Reading, MA: Addison-Wesley Longman)
- Van Ness H C 1969 *Understanding Thermodynamics* (New York: McGraw-Hill Book Company)

IOP Publishing

Thermodynamics and Statistical Mechanics
An introduction for physicists and engineers
Samya Zain

Chapter 8

Heat transfer-to-power cycle I

We introduced the heat transfer-to-power cycle in section 6.3. In the next two chapters we will discuss these cycles in detail. A device that utilizes the heat transfer-to-power cycle is called a '**heat engine**' and it uses the fact that heat spontaneously flows from a hot object to a cold object to perform mechanical work.

8.1 Heat transfer-to-power cycle

A basic heat engine, introduced in section 6.3.2 is nothing other than a simple piston-and-cylinder system filled with a working fluid. A working fluid can be a gas or a vapor-gas mixture. A heat engine made up of a piston-and-cylinder device is also called a '***reciprocating***' engine. Despite its simplicity, a piston-and-cylinder device is one of the most important inventions of all times and it has left its mark on all aspects of modern life from transportation, from car engines and airplane engines to electric power generation that powers our homes.

For a heat engine to do useful work, the working substance in the cylinder must operate in a cycle, which means it must pass through a series of thermodynamic processes repeatedly that return it to the initial state. Most power-producing devices operate on cycles and most practical engines go through four cycles defined by the motion of the piston as it moves back and forth in the cylinder. When the gas is heated the piston moves outwards, whereas when gas is cooled the piston moves into the cylinder. A heat transfer-to-power cycle during which a net amount of useful (mechanical) work is produced is called a '***power cycle***', and a power cycle during which the working fluid remains a gas throughout the cycle is called a '*gas power cycle*'.

The cycles encountered in real life are difficult to analyze because of the complications due to drag forces, such as friction or the insufficient time available to set up the equilibrium conditions. So in order to study power cycles we approximate the actual cycle to something called an '***ideal cycle***'. Comparison of an actual cycle versus an ideal cycle can be seen in figure 8.1. The ideal cycles, shown

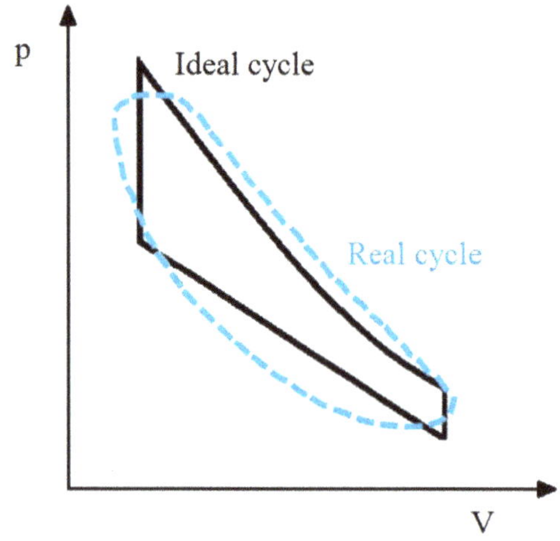

Figure 8.1. The analysis of complex processes are reduced to a manageable level by utilizing certain idealizations.

in figure 8.1 deviate somewhat from the real cycles but they still retain the general characteristics of the actual cycles, which allows us to perform calculations for the power cycles. The main difference is that ideal cycles are made up totally of internally reversible processes, which allows them to avoid all the complexities that internal irreversibilities and frictional forces add to the real cycles.

Recall that heat engines are designed for converting thermal energy to work, and their performance is expressed in terms of the thermal efficiency (η_{th}). From conservation of energy, the mechanical work done (W) during the cycle is written as:

$$Q_H - Q_C = W \tag{8.1}$$

where Q_H is heat added at the high temperature (T_H) of the reservoir (source) and Q_C is the heat removed at the cold temperature (T_C) of the sink. Both T_H and T_C are measured in kelvin. The efficiency (η) of the engine is given as:

$$\eta_{th} = \frac{\text{What you want out}}{\text{Input cost}} = \frac{W_{out}}{Q_H} = 1 - \frac{Q_C}{Q_H} = 1 - \frac{T_C}{T_H} \tag{8.2}$$

8.1.1 Idealizations and simplifications for engine operation

In this chapter we discuss the operation of various cycles and their practical applications. But before we go into the details of engines, let us discuss a few important idealizations and simplifications we employ to do so.

First of all, we know from chapter 6, the most efficient cycle is the Carnot cycle. A Carnot cycle is a totally reversible cycle with the highest thermal efficiency of all heat engines operating between the same temperature difference. Then the question is, why even bother developing an ideal cycle? The simple answer to this question is, we need to do this because of the hardware limitations. Most cycles in practice differ from the Carnot cycle, which means we cannot use the Carnot cycle as a model for our calculations. Additionally, different work-producing devices operate on different ideal cycles, so we have to be very careful when defining what we mean when we say the '*ideal cycle*'.

One thing we can say about ideal cycles is that like Carnot cycles, ideal cycles are internally reversible, but, unlike Carnot cycles, ideal cycles may not be externally reversible. That is, ideal cycles may involve irreversibilities external to the system, such as heat transfer through a finite temperature difference. Therefore, the thermal efficiency of an ideal cycle is less than that of a totally reversible cycle operating between the same temperature limits. There are some commonly employed simplifications in the analysis of power cycles,

1. An ideal cycle is completely frictionless, which means that the working fluid does not experience any pressure gain or loss as it flows through pipes or heat exchangers.
2. All expansion and compression processes occur in a state of quasi-static equilibrium.
3. All pipes connecting the components of a system are well insulated and no heat transfer occurs through them.
4. The changes in kinetic and potential energies of the working fluid are negligible.

8.1.1.1 Cold-air-standard assumptions

Actual gas power cycles are rather complex, hence in order to reduce complexities during calculations we utilize the following approximations, called the '***cold-air-standard assumptions***'. These provide us with some simplification of the inherent complications in our system without introducing major errors in our analysis. Cold-air-standard assumptions are:

1. All cyclic processes are internally reversible.
2. The combustion process is a heat-addition process from an external source at temperature (T_H).
3. The exhaust process is a heat-rejection process that restores the working fluid to its initial state at temperature (T_C).
4. When the working fluid used in our system is air, it always behaves as an ideal gas and is assumed to have a constant specific heat value determined at room temperature (25 °C).

8.1.2 Reciprocating engine

The basic components of a reciprocating engine are shown in figure 8.2 and include:

1. **The piston:** The piston moves in the cylinder between two fixed positions, top dead center (TDC) and bottom dead center (BDC).
    (a) The top dead center (TDC), is the position of the piston when it forms the smallest volume in the cylinder.
    (b) The bottom dead center (BDC) is the position of the piston when it forms the largest volume in the cylinder.

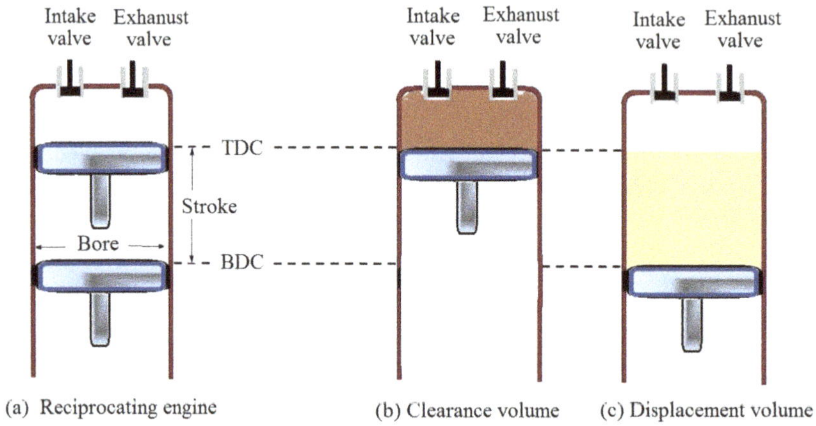

Figure 8.2. (a) Reciprocating engine. (b) Clearance volume. (c) Displacement volume.

The distance between the TDC and the BDC is the largest distance that the piston can travel in one direction, and it is called the 'stroke of the engine'. The diameter of the piston is called the 'bore'. The minimum volume formed in the cylinder when the piston is at TDC is called the 'clearance volume'. The volume displaced by the piston as it moves between TDC and BDC is called the 'displacement volume'.

2. **The intake valve:** The air or air–fuel mixture is drawn into the cylinder through the intake valve.
3. **The exhaust valve:** The combustion products are expelled from the cylinder through the exhaust valve.

The ratio of the maximum volume formed in the cylinder to the minimum volume is called the compression ratio (r) of the engine:

$$r = \frac{V_{\text{max}}}{V_{\text{min}}} = \frac{V_{\text{BDC}}}{V_{\text{TDC}}} \tag{8.3}$$

Reciprocating engines are classified, depending on how the combustion process in the cylinder is initiated, as,

1. **Spark-ignition (SI)-engines:** In SI-engines, the combustion of the air–fuel mixture is initiated by means of a spark plug. For example as done in an Otto cycle, explained in detail in section 8.2.
2. **Compression-ignition (CI)-engines:** In CI-engines, the air–fuel mixture is self-ignited as a result of compression. Compression is achieved by compressing the working fluid (in this case the air–fuel mixture) above its self-ignition temperature, as done in a Diesel cycle, explained in detail in section 8.3.

8.1.3 Mean effective pressure (MEP)

Engineers describe the performance of engines by using a term called '***mean effective pressure***' (MEP). MEP is used as a parameter to compare the performance of

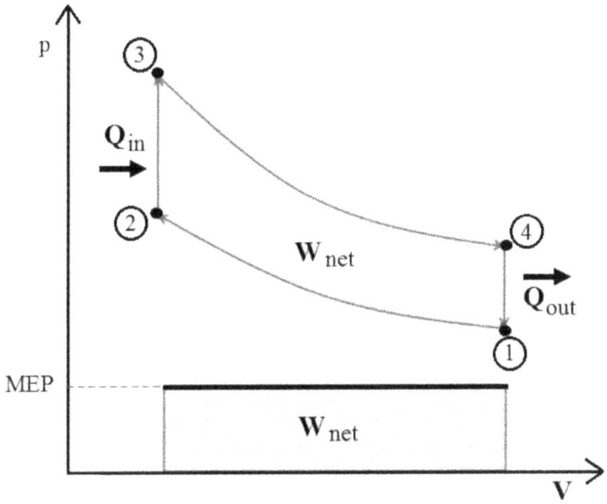

Figure 8.3. MEP—Otto Cycle. The net work output of a cycle is equivalent to the product of the MEP and the displacement volume.

engines of equal size. Since MEP does not depend on the engine displacement it is a very important measure of the useful work done by the engine. Work done during a cycle (W_{net}) is defined as the product of MEP, the piston area and the stroke of the engine. The product of the piston area and the stroke gives us the displacement volume, i.e., W_{net} = MEP × (piston area × stroke) = MEP × (displacement volume), as seen if figure 8.3. Hence, mathematically, MEP can be defined as,

$$\text{MEP} = \frac{\text{Useful work per cycle } (W_{\text{net}})}{\text{Displacement volume}} = \frac{W_{\text{net}}}{V_{\text{max}} - V_{\text{min}}} \tag{8.4}$$

For two engines of equal displacement volume, the one with a higher MEP will produce the greater net work and if the engines run at the same speed, the greater power. Hence the engine with larger MEP value delivers more useful work per cycle and thus performs better than the one with lower MEP value. Please note, even though MEP contains the word 'pressure', it is not an actual pressure within the engine, it is just a measure of the work developed in one complete cycle.

There are several types of MEP, for example, indicated mean effective pressure (IMEP), brake mean effective pressure (BMEP) and friction mean effective pressure (FMEP). These are defined by the location of measurement and method of calculation. IMEP is the mean effective pressure calculated by the work done during the cycle. However, IMEP does not take the engine efficiency into account. BMEP is the actual output of the internal combustion engine at the crankshaft and it takes into account the engine efficiency. FMEP indicates the MEP of the engine lost through friction and it is numerically the difference between IMEP and BMEP. From all of these we can calculate the mechanical efficiency of any engine as,

$$\eta_{\mathrm{mech}} = 1 - \frac{\mathrm{FMEP}}{\mathrm{IMEP}}$$

8.2 Otto engine cycle

An Otto cycle is used in gasoline powered car engines. It is the idealized thermodynamic *heat transfer-to-power cycle*, also called the '*spark-ignition (SI) reciprocating engine*' or simply the '*gasoline engine*'.

German engineer Nikolaus Otto created the Otto cycle and built the first successful four-stroke engine in 1876 CE based on the cycle proposed by Frenchman Beau de Rochas in 1862 CE. Otto found a way to layer the fuel mixture into the cylinder which caused the fuel to burn in a specific way that he called '*layered charge*'. This resulted in a controlled combustion rather than the explosion which destroyed all the other engines developed at the time. The Otto engine also had a longer stroke of the piston in the cylinder, which made Otto's device more efficient.

In an Otto engine, there is no ' hot reservoir' connected to the engine. The thermal energy in this engine is produced by burning the fuel inside the cylinder itself. The burning of fuel causes the working fluid (gas) to get a high temperature and pressure, exactly as if the gas had absorbed the heat from an external source. The first person to utilize Otto cycle to power cars was Karl Benz in 1885 CE. However, Henry Ford was the first to mass produce cars with engines that were based on the Otto cycle when he introduced his Model T to the public in 1908 CE.

The Otto cycle consists of four processes, two at constant volume during which heat is transferred to and from the working fluid that are connected by two adiabatic steps, as seen in the pV-diagram of the Otto cycle in figure 8.4. The steps are briefly explained as follows:

1. **At point (1):** Otto cycle starts at point 1 of figure 8.5. At point 1, the piston is at the point of maximum withdrawal from the cylinder, the bottom dead center (BDC). The cylinder which is full of air at about 1 atm at this point. In an actual engine, a particular amount of gasoline vapor is mixed in air but we will ignore that here and consider only an idealized process with an ideal gas.
2. **From point (1) to (2): adiabatic compression step:** The piston is carried into the cylinder by the inertia of a fly-wheel, which provides the work (W_{in}). During this process the air in the cylinder is compressed reversibly and adiabatically, ($Q_{12} = 0$). The work is given as,

$$W_{\mathrm{in}} = -\Delta U_{12} = -c_V(T_2 - T_1)$$

3. **From point (2) to (3): isochoric step:** The piston reaches the limit of its stoke at point 2 and is at the top dead center (TDC) position. Heat (Q_{in}) is added quickly to increase the pressure and temperature while keeping the volume constant. The piston remains in the minimum limiting position ($V_2 = V_3$

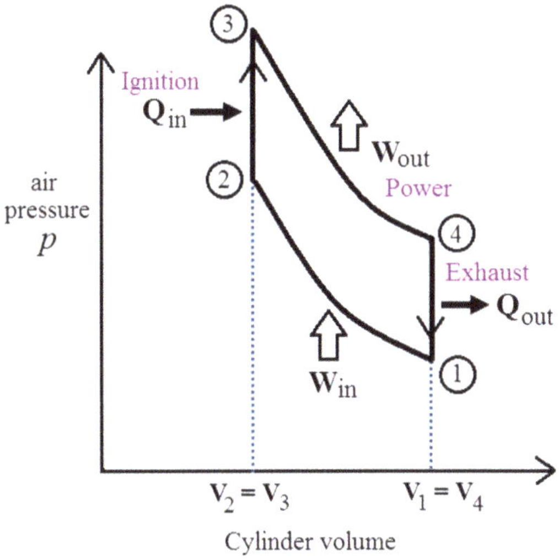

Figure 8.4. A pV-diagram of the Otto Cycle.

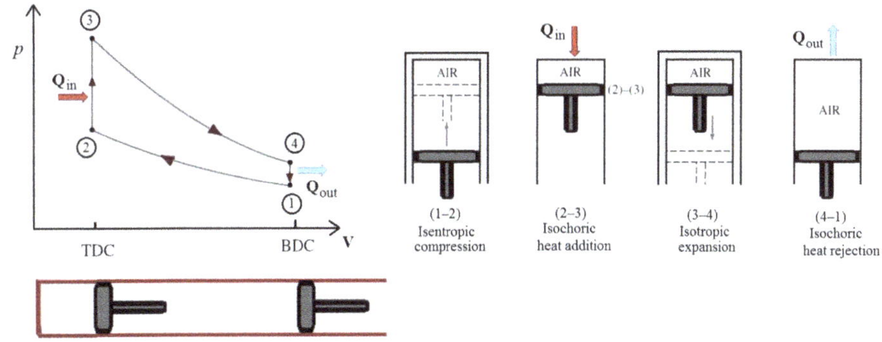

Figure 8.5. Otto cycle for an ideal gasoline engine.

which implies $W_{23} = 0$), but both pressure and temperature are increased from P_2 and T_2 to much higher values at P_3 and T_3.

$$Q_{23} = \Delta U_{23} = +c_V(T_3 - T_2)$$

4. **From point (3) to (4): adiabatic expansion step:** During this step the gas expands against the piston and does useful work (W_{out}) in a reversible and adiabatic process, given as,

$$W_{out} = -\Delta U_{34} = -c_V(T_4 - T_3)$$

5. **From point (4) to (1): isochoric step:** Heat is quickly removed from the working fluid (air) such that $V_1 = V_4$ which implies $W_{41} = 0$, which allows the cycle to get back to point 1. In a real engine exhaust valves open at point (4) and hot gases rush out of the cylinder.

$$Q_{41} = \Delta U_{41} = +c_V(T_1 - T_4)$$

Table 8.1. Otto cycle in a nutshell.

		Heat	Work
Point 1 to 2	Adiabatic	$Q_{12} = 0$	$W_{12} = -\Delta U_{12}$ $W_{12} = -c_V(T_2 - T_1)$
Point 2 to 3	Const V	$Q_{23} = \Delta U_{23}$ $Q_{23} = +c_V(T_3 - T_2)$	$W_{23} = 0$
Point 3 to 4	Adiabatic	$Q_{34} = 0$	$W_{34} = -\Delta U_{34}$ $W_{34} = -c_V(T_4 - T_3)$
Point 4 to 1	const V	$Q_{41} = \Delta U_{41}$ $Q_{41} = +c_V(T_1 - T_4)$	$W_{41} = 0$

Table 8.2. Otto Cycle.

$T_1 = 140\ °\text{F} = 333\ \text{K}$	$P_1 = 1\ \text{atm} = 1.01 \times 10^5\ \text{Pa}$
$T_2 = ?$	$P_2 = 15\ \text{atm} = 1.52 \times 10^6$
$T_3 = 3940\ °\text{F} = 2444\ \text{K}$	$P_3 = ?$
$T_1 = ?$	$P_4 = ?$

These steps are summarized in table 8.1. Please note that an actual engine will go through an extra step during the cycle called a point 0, that does not produce any work. The purpose of this extra step is to push the working fluid from cylinder (point $1 \to 0$) and to draw in fresh air and go back to step 1. The net result is to return system to its initial state, which we will see in section 8.2.2.

The thermal efficiency for the Otto cycle, defined as the ratio of net useful work extracted by the engine during one cycle to the input heat, is given as:

$$\eta_{\text{Otto}} = \frac{\text{Work}_{\text{cycle}}}{Q_{\text{in}}} = \frac{W_{\text{net}}}{Q_{23}} \tag{8.5}$$

where Q_{23} is the heat added during the isochoric step (2) to (3).

Worked example 8.1. Let's look at an example for an Otto Cycle (table 8.2). We will assume that the following values are given: at point (1), the pressure is $P_1 = 1$ atm and the temperature is $T_1 = 140$ °F; at point 2, pressure $P_2 = 15$ atm and at point 3, the temperature is $T_3 = 3940$ °F. Hint: Assume air behaves like an ideal gas and take heat capacity $C_V = 5200\ \text{J K}^{-1}$, $\gamma = C_P/C_V = 7/5 = 1.4$.

Solution: In order to solve for the values of heat, internal energy and the work done, it is clear from table 8.2 we have to find the values of temperatures at all steps.

1. Let's first find T_2: From point (1) to (2), the process is an adiabatic, reversible compression. From chapter 4, equation (4.57) we know,

$$\frac{T_2}{T_1} = \left(\frac{P_2}{P_1}\right)^{\frac{\gamma-1}{\gamma}}$$

Putting values, we solve for T_2 and get = 722 K.

2. To find T_4, we know that the process from point (3) to (4) is an adiabatic, reversible expansion. From chapter 4 equation (4.65).

$$\frac{T_4}{T_3} = \left(\frac{V_3}{V_4}\right)^{\gamma-1}$$

But from step (1) to step (2),

$$\frac{T_1}{T_2} = \left(\frac{V_2}{V_1}\right)^{\gamma-1}$$

Since both are constant volume steps, we can say $\frac{V_3}{V_4} = \frac{V_2}{V_1}$, which means we can write,

$$T_4 = T_3\left(\frac{T_1}{T_2}\right)$$

Putting values we get $T_4 = 1127$ K.

3. Since we have all the values of temperature we can use the First law to find heat (Q), work (W), and energy (ΔU) as shown below:
   (a) **Step (1) → (2): Adiabatic compression step**
      i. Heat: $Q_{12} = 0$
      ii. Internal energy: $\Delta U_{12} = C_V(T_2 - T_1) = 2052$ kJ.
      iii. Work done: $W_{12} = Q + W = W = 2052$ kJ
   (b) **Step (2) → (3): Constant Volume step**
      i. Work done: $W_{23} = 0$.
      ii. Internal energy: $\Delta U_{23} = C_V(T_3 - T_2) = 9084$ kJ.
      iii. Heat: $Q_{12} = Q + W = Q = 9084$ kJ.
   (c) **Point (3) → (4): Adiabatic expansion step**
      i. Heat: $Q_{34} = 0$
      ii. Internal energy: $\Delta U_{34} = C_V(T_4 - T_3) = -6947$ kJ.
      iii. Work done: $W_{34} = -6947$ kJ
   (d) **Point (4) → (1): Constant volume step**
      i. Work done: $W_{41} = 0$
      ii. Internal energy: $\Delta U_{41} = C_V(T_1 - T_4) = -4188$ kJ.
      iii. Heat: $Q_{41} = -4188$ kJ.

   Next step is to summarize all calculated values in a table for easy reference, as I have done in table 8.3. Now that we have all

Table 8.3. Otto cycle Example 8.1 summary.

		Heat	Internal energy	Work
Point 1 to 2	Adiabatic compression	$Q_{12} = 0$	$U_{12} = 2052$ kJ	$W_{12} = 2052$ kJ
Point 2 to 3	Constant volume	$Q_{23} = 9084$ kJ	$U_{23} = 9084$ kJ	$W_{23} = 0$
Point 3 to 4	Adiabatic expansion	$Q_{34} = 0$	$U_{34} = -6947$ kJ	$W_{34} = -6947$ kJ
Point 4 to 1	Constant volume	$Q_{41} = -4188$ kJ	$U_{41} = -4188$ kJ	$W_{41} = 0$

the values we can find the net work done, internal energy, heat and the total efficiency of the cycle.

(e) **Net**

i. Heat: $Q_{\text{net}} = Q_{23} + Q_{41} = 4852$ kJ
ii. Work done: $W_{\text{net}} = -4895$ kJ
iii. Internal energy: $\Delta U_{\text{net}} = 0$
iv. Efficiency: $\eta = \left|\frac{W_{\text{net}}}{Q_{\text{in}}}\right| = 0.54$

Hence efficiency $\eta = 54\%$, which means that 54% of the heat added went toward useful work and 46% of the heat added leaves the system as $Q_{\text{out}} = Q_{41} = -4188$ kJ.

8.2.1 Thermal efficiency for a real Otto cycle

We know that real engines do not execute the idealized cycle discussed above. The compression and expansion steps are not completely reversible, nor are they completely adiabatic, hence the useful work output is also considerably lower than what we calculate for the idealized cycle. So how do we calculate engine efficiency (η) for a real engine?

One way to calculate engine efficiency (η) is to estimate the amount of work done in each step individually. In steps that produce useful work, we define efficiency as the ratio of actual useful work to the reversible work done by the engine, as,

$$\eta_1 = \frac{\text{Actual Work Produced}}{\text{Reversible Work}} = \frac{W_{\text{prod}}}{Q_{\text{rev}}} < 1 \tag{8.6}$$

For the steps that require work input, the reversible work is the minimum necessary work. Hence irreversibilities only serve to increase the work required,

$$\eta_2 = \frac{\text{Reversible work}}{\text{Work required}} = \frac{W_{\text{rev}}}{W_{\text{req}}} < 1 \tag{8.7}$$

For the compression and expansion processes, experience shows that an efficiency of about 0.75 is a reasonably good estimate for the maximum engine efficiency (η). We cannot hope for any better.

Worked example 8.2. Suppose for a real engine we have a situation where the W_{rev} is 3500 J, W_{req} is 11 850 J and $Q_{23} = 15\,500$ J. Also given for the engine are $T_1 = 600$ K, $T_2 = 1300$ K, $T_3 = 4400$ K, and $T_4 = 2030$ K. Then for a real engine we will get,

1. **Point (1) → (2):** Adiabatic compression step, work is required:

$$\eta_2 = \frac{W_{\text{rev}}}{W_{\text{req}}}$$
$$0.75 = \frac{-3500}{W_{12}} \Longrightarrow W_{12} = -4670 \text{ J}$$

   and from the First law of thermodynamics:

$$\begin{aligned} \Delta U &= Q_{12} - W_{12} \\ C_V \Delta T &= Q_{12} - W_{12} \\ 5(1300 - 600) &= Q_{12} - 4670 \Longrightarrow Q_{12} = -1170 \text{ J} \end{aligned}$$

   This is the amount of heat transferred from the gas to the surroundings, not adiabatic anymore.

2. **Point (3) → (4):** Adiabatic expansion step, where engine does work:

$$\eta_1 = \frac{W_{\text{prod}}}{W_{\text{req}}}$$
$$0.75 = \frac{W_{34}}{11\,850} \Longrightarrow W_{34} = 8900 \text{ J}$$

   and from the First law of thermodynamics:

$$\begin{aligned} \Delta U &= Q_{34} - W_{34} \\ C_V \Delta T &= Q_{34} - W_{34} \\ 5(2030 - 4400) &= Q_{34} - 8900 \Longrightarrow Q_{34} = -2950 \text{ J} \end{aligned}$$

   Again this is the amount of heat transferred from the gas to the surroundings; note this value is even bigger than the one in steps 1 → 2. It is because the gas temperatures are higher.

3. Then the net work done by the engine:

$$\begin{aligned} W_{\text{engine}} &= W_{12} + W_{34} \\ &= -4670 + 8900 \\ &= 4230 \text{ J} \end{aligned} \tag{8.8}$$

   and efficiency is given by:

$$\eta = \frac{\text{Net work out}}{\text{Heat input}} = \frac{W_{\text{engine}}}{Q_{23}} = \frac{4230}{15\,500} = 0.27 \tag{8.9}$$

Then thermal efficiency will be given as $\eta = 27\%$. The irreversibilities which we have assumed to exist in our engine by taking an efficiency of 0.75 for each step of the cycle have reduced the thermal efficiency to 27% for a real engine operating on the Otto cycle. This is a reasonable value for a real engine, well-tuned and operating under conditions for which it was designed.

8.2.2 Internal combustion engine

The Otto cycle defines a set of processes used by SI internal combustion engines for both two-stroke and four-stroke cycles. A schematic of an internal combustion engine is shown in figure 8.6. Most gasoline engine cars use the basic principles of the Otto cycle and use four-stroke engines. On the other hand, motorcycles and lawn mowers employ the Otto cycle and use two-stroke engines.

The working substance for these engines is a mixture of air and vaporized gasoline. Both the two-stroke and four-stroke engines proceed in a similar manner. At the beginning of each cycle, they both take in fuel and air, compress it and cause it to react, thereby adding heat by converting chemical energy into thermal energy inside the cylinder, expand and then eject the combustion products at the end of the cycle and finally at the beginning of the next cycle replace them with a new charge of fuel and air.

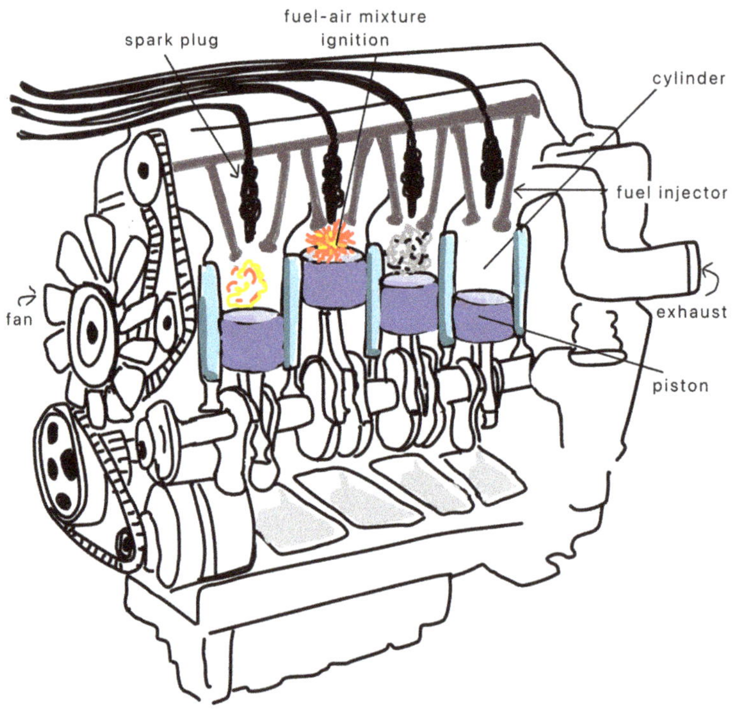

Figure 8.6. Schematic of an internal combustion engine.

8.2.2.1 *Efficiency of gasoline engine*

$$\text{Efficiency of Gasoline Engine} = \frac{\text{Useful work produced during cycle}}{\text{Heat absorbed during ignition step}}$$

$$\eta = \frac{W_{\text{out}}}{Q_{12}}$$

$$\eta = 1 - \left(\frac{V_1}{V_2}\right)^{\gamma-1}$$

where $\frac{V_1}{V_2}$ is called the compression ratio and γ is the specific heat ratio, also called adiabatic exponent ($\equiv C_P/C_V$).

For air, $\frac{V_1}{V_2} = 8$ and $\gamma = 7/5$; which gives efficiency, $\eta = 1 - \left(\frac{1}{8}\right)^{(7/5)-1} = 0.56$.

Hence, $\eta = 56\%$, which is pretty good as far as efficiencies go.

Another way of expressing efficiency:

We also know $TV^{\gamma-1}$ = constant, which implies:

$$\frac{T_2}{T_1} = \left(\frac{V_1}{V_2}\right)^{\gamma-1} \tag{8.10}$$

Then,

$$\eta = 1 - \left(\frac{T_1}{T_2}\right) = 1 - \left(\frac{T_4}{T_3}\right) \tag{8.11}$$

8.2.2.2 *Four-stroke engines*

In most SI engines, the piston executes four complete strokes (two mechanical cycles) within the cylinder, and the crankshaft completes two revolutions for each thermodynamic cycle. These engines are called '*four-stroke*' internal combustion engines. A schematic of each stroke as well as a pV-diagram for an actual four-stroke SI engine is given in figure 8.7. Steps followed in the internal combustion (four-stroke cycles) engine in cars include the following,

- Step 0: Initially, both the intake and the exhaust valves are closed, and the piston is at its lowest position, bottom dead center (BDC), at maximum

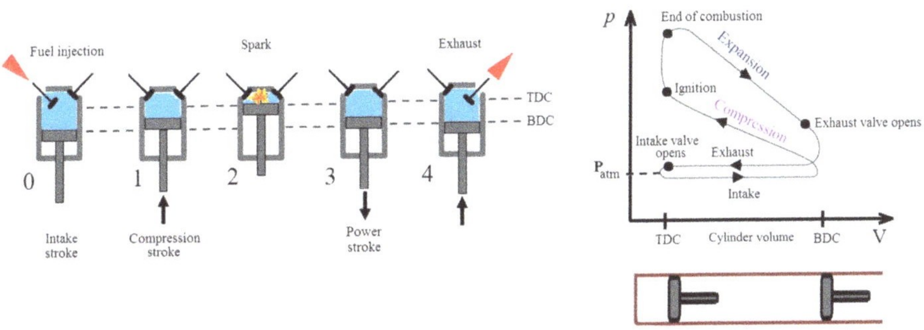

Figure 8.7. Otto cycle for a real gasoline engine.

volume. The mixture of fuel and air is injected into a cylinder. This is called the '*intake stroke*', as seen in figure 8.7(a).

- Step 1: During the compression stroke, the piston moves upwards, and in the process compresses the air–fuel mixture, as seen in figure 8.7(b).
- Step 2: Just before the piston reaches its highest position, top dead center (TDC), of minimum volume, a spark plug ignites the fuel–air mixture which raises the temperature and pressure of the mixture while keeping the volume constant.
- Step 3: As a result the gas mixture expands adiabatically; this is called the expansion or '*power stroke*', as seen in figure 8.7(d). During this step high pressure gases push the piston down, which forces the crankshaft to rotate and produce useful work output, in this case the forward or backwards motion of a car.
- Step 4: At the end of the power stroke, the piston is at its lowest position and this completes the first mechanical cycle. The cylinder at this time is filled with combustion products.
- Step 5: The piston moves upward one more time and in the process expels the hot exhaust gases through the exhaust valve, as shown in figure 8.7(e). This is called the '*exhaust stroke*'.
- Step 0: The piston moves down a second time and draws in fresh new air–fuel mixture through the intake valve, the '*intake stroke*', at the lower temperature and pressure.
- and the cycle is reset to begin again.

Note, the pressure in the cylinder is slightly above the atmospheric value during the exhaust stroke and slightly below during the intake stroke.

8.2.2.3 Two-stroke engines

Two-stroke engines have only two stokes, the '*power stroke*' and the '*compression stroke*'. In these engines, the crank-case is sealed and the intake and exhaust valves are replaced by openings in the lower portion of the cylinder wall, as shown in figure 8.8.

During the end of the power stroke, the piston uncovers the exhaust port (seen on the left side of figure 8.8) which allows the exhaust gases to be partially expelled from the cylinder. This then allows the fresh air–fuel mixture to rush in at the intake port

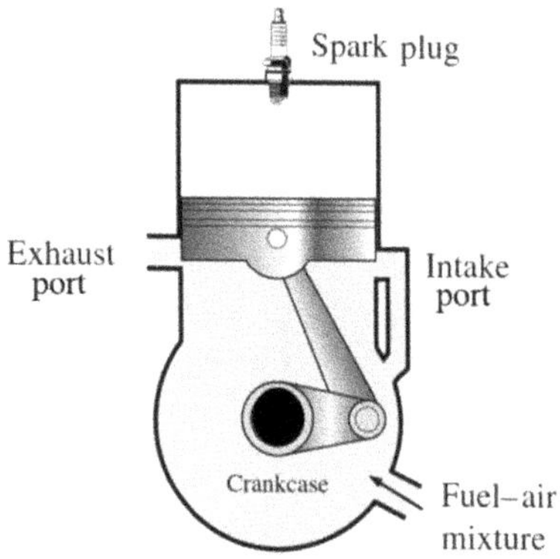

Figure 8.8. Schematic of a two-stroke reciprocating engine.

(seen on the right side of figure 8.8) and drive most of the remaining exhaust gases out of the cylinder. This mixture is then compressed as the piston moves upward during the compression stroke, which is then ignited by a spark plug.

There are many advantages to using two-stroke engines rather than four-stroke engines. Two-stroke engines are relatively simple and inexpensive to build and operate. They also use a lot less fuel and have high power-to-weight and power-to-volume ratios which makes them perfect for applications that require small size and small weights. A well-designed two-stroke engine can provide significantly more power than its four-stroke counterpart because two-stroke engines produce power on every engine revolution instead of every other one. Hence they are used in most motorcycles, chain saws and lawn mowers. However, two-stroke engines are generally less efficient than four-stroke engines because of the placement limitations of the exhaust and intake ports in the cylinder, the exhaust gases are not completely expelled in each cycle and additionally the new intake of air–fuel mixture is sometimes exhausted with the expulsion of exhaust gases.

8.3 Diesel engine

The Diesel engine was first proposed by Rudolph Diesel in the 1890s. It is a *heat transfer-to-power* engine that uses a Diesel cycle to convert heat input into mechanical work. The Diesel cycle is known as a constant pressure cycle as pressure is kept constant during two of the four steps in this cycle. The operation of a Diesel engine is shown in figure 8.9. A Diesel engine is also called a *compression ignition* CI-engine and is very similar to the SI engine discussed in section 8.2. The difference between SI and CI engines is the method of initiating combustion. A Diesel engine uses a high compression to ignite the fuel, whereas an SI engine uses a spark plug for initiating ignition.

In SI engines, the air–fuel mixture is compressed to a temperature just below its auto-ignition temperature and the combustion process is started by firing a spark plug. In CI engines the air is compressed to a temperature above the auto-ignition temperature of the fuel, and combustion starts immediately on contact as the fuel is

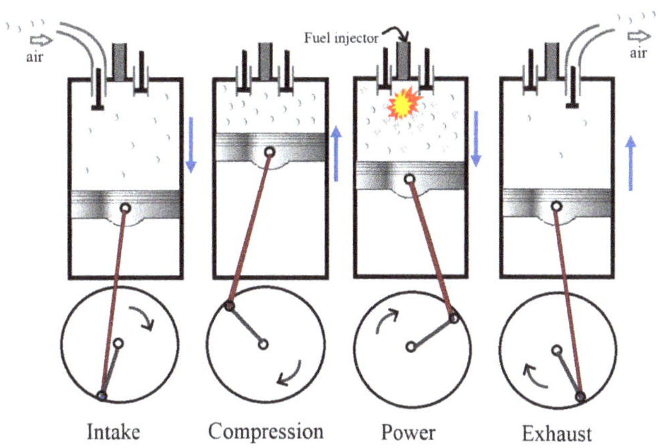

Figure 8.9. Operation of a Diesel engine.

injected into this hot air. The spark plug and carburetor in gasoline engines are replaced by fuel injectors in Diesel engines.

The air in the Diesel engine is compressed adiabatically, which raises the temperature of air in the cylinder above the ignition temperature of the fuel–air mixture formed by injecting fuel once the air is compressed and the piston approaches top dead center (TDC). Please note the combustion process in Diesel engines takes place over a longer time interval as compared to the SI engines, which makes the reaction time for CI engines comparatively slower than their SI counterparts.

The steps for a diesel cycle include two isobaric (constant pressure) and two isentropic processes (constant entropy), as shown in figure 8.10, and are described as follows,

1. **Step (0): air injection process:** The air is injected adiabatically in step (0), as shown in figure 8.10.
2. **Step (1) to (2): isentropic compression:** Step 1 to 2 is the isentropic compression of the air. Energy is transferred in as work (W_{in}) done by the piston in compressing the air. The pressure at the end of compression, in Diesel engines, is approximately 35 atm.
3. **Process (2) to (3): isobaric heat addition:** Process 2 to 3 is a reversible isobaric (constant pressure) process. Energy enters the system as heat (Q_{in}) by the combustion of the fuel and is mathematically given as,

$$Q_{in} = c_P(T_3 - T_2)$$

Please keep in mind that the combustion in a CI engine is heterogeneous, whereas the combustion is homogeneous in SI engines. In heterogeneous combustion, the solid and gas phase of the mixture interact to enhance the transfer of reactants to their final products. Heterogeneous combustion is

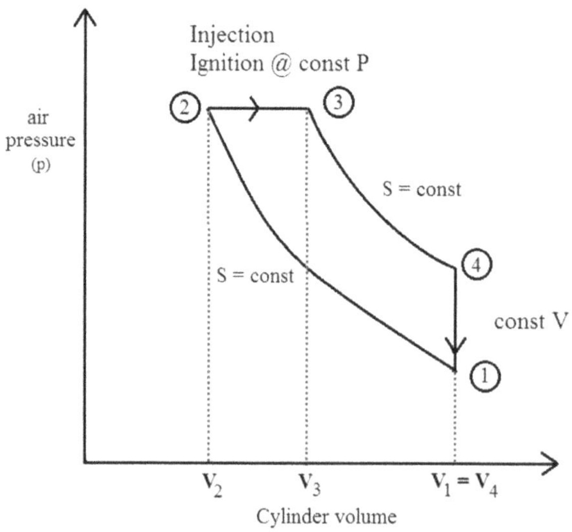

Figure 8.10. pV-diagram for a Diesel engine.

noisier than homogeneous combustion due to the high rate of the in-cylinder pressure. This noise is called 'diesel knocking' and is the major contributor to poor sound quality in Diesel powered trucks or cars.

4. **Process (3) to (4): isentropic expansion:** Process 3 to 4 is the isentropic expansion of the working fluid, also called the *power stroke*. During this process, (W_{out}) is done by the expanding working fluid, which pushes the piston which turns the crankshank and produces useful work for the Diesel cycle. This work is also called the '*gross work*'. Some portion of W_{out} in this case is used in the next cycle to compress the next charge of air.

   The net work produced (W_{net}) is represented by the area enclosed by the cycle on the pV-diagram, as shown in figure 8.10.

$$W_{net} = Q_{in} - Q_{out}$$

5. **Process (4) to (1): constant volume heat rejection:** Process 4 to 1 is the reversible constant volume process in which some energy flows out of the system as heat (Q_{out}) done by venting the air and is mathematically given as,

$$Q_{out} = c_V(T_4 - T_1)$$

Diesel engines have some major advantages over other engines. One, in Diesel engines air is compressed during the compression stroke, before the fuel mixture is introduced into the system. This eliminates the possibility of auto-ignition of the fuel and hence Diesel engines can be designed to operate at much higher compression ratios. Additionally, not having to deal with auto-ignition issues, fuels that can be used in these engines can be less refined and hence much less expensive.

Diesel engines also have the lowest fuel consumption of any large internal combustion engine. Due to their higher efficiencies and lower fuel costs Diesel engines are very attractive and are extensively used where large amounts of power are required, such as in commercial trucks, trains and cargo ships. For a given comparative ratio, efficiency for a Diesel engine is more than that for the Otto cycle. $\eta_{Diesel} \approx 40\%$ versus $\eta_{Otto} \approx 20\% - 30\%$, hence,

$$\eta_{Diesel} > \eta_{Otto} \tag{8.12}$$

when both cycles operate on the same compression ratio. pV diagrams for Diesel versus Otto cycle are shown in figure 8.11.

Worked example 8.3. Assume an ideal air-standard Diesel cycle engine has a compression ratio of 18 and a cutoff ratio of 2. At the beginning of the compression process, the working fluid is at 100 kPa, 25 °C (300 K). Determine given nominal specific heat capacity for air at 300 K, $c_p = 1.005$ kJ kg^{-1} K^{-1}, $c_V = 0.717$ kJ kg^{-1} K^{-1}, $R = 0.287$ kJ kg^{-1} K^{-1}, and $\gamma = 1.4$. Heat capacity for fuel at 300 K, $c_p = 1.121$ kJ kg^{-1} K^{-1}, and $c_V = 0.834$ kJ kg^{-1} K^{-1}.

(a) The temperature and pressure of the air at the end of each step

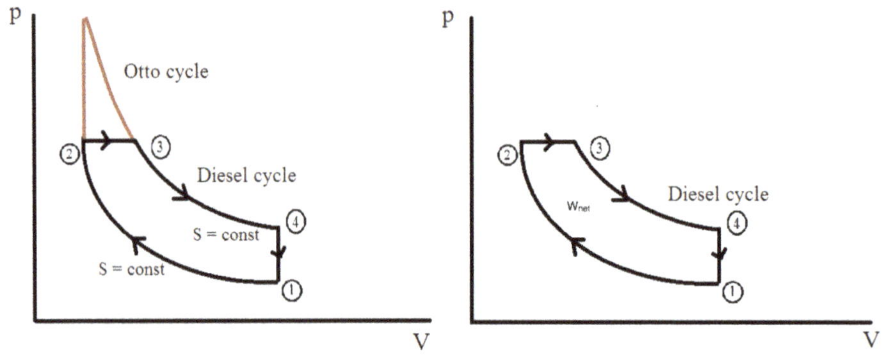

Figure 8.11. pV-diagram for a Diesel Cycle versus Otto Cycle.

(b) The net work output per cycle [kJ kg^{-1}], and
(c) The thermal efficiency.

Solution: Given values:

Point 1; $T_1 = 300$ K, $P_1 = 100$ kPa.

Compression ratio $(r) = \frac{V_1}{V_2} = 18$.

Cutoff ratio $(r_c) = \frac{V_3}{V_2} = 2$.

For adiabatic processes, $PV^\gamma = \text{const}$ and $TV^{\gamma-1} = \text{const}$.

1. **Point 1 to 2: adiabatic compression:** From $TV^{\gamma-1} = \text{const}$ we can write,

$$\frac{T_2}{T_1} = \left(\frac{V_1}{V_2}\right)^{\gamma-1} = r^{\gamma-1} = 18^{1.4-1} = 3.2$$
$$\Longrightarrow T_2 = 811\ K$$

Also from $PV^\gamma = \text{const}$ we can write,

$$\frac{P_2}{P_1} = \left(\frac{V_1}{V_2}\right)^{\gamma} = r^\gamma = 18^{1.4} = 57.2$$
$$\Longrightarrow P_2 = 5720\ \text{kPa}$$

2. **Point 2 to 3: constant pressure expansion:** For a constant pressure process, $P_2 = P_3 = 5720$ kPa. We can use the ideal gas law $(pV = nRT)$:

$$\frac{P_3V_3}{T_3} = \frac{P_2V_2}{T_2}$$
$$T_3 = \frac{V_3}{V_2}T_2 = r_cT_2 = 2(811)$$
$$\Longrightarrow T_3 = 1622\text{K}$$

Heat during fuel injection,

$$Q_{\text{in}} = c_p(T_3 - T_2) = 1.121(1622 - 811) = 909\ \text{kJ kg}^{-1}$$

3. **Point 3 to 4: adiabatic expansion:** Again using equations for adiabatic processes

$$\frac{T_4}{T_3} = \left(\frac{V_3}{V_4}\right)^{\gamma-1} = \left(\frac{r_c}{r}\right)^{\gamma-1} = \left(\frac{2}{18}\right)^{1.4-1} = 0.42$$
$$\Longrightarrow T_4 = 674\ \mathrm{K}$$

also from PV^γ = const we can write,

$$\frac{P_4}{P_3} = \left(\frac{V_3}{V_4}\right)^{\gamma} = \left(\frac{r_c}{r}\right)^{\gamma} = \left(\frac{2}{18}\right)^{1.4} = 0.046$$
$$\Longrightarrow P_4 = (4866)(0.046)\ \mathrm{kPa} = 224\ \mathrm{kPa}$$

4. **Point 4 to 1: constant volume exhaust:**

$$Q_{\text{out}} = c_V(T_4 - T_1) = 0.834(674 - 300) = 311\ \mathrm{kJ\ kg^{-1}}$$

5. Net work output per cycle will be calculated as,

$$W_{\text{net}} = Q_{\text{in}} - Q_{\text{out}} = 909 - 311 = 598\ \mathrm{kJ\ kg^{-1}}$$

6. Thermal efficiency,

$$\eta = \frac{W_{\text{net}}}{Q_{\text{in}}} = \frac{Q_{\text{in}} - Q_{\text{out}}}{Q_{\text{in}}} = 1 - \frac{311}{909} = 0.65$$

This calculated efficiency (65%) is unrealistically high. It is worth remembering that in this idealized analysis we have ignored many loss effects that exist in practical heat engines.

Concept questions 8.1

1. Humans used fire to raise temperatures of objects (by adding heat) much earlier in history than using mechanical work to raise temperatures of objects like those used in steam engines. Could you think of a reason why?
2. A heat engine that in each cycle does positive work and loses energy as heat, with no heat energy input, would violate:
   (a) the Zeroth law of thermodynamics.
   (b) the First law of thermodynamics.
   (c) the Second law of thermodynamics.
   (d) the Third law of thermodynamics.
   (e) Newton's second law.

Exercises 8.1

1. In an air Otto cycle, assume the maximum temperature of the cycle is 1100 °C and the compression begins at 35 °C and 1 atm. Given that the compression ratio is 7 and c_V is 0.718 kJ kg^{-1} K^{-1}, find,
   (a) The various temperatures at all points in the cycle,
   (b) The pressure values at all points in the cycle,
2. Consider the Otto cycle given in Q1. and using the calculated values, find
   (a) The heat input per kg of air,
   (b) The heat rejected per kg of air,
   (c) Net work done per kg of air,
   (d) The efficiency of the cycle.
3. In an regular air Diesel cycle, suppose the compression ratio is 15. Compression begins at 1 atm, and 40 °C. The heat added is 1675 kJ kg^{-1}. Given $c_p = 1$ kJ kg^{-1} K^{-1}, $c_V = 0.718$ kJ kg^{-1} K^{-1} and for air, $R = 287$ J kg^{-1} K^{-1}. Find,
   (a) The various temperatures at all points in the cycle,
   (b) The pressure values at all points in the cycle,
   (c) The volume per kg values at all points in the cycle,
   (d) Work done per kg of air,
   (e) The cycle efficiency.

8.4 The ideal cycle for a gas-turbine engine: Brayton cycle

The Brayton cycle, also known as the Joule cycle, is named after the American engineer George Bailey Brayton, who first proposed it. Brayton originally developed the Brayton cycle to be used in piston engines and the first Brayton engine was a two-stroke engine which produced power. In 1875 CE Brayton engines were used to power the world's first submarines. In 1890 CE Brayton developed a four-stroke engine that was later used by Rudolf Diesel in his early Diesel engines.

Today, the Brayton cycle is used for gas turbines where both the compression and expansion processes take place in rotating machinery. Gas turbines operate on basically two types of cycles: one, which is open to the atmosphere, called an open gas-turbine cycle, that uses an internal combustion chamber; and the other, that is a closed to the atmosphere, called the closed gas-turbine cycle, which uses a heat exchanger, as shown in figure 8.12.

1. **Open gas-turbine cycle:** In an open cycle, fresh air is drawn into the compressor, where both its temperature and pressure are raised. From here the high-pressure air goes to the combustion chamber, where the fuel is burned at constant pressure. These high-temperature gases enter the turbine, where they are allowed to expand to the atmospheric pressure and in turn produce power. This cycle is classified as an open gas-turbine cycle because the exhaust gases leaving the turbine are ejected to the atmosphere.
2. **Closed gas-turbine cycle:** In a closed gas-turbine cycle, the compression and expansion processes are the same as those in the open gas-turbine cycle. However, the combustion process of the open cycle is replaced by a

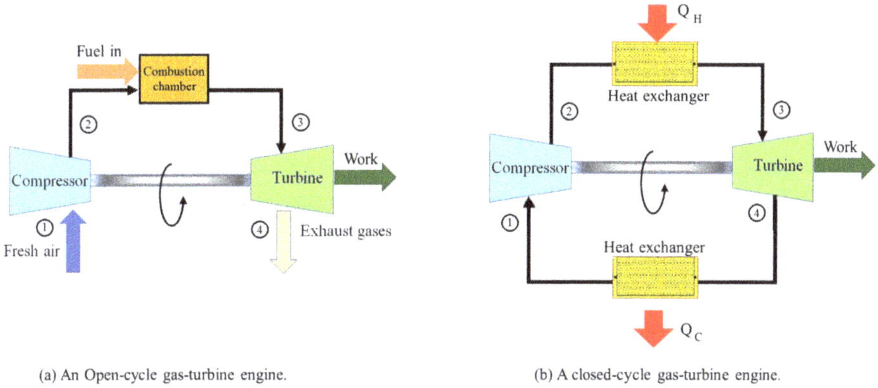

Figure 8.12. Open-cycle and a closed cycle gas-turbine engine.

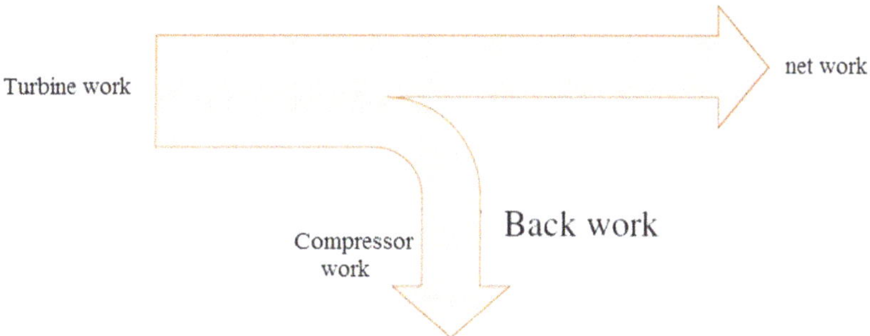

Figure 8.13. BWR is defined as the fraction of the turbine work used to drive the compressor.

constant-pressure heat-addition process from an external source and the exhaust process in the open cycle is replaced by a constant pressure heat-rejection process to the atmospheric air. The ideal cycle working fluid undergoes in the closed loop is called the '*idealized air standard Brayton cycle*' or simply the '***Brayton cycle***'.

There are many advantages to using the Brayton cycle. It is simple in construction so it is easy to maintain, and it can use many different kinds of fuel. It can also handle large volumes of gas and has a small weight-to-power ratio.

8.4.1 Back work ratio

The fraction of the turbine work used to drive the compressor is called the '**back work ratio**' (BWR). It is shown in figure 8.13 and is mathematically expressed as,

$$\text{Back Work Ratio} \equiv \frac{\text{Compressor Work}}{\text{Turbine Work}} \tag{8.13}$$

8.4.2 Major applications of gas-turbine engines

The two major application of gas-turbine engines are in aircraft propulsion and electric power generation. When gas-turbine engines are used for aircraft propulsion, the gas turbine produces just enough power to drive both the compressor and a small generator to power the auxiliary equipment. The high-velocity exhaust gases are responsible for producing the necessary thrust to propel the aircraft. When gas-turbine engines are used in stationary power plants to generate electricity as a stand-alone unit or in conjunction with steam power plants, the exhaust gases of the gas turbine are used as the heat source for the steam. The gas-turbine cycle can also be executed as a closed cycle for use in nuclear power plants.

In steam power plants the BWR is only a few percent, whereas in gas-turbine power plants the BWR is very high because more than half of the turbine work output is used to drive the compressor. This situation in gas-turbine power plants is not ideal by any standards. But it gets even worst when isentropic efficiencies of the compressor and the turbine are low. This is not surprising because the working fluid in a steam power plants is a liquid instead of a gas, and the steady-flow work is proportional to the specific volume of the working fluid. In order to get a high BWR, a power plant must have a large turbine to provide the additional power requirements of the compressor. Therefore, the turbines used in gas-turbine power plants are larger than those used in steam power plants of the same net power output.

8.4.3 Ideal Brayton cycle

In the ideal cycle a working fluid, usually air, undergoes a closed gas-turbine loop is called the '*Brayton cycle*'. The Brayton cycle is made up of four internally reversible processes; two are isentropic and two are isobaric processes and are shown in figure 8.14. Steps in a Brayton cycle are discussed briefly here, and are summarized in table 8.4.

1. **Step 1 to 2: isentropic compression:** During this step air is drawn into the compressor, and its pressure is increased from atmospheric pressure to P_2.

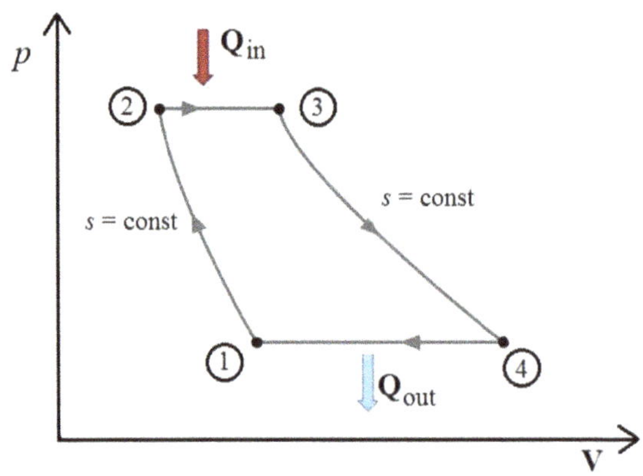

Figure 8.14. pV-diagram for the ideal Brayton cycle.

Table 8.4. Process in the ideal versus actual Brayton cycles.

Brayton Cycle	Ideal	Actual
Step 1 → 2—compression in a compressor	isentropic	adiabatic
Step 2 → 3—heat addition	isobaric	isobaric
Step 3 → 4—expansion in a turbine	isentropic	adiabatic
Step 4 → 1—heat rejection	isobaric	isobaric

2. **Step 2 to 3: isobaric step:** This is a constant-pressure process ($P_2 = P_3$), since the chamber is open to flow in and flow out. The compressed air moves through the combustion chamber. In the combustion chamber fuel is burned which heats the working fluid (air). Added heat is mathematically given as,

$$Q_{\text{in}} = Q_{32} = H_3 - H_2 = \dot{m}c_p(T_3 - T_2) \tag{8.14}$$

where c_p is the specific heat capacity at constant pressure. The highest temperature (T_3) in the cycle occurs at the end of the combustion process. Please note that T_3 is limited by the maximum temperature that the turbine blades can tolerate.

3. **Step 3 to 4: isentropic expansion:** In this step heated and pressurized working fluid (air) expands as it passes through a turbine and useful work (W_{net}) is done by the turbine and is used to drive the compressor. The useful work is mathematically given as,

$$W_{\text{net}} = (\text{Turbine Work}) - (\text{Compressor Work})$$
$$W_{\text{net}} = W_T - W_C$$

4. **Step 4 to 1: isobaric process:** Isobaric process is a constant pressure step, that is ($P_4 = P_1$). Heat (Q_{out}) is rejected to the atmosphere and is given as,

$$Q_{\text{out}} = |-Q_{14}| = H_4 - H_1 = \dot{m}c_P(T_4 - T_1) \tag{8.15}$$

The thermal efficiency of the ideal Brayton cycle under the cold-air standard assumptions, introduced in section 8.1.1.1 is,

$$\begin{aligned}\eta_{\text{Brayton}} &= \frac{W_{\text{net}}}{Q_{\text{in}}} \\ &= 1 - \frac{Q_{\text{out}}}{Q_{\text{in}}} \\ &= 1 - \frac{\dot{m}c_P(T_4 - T_1)}{\dot{m}c_P(T_3 - T_2)} \\ &= 1 - \frac{(T_4 - T_1)}{T_3 - T_2}\end{aligned}$$

Then, in terms of the **pressure ratio** ($r_p \equiv \frac{P_2}{P_1} = \frac{P_3}{P_4}$) and the specific heat ratio (γ), knowing, steps 1–2 and 3–4 are isentropic, and $P_2 = P_3$ and $P_4 = P_1$ we can say,

$$\frac{T_2}{T_1} = \left(\frac{P_2}{P_1}\right)^{(\gamma-1)\gamma} = \left(\frac{P_3}{P_4}\right)^{(\gamma-1)\gamma} = \frac{T_3}{T_4}$$

We can simplify the thermal efficiency of the ideal Brayton cycle as,

$$\begin{aligned} \eta_{\text{Brayton}} &= 1 - \frac{T_1(T_4/T_1 - 1)}{T_2(T_3/T_2 - 1)} \\ &= 1 - \frac{1}{r_p^{\frac{(\gamma-1)}{\gamma}}} \end{aligned} \tag{8.16}$$

$$\eta_{\text{Brayton}} = 1 - \frac{1}{r_p^{\frac{(\gamma-1)}{\gamma}}} \tag{8.17}$$

8.4.4 Real gas-turbine cycles

The real gas-turbine cycle differs from the ideal Brayton cycle in many respects, for example,

1. The ideal Brayton cycle assumes no pressure change. In reality some pressure drop during the heat-addition and heat rejection processes is inevitable.
2. Irreversibilities in a real Brayton cycle introduce two major errors:
   (a) The actual work input required (to compressor) is more than calculated for in an ideal Brayton cycle.
   (b) The actual work output from the turbine is less than what is calculate for an ideal Brayton cycle.

8.4.5 Brayton cycle with regeneration

In gas-turbine engines the temperature of the normally rejected exhaust gas leaving the turbine (T_4) is much higher than the temperature of those exiting the compressor (T_2). This means high-pressure air leaving the compressor at T_2 can be heated if we can somehow transfer the normally rejected heat at T_4 to the air. This transfer is usually achieved by using a heat exchanger known as a '***regenerator***', sometimes also called the '***recuperator***'. A schematic of a gas-turbine engine with a regenerator is shown in figure 8.15.

Since less energy is rejected from the cycle (Q_{out} decreases), the thermal efficiency of the Brayton cycle increases as a result of regeneration. Under the cold-air-standard assumptions, the thermal efficiency of an ideal Brayton cycle with regeneration is given as,

$$\eta_{\text{regen}} = \frac{W_T - W_C}{Q_{\text{in}}}$$

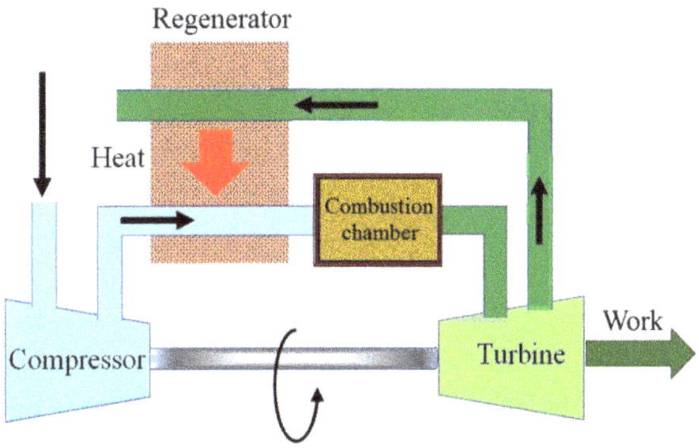

Figure 8.15. A gas-turbine engine with regenerator.

For an ideal regenerator $Q_{\text{in}} = W_{\text{T}}$, therefore,

$$\eta_{\text{regen}} = 1 - \frac{W_C}{W_T} = 1 - \frac{\dot{m}c_P(T_2 - T_1)}{\dot{m}c_P(T_3 - T_4)} = 1 - \frac{T_1}{T_3}\frac{(T_2/T_1 - 1)}{(1 - T_4/T_3)}$$

using isentropic relations,

$$\frac{T_2}{T_1} = \left(\frac{P_2}{P_1}\right)^{(\gamma-1)\gamma} = r_p^{\frac{(\gamma-1)}{\gamma}}$$

and

$$\left(\frac{P_3}{P_4}\right)^{(\gamma-1)\gamma} = \frac{T_3}{T_4} = r_p^{\frac{(\gamma-1)}{\gamma}} \Longrightarrow \frac{T_4}{T_3} = r_p^{-\frac{(\gamma-1)}{\gamma}}$$

Put these values in the above equation and calculate η_{regen} as,

$$\eta_{\text{regen}} = 1 - \left(\frac{T_1}{T_3}\right)\frac{r_p^{\frac{(\gamma-1)}{\gamma}} - 1}{1 - r_p^{-\frac{(\gamma-1)}{\gamma}}}$$

which can be simplified to,

$$\eta_{\text{regen}} = 1 - \left(\frac{T_1}{T_3}\right)r_p^{\frac{(\gamma-1)}{\gamma}} \tag{8.18}$$

Please note that the main difference between the ideal Brayton cycle (equation (8.17)) and Brayton cycle with regeneration (equation (8.18)) is that equation (8.18) has an additional ratio term (T_1/T_3), where T_1 is the lowest temperature and T_3 is the highest temperature during the cycle. These ratios are further related to the pressure ratios. This implies that the thermal efficiency for a Brayton cycle with regeneration depends on both the pressure ratio and the temperature ratio and the thermal efficiency of a Brayton cycle with regeneration decreases with the increase in pressure ratio. When r_p is kept constant then η_{regen} increases when the ratio T_1/T_3

decreases, while if the ratio T_1/T_3 is kept constant, η_{regen} increases when r_p decreases. This means not all combinations of pressure and temperature ratios will increase the thermal efficiency of a Brayton cycle with regeneration and we have to be careful when choosing the combinations of pressure and temperature ratios.

Worked example 8.4. Consider an ideal Brayton cycle in which air is used as a working fluid and the minimum temperature (T_1) and maximum temperature (T_3) are 300 K and 1300 K, respectively. Suppose the pressure ratio across the compressor is 6:1, and values of enthalpies and pressure ratios at minimum and maximum temperatures are, $H_1 = 300.19$ kJ kg^{-1}; $P_{r_1} = 1.386$ and $H_3 = 1395.97$ kJ kg^{-1}; $P_{r_3} = 330.9$. Determine:

a. The net work output per unit mass.
b. The thermal efficiency of the cycle.

Assume: Air is an ideal gas with variable specific heats and the process occurs at air-standard assumptions and $\gamma = 1.4$.

Solution:

a. The net work output per unit mass is calculated using the formula,

$$W_{\text{net}} = Q_{\text{in}} - Q_{\text{out}}$$

where $Q_{\text{in}} \equiv H_3 - H_2$ and $Q_{\text{out}} \equiv H_4 - H_1$. We are given the values of $H_1 = 300.19$ kJ kg^{-1} and $H_3 = 1395.97$ kJ kg^{-1}, however, we do not know values of H_2 and H_4. We can calculate these, using the compression ratio $r_p = 6$, and knowing $P_2/P_1 = P_3/P_4 = 6$ and $P_{r_1} = 1.386$ and $P_{r_3} = 330.9$ as,

$$P_{r_2} = \left(\frac{P_2}{P_1}\right)P_{r_1} = 6(1.386) = 8.316$$

$P_{r_2} = 8.3$ corresponds to the following values, temperature (in this case T_2) 500 K and $H_2 = 503$ kJ kg^{-1}.[1]

$$P_{r_4} = \left(\frac{P_4}{P_3}\right)P_{r_3} = \frac{1}{6}(330.9) = 55.15$$

$P_{r_4} = 55.15$ corresponds to the following values; temperature (in this case T_4) 840 K and $H_4 = 860$ kJ kg^{-1}.

$$Q_{\text{in}} \equiv H_3 - H_2 = (1395.97 - 503)\ \text{kJ kg}^{-1} = 892.97\ \text{kJ kg}^{-1}$$
$$Q_{\text{out}} \equiv H_4 - H_1 = (860 - 300.19)\ \text{kJ kg}^{-1} = 559.81\text{kJ kg}^{-1}$$

Put these values in the above equation,

$$W_{\text{net}} = Q_{\text{in}} - Q_{\text{out}} = (892.97 - 559.81) = 333.2\ \text{kJ kg}^{-1}.$$

[1] These values are taken from the properties of ideal gases tables, called steam tables, in engineering textbooks.

b. The thermal efficiency of the cycle.

$$\eta = \frac{W_{\text{out}}}{Q_{\text{in}}} = \frac{333.2}{892.97} = 37.9\%.$$

8.4.6 Jet engines

Gas-turbine engines are comparatively light and compact and they have a high power-to-weight ratio, which makes them perfect to be used in aircraft. Aircraft gas-turbines operate on an open gas-turbine cycle called a '***jet-propulsion cycle***'. In an aircraft, the engine called a '*jet engine*' is used to convert fuel, usually kerosene, into a force named '*thrust*'. The pV diagram for a jet engine is shown in figure 8.16. Jet engines are also called '***reaction engines***' because thrust in the engine is produced by emitting a jet of fluid behind the engine, in the opposite direction to motion, at very high speeds, which pushes the airplane forward. This is accomplished in reality by either somewhat accelerating a large mass of fluid, as is done in a propeller-driven engine, or by accelerating a small mass of fluid by a lot, as done in a jet or in a turbojet engine or both as done in a turboprop engine.

To understand jet engines, we have to compare them to regular piston engines. A piston engine makes power in steel cylinders where a mixture of fuel and air is added. The pistons in these cylinders raise the temperature of the mixture by compressing it until it ignites, either with help of a spark plug, as done in a gas engine (SI engines, discussed in section 8.2.2), or spontaneously, as done in a Diesel engine (CI engines, discussed in section 8.3). In regular car engines, the burning fuel and air mixture explodes, expands and hence pushes the piston back. The piston drives the crankshaft that powers the car. As a result during these cycles we get four steps, intake, compression, combustion, and exhaust. There are a few problems in this system that include,

1. First, the piston is driven only during one of the four steps, that is during the combustion step. This means power is only made during combustion.

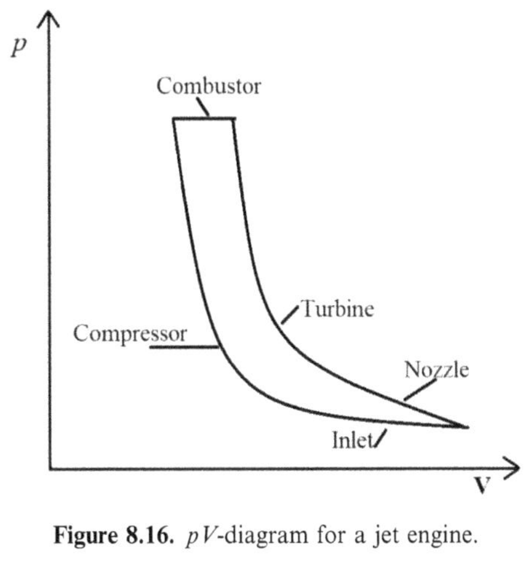

Figure 8.16. pV-diagram for a jet engine.

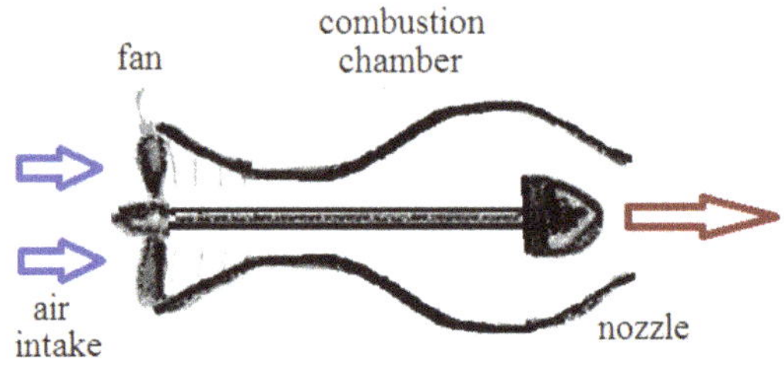

Figure 8.17. Jet engine.

The rest of the time the engine is not supplying any power to the car. This is not ideal; we want maximum output power.

2. Second, power is directly proportional to the size of cylinder and the distance the piston moves. Larger sized cylinders will produce more power, however, this also increases the weight of the car. Hence the power-to-weight ratio becomes an issue.

A jet engine on the other hand burns a mixture of fuel and air to release energy that powers the engine, but instead of using cylinders that go through four steps, it uses one long metal tube that carries out the steps in a single sequence, as shown in figure 8.17. This means that, unlike a piston-and-cylinder engine, a jet engine produces maximum power all the time. Jet engines are also designed to take in a large amount of air and burn it with large amount of fuel to produce a lot of power. The power of a jet engine is equivalent to that of 28 Formula-1 racing cars. Please keep in mind that the net work output of a jet propulsion cycle is zero. The gases that exit the turbine at relatively high pressures are accelerated in a nozzle to provide the thrust to propel the aircraft.

8.4.6.1 Properties of a jet engine

The thrust of a typical jetliner engine is about 510 000 N. Thrust is the unbalanced force caused by the difference in the momentum of the low-velocity air entering the engine and the high-velocity exhaust gases leaving the engine. Thrust developed by the engine is determined from Newton's second law of motion ($\Sigma\vec{F} = m\vec{a}$), where $\vec{F}$ is the sum of forces acting on an object, m is the mass and $\vec{a}$ is the acceleration produced as a result. Assuming the pressures at the inlet and the exit of a turbojet engine are identical, the net thrust developed by the engine is calculated using the equation (8.19).

$$\begin{aligned} F_{\text{thrust}} &= (\dot{m}v)_{\text{exit}} - (\dot{m}v)_{\text{in}} \\ &= \dot{m}(v_{\text{exit}} - v_{\text{in}}) \end{aligned} \tag{8.19}$$

where v_{exit} is the exit velocity of the exhaust gases relative to the airplane, v_{in} is the inlet velocity of the air relative to the airplane and $\dot{m}$ is the mass flow rate of air through the engine.

For a plane cruising at a particular altitude with constant speed, the thrust is used to overcome air drag, and the net force acting on the body of the aircraft is zero.

Air is thinner at higher altitudes and hence it exerts a smaller drag force on a plane. Commercial airplanes use this fact to save on fuel consumption by flying at higher altitudes during long trips.

In reality, the thrust is mathematically given as,

$$\begin{aligned} F_{\text{thrust}} &= (\text{Nozzle Gross Thrust}) - (\text{Drag of the Intake}) \\ &= (\dot{m}_{\text{air}} + \dot{m}_{\text{f}})v_j - (\dot{m}_{\text{air}} V) \end{aligned} \tag{8.20}$$

where $\dot{m}_{\text{air}}$ is the rate of flow of air through the engine, $\dot{m}_{\text{f}}$ is the rate of flow of fuel entering the engine, v_j is the speed of jet (basically the exhaust plume) and V is the true air speed of the plane. The power developed from the thrust of the engine is called the **propulsive power** ($\dot{W}_P$), mathematically given by the equation,

$$\begin{aligned} \dot{W}_P &= Fv_{\text{plane}} \\ &= \dot{m}(v_{\text{exit}} - v_{\text{in}})v_{\text{plane}} \end{aligned} \tag{8.21}$$

Propulsive efficiency (η) is a measure of how efficiently the thermal energy released during the combustion process is converted to propulsive energy and is given by,

$$\begin{aligned} \eta &= \frac{\text{Propulsive power}}{\text{Energy input rate}} \\ &= \frac{\dot{W}_{\text{P}}}{\dot{Q}_{\text{in}}} \end{aligned} \tag{8.22}$$

The remaining energy released becomes the kinetic energy of the exhaust gases relative to a fixed point on the ground and causes an increase in the enthalpy of the gases leaving the engine.

8.4.6.2 Main parts of a jet engine

The simplest type of jet engine is a '*turbojet*' engine, as shown in figure 8.17. A jet engine is divided into two main parts, a cold section and a hot section. The cold section is made-up of the '**air inlet**', which is basically a duct that carries air to the engine, a number of '**turbofans**', which play the role of a compressor in a jet engine, and a '**shaft**', which connects the turbine to the compressor.

The hot section of the jet engine is made-up of the following parts: '**combustor**', which is equivalent to the combustion chamber in the other engines and where fuel is burned; the '**turbine**', which is a series of bladed discs that act like windmill blades and extract energy from the hot gases; and the '**exhaust**', also called the propelling nozzle, where the exhaust gases from the turbine pass through the propelling nozzle and produce a high velocity jet. The turbine blades, compressor, and the turbofans are all connected to a long axle. As the turbine blades spin, they also turn the compressor and the fan. A schematic of a jet engine is shown in figure 8.18.

In the operation of a jet engine the following sequence of events occur: first, the air is drawn in at the front of a tube, and the air intake carries the air into the engine. Suppose a jet is flying at about 600 mph (or 100 km h^{-1}). The inlet sucks in air, and slows it down by about 60% or to about 240 mph (400 km h^{-1}). The air then

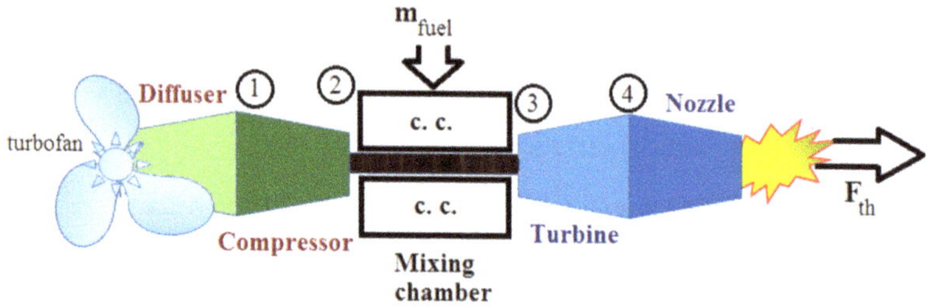

Figure 8.18. Schematic of a jet engine.

encounters multiple turbofans that squeeze the air. As the air moves through the turbofans, its pressure increases about eight times, which in turn increases its temperature. Air then gets to the combustor, where it is mixed with fuel, usually kerosene supplied from a fuel tank in the plane's wing, and combusted. This gives off extremely hot exhaust gases (at almost 900 °C) and produces a huge increase in temperature. The turbine extracts energy from the hot gases, the turbine gains energy and the gas loses the same amount of energy and the temperature of the gas goes down. The exhaust gases exit the engine through a tapering **exhaust** at the back of the tube. As the gases pass through the exhaust nozzle, it speeds up to about 2100 km h^{-1}; think of the equation of continuity ($A_1 v_1 = A_2 v_2$), from chapter 2, where the A's are the cross-sectional areas of the pipe and the v's are the speeds of the fluid at points 1 and point 2. This backward thrust powers the jet engine and propels the jet plane forwards at great speeds.

8.4.7 Nuclear power plants

An atom consists of neutrons and positively charged protons bound together by very strong nuclear forces in the nucleus, and negatively charged electrons orbiting around it. The nuclear forces are much larger than the forces that bind the electrons to the nucleus. The tremendous amount of energy associated with the strong bonds within the nucleus of the atom itself is called nuclear energy.

Nuclear power plants use nuclear fission to generate electricity. The best known fission reaction involves the splitting of the uranium atom (the U-235 isotope) into other elements and is commonly used to generate electricity in nuclear power plants, to power nuclear submarines and aircraft carriers and many other applications. When a U-235 atom absorbs a neutron and splits during a fission process, it produces a cesium-140 atom, a rubidium-93 atom, 3 neutrons, and energy. The complete fission of 1 kg of U-235 releases 6.73×10^{10} kJ of heat, which is more than the heat released when 3000 tons of coal are burned. Therefore, for the same amount of fuel, a nuclear fission reaction releases several million times more energy than a chemical reaction. Nuclear power provides the world with around 11% of its total electricity. Typical nuclear power plants achieve efficiencies around 33%–37%, which is comparable to fossil fueled power plants.

Nuclear power plants generate electricity by utilizing nuclear reactors with the Rankine cycle, where the heat generated by the reactor converts water into steam,

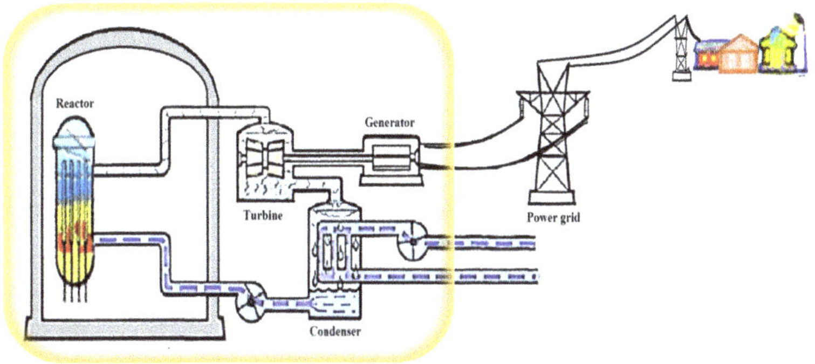

Figure 8.19. A boiling water nuclear reactor in combination with the Rankine cycle forms the basis of a nuclear power plant.

which spins a turbine and a generator. A nuclear power plant consists of the following main parts, as shown in figure 8.19:

1. **Nuclear reactor:** The nuclear reactor is the main component of a nuclear power plant. The reactor is to the nuclear power plant what the boiler is to a coal plant, that is, it is the heat source for the nuclear power plant. The reactor contains the nuclear fuel and is where the reaction happens and is also where waste products are deposited at the end.
2. **Heat rejection:** When the nuclear power plant is equipped with cooling towers the power plant rejects waste heat to the surrounding air. Other nuclear power plants are built near bodies of water, such as lakes and rivers so they can reject waste heat into the river or lake, instead of having cooling towers. Whether the power plant is built with cooling towers or not is a simple matter of the physical location of the power plant.
3. **Steam generators:** Most power plants in the world use pressurized water reactors, which use two loops of circulating water to produce steam. The first loop carries the extremely hot liquid water to a heat exchanger, where liquid water is heated up so it becomes steam. The steam is then sent to the turbine. The second most common reactors used are called boiling-water reactors. In these reactors the water is heated directly in the core to steam, as seen in figure 8.19.
4. **Turbine:** Steam from the steam generator is then sent to the turbine. It travels very quickly and at high pressures and goes through one or more turbines. In the turbine the steam loses energy and condenses back to liquid water.
5. **Generator:** The rotation of the turbines is used to spin an electric generator, which produces electricity that is sent out to the electrical grid and powers the community and runs the factories.

Worked example 8.5. Suppose you wish to drive a car on nuclear power and the car comes with 10 g (0.4 ounces) of U-235 fuel. Determine under average driving conditions how long will you have to drive the car before you have to replace the 10 g of U-235 fuel? Assume the capacity of your car's fuel tank is 14 gallons (about 50 L) and you use up about a gallon (about 4 L) of gasoline a day and the heating value of gasoline is 44 000 kJ kg^{-1}.

Solution: Suppose the density of gasoline is 0.75 kg L^{-1}, then you use,

$$m_{\text{gasoline}} = \rho V = (0.75 \text{ kg L}^{-1})(4 \text{ kg L}^{-1}) = 3 \text{ kg/day}$$

Given that the heating value of gasoline is 44 000 kJ kg^{-1}, the energy supplied to the car per day is,

$$\begin{aligned} E &= (m_{\text{gasoline}})(\text{Heating value}) \\ &= (3 \text{ kg/day})(44\,000 \text{ kJ kg}^{-1}) \\ &= 132\,000 \text{ kJ kg}^{-1} \end{aligned}$$

The heat released by the complete fission of 10 g of U-235 is,

$$(0.01\text{kg})(6.73 \times 10^{10} \text{ kJ kg}^{-1}) = 6.73 \times 10^{8} \text{ kJ}.$$

which is enough to run the car for,

$$\begin{aligned} \text{Number of days} &= \frac{\text{Energy content of fuel}}{\text{Daily energy use}} \\ &= \frac{6.73 \times 10^{8} \text{ kJ kg}^{-1}}{132\,000 \text{ kJ kg}^{-1}} \\ &= 5098 \text{ days} \equiv 14 \text{ years!} \end{aligned}$$

Concept questions 8.2

1. How does the Brayton cycle work?
2. The molecules in a combustion chamber of a rocket engine are in a state of random motion. When some of these molecules are expelled through a jet nozzle, what happens to the temperature (lower, higher or remains the same) of the molecules left in the chamber? Assume $Q = 0$ during this process.
3. A coal-fired power plant needs coal to run. About how many tons of coal are required to fuel a 1 GigaWatt plant for a day? How much fuel will be needed in a nuclear plant?
4. In a coal-fired power plant, we burn coal to boil water to produce steam,. The steam travels through the turbine and rotates a large shaft. The rotating shaft turns massive coils of copper conductor. Why does this induce an electrical current?

Exercises 8.2

1. In a Brayton cycle based power plant, the air at the inlet is at 27 °C, 0.1 MPa. The pressure ratio is 6.25 and the maximum temperature is 800 °C. Find,
   (a) the compressor work per kg of air,
   (b) the turbine work per kg or air,

(c) the heat supplied per kg of air, and
(d) the cycle efficiency.

2. Consider an ideal Brayton cycle in which air is used as a working fluid into the compressor at room temperature and 100 kPa. The maximum temperature in the cycle is 1100 °C and the air flow rate is 10 kg s^{-1}. Find, the compressor work, the turbine work and the thermal efficiency of the cycle.

Further reading

- Cengel Y and Boles M 2006 *Thermodynamics: An Engineering Approach* 5th edn (New York: McGraw-Hill Higher Education)
- Resnick R, Halliday D and Krane K 2002 *Physics* vol 1 5th edn (New York: Wiley)
- Schroeder D 2000 *Thermal Physics* (Reading, MA: Addison-Wesley Longman)

Chapter 9

Heat transfer-to-power cycle II

The ideal Otto and Diesel cycles discussed in chapter 8, are made up of entirely internally reversible processes and are therefore totally internally reversible cycles. However, real Otto and Diesel cycles involve heat transfer through a finite temperature difference during the non-isothermal heat-addition and heat-rejection processes, which are irreversible. Therefore, the thermal efficiency of an Otto or Diesel engine is always less than the efficiency of an ideal Carnot engine operating between the same temperature limits.

For a heat engine operating between a heat source at T_H and a heat sink at T_C to be totally reversible, the temperature difference between the working fluid and the heat source (or sink) should never exceed a differential amount (dT) during any heat-transfer process. That is, both the heat-addition (T_H) and heat-rejection (T_C) processes during the cycle must take place isothermally just like what happens in a Carnot cycle.

There are two other cycles that involve isothermal heat-addition at T_H and isothermal heat-rejection at T_C. These cycles are called the **Stirling cycle** and the **Ericsson cycle**. In the Stirling cycle, the two isentropic processes of the Carnot cycle are replaced by two constant-volume regeneration processes, whereas in the Ericsson cycle, the two isentropic processes in the Carnot cycle are replaced by two constant-pressure regeneration processes. These are discussed in this chapter.

9.1 Stirling engines

Early steam engines often exploded, which was a major concern for people working in close proximity to the engines. Some people were maimed or even killed as a result of these explosions. Robert Stirling was a minister in the Church of Scotland and an inventor. He was deeply concerned about the effect steam engines were having on the people. He invented an engine he called the 'air engine' that could not explode and which produced more power than the steam engines of the time. Robert Stirling patented his air engine in 1816 CE and continued to improve upon his basic idea.

doi:10.1088/978-0-7503-3083-1ch9

Engines that followed eventually became known as 'hot air engines', which is what they were called until the 1940s.

The Stirling engine is classified as an '*external combustion engine*', because all heat transfers to and from the working fluid take place through a solid boundary, called the '*heat exchanger*'. This isolates the combustion process from any contaminants that may be produced in the engine and makes it safer for the public. Recall from chapter 8 that in the internal combustion engine, heat input is provided within the body of the working fluid in the combustion chamber.

9.1.1 Regenerator

The 'regenerator', or 'economizer' as Robert Stirling called it, is what makes the Stirling cycle different from other cycles. A 'regenerator' is an internal heat exchanger and temporary heat storage placed between the hot and cold reservoirs in a way that the working substance is forced to pass through it first in one direction, taking heat from the working fluid in one direction, and then forced to pass in the opposite direction where it returns heat to the working fluid returning it in the other direction.

The basic function of a regenerator is to retain heat within the system and recycle it at temperatures between T_H and T_C. This heat without the 'regenerator' would be ejected and lost to the environment. The advantage of this is that it allows the thermal efficiency of the Stirling cycle to approach the limiting Carnot efficiency, mathematically given as:

$$\eta_{\text{Stirling}} = 1 - \frac{T_C}{T_H} \tag{9.1}$$

The regenerator, which can be a wire or a mesh, is assumed to be a reversible heat transfer device, used for the temporary storage of heat energy. Usual regenerators are simple meshes made up of simple metals or ceramic or foam. The main qualifications for a 'regenerator' are that it should have a high surface area, a high heat capacity, high thermal mass (mass times specific heat), low conductivity and low fluid friction. The mass of the working fluid contained within the regenerator at any instant is considered negligible.

9.1.2 Steps of a Stirling cycle

In the Stirling cycle, the compression and expansion isentropic processes of the Carnot cycle are replaced by two constant-volume regeneration processes. The Stirling cycle is made up of four reversible processes:

1. **Step 1 to 2: isothermal expansion**. This step is an isothermal (constant T) expansion step, where heat ($Q_{in} = Q_H$) is added from an external source.
2. **Step 2 to 3: isochoric regeneration I**. This is a constant-volume step. This is the first regeneration step. Internal heat during this step transfers from the working fluid to the regenerator.

3. **Step 3 to 4: isothermal compression**. This is a compression step, performed at constant temperature, in which heat ($Q_{out} = Q_C$) is rejected to the external sink.
4. **Step 4 to 1: isochoric regeneration II**. This is the second regeneration step. During this step, internal heat transfers back from the regenerator to the working fluid.

9.1.3 Actual Stirling engine

In theory, the Stirling engine is the simplest of all heat engines. However, the actual Stirling cycle requires innovative hardware to be made a reality. They are usually bulky and complicated to build, so in this text I will avoid all the engineering issues and explain the execution of the Stirling cycle in a closed system by an ideal engine.

When compared to a piston-and-cylinder system, a Stirling engine has two pistons within one cylinder with a regenerator in the middle. The regenerator divides the cylinder into two chambers, the left and the right chamber. The two pistons in a Stirling engine are called '**power piston**' and '**cold piston**', as shown on the left and right side of in figure 9.1, respectively, with a regenerator mesh in the middle.

Initially the left chamber has the entire working fluid, usually a gas, at a high pressure and high temperature (T_H). The Stirling engine undergoes the following steps while executing the Stirling cycle:

1. **Step (1) to (2): power stroke**—isothermal expansion (T_H). The power piston is pushed outwards (to the left) by the expanding gas, thus doing work, and the

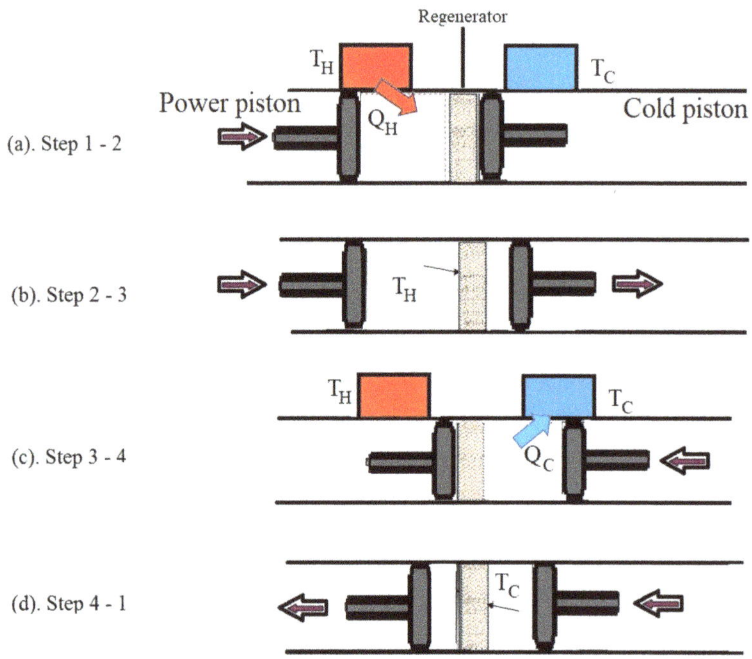

Figure 9.1. Working principles of a Stirling engine with two pistons.

gas pressure drops. The gas absorbs heat from a source at T_H. While the cold piston in the cylinder remains at rest all the way inwards, as shown in figure 9.1(a). Assuming the working fluid is an ideal gas, the work done during this step is given as,

$$W_{12} = -nRT_H \ln\left(\frac{V_2}{V_1}\right)$$

2. **Step (2) to (3): isochoric heat removal**—transfer of gas to the cold cylinder at constant volume ($Q_C = Q_{out}$). During step 2 to 3, the power piston moves into the cylinder (to the right) and the cold piston moves outwards (toward the left), that is, both pistons move to the right at the same rate to keep the total volume constant. The entire working fluid (gas) passes from the hot part of the cylinder through to the cold cylinder. On its way working fluid (gas) passes through the regenerator, where it cools, and transfers heat to the regenerator for use in the next cycle, as shown in figure 9.1(b).
3. **Step (3) to (4): compression stroke**—isothermal contraction (T_C). In this step the cold piston moves inwards (to the left) isothermally, which compresses the working fluid (gas) to its original volume. During this process, work ($W_{in,compression}$) is done on the gas and the gas gives up $Q_{H,in}$ heat to the cold reservoir. The power piston remains at rest all the way in, as shown in figure 9.1(c). The net work output is determined as,

$$W_{out} = \eta Q_{in} = mT_H(S_2 - S_1), \tag{9.2}$$

where the thermal efficiency of this totally reversible cycle is determined from $\eta = 1 - T_C/T_H$ and S_1 and S_1 are entropies during step 1 and step 2, respectively.
4. **Step (4) to (1): isochoric heat absorption**—transfer of gas to the hot cylinder at constant volume. During this step, the gas passes back through the regenerator on its way to the hot chamber where it recovers much of the heat ($Q_{H,in}$) transferred to the regenerator in step 2 to 3. The gas heats up during this passage, as shown in figure 9.1(d). The amount of heat transferred in the regenerator is,

$$Q_{regen} = Q_{41,in} = mc_V(T_1 - T_4) \tag{9.3}$$

This completes the cycle. The pV diagram for a Stirling cycle is shown in figure 9.2.

9.1.4 Efficiency of a Stirling engine

We know that the efficiency of engines is defined as,

$$\eta = \left|\frac{W_{net}}{Q_{in}}\right| \tag{9.4}$$

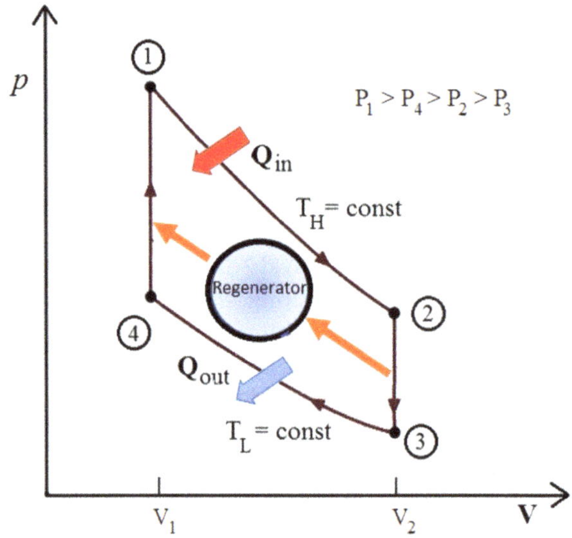

Figure 9.2. *pV*-diagram for a Stirling cycle with a regenerator.

Using the ideal gas law for the working fluid, the net work for Stirling engine is calculated as,

$$\begin{aligned}W_{\text{net}} &= W_{\text{out,expansion}} + W_{\text{in,compression}} \\ &= -\int_{\text{expansion}} PdV - \int_{\text{compression}} PdV \\ &= -\int_{\text{expansion}} \frac{nRT_{\text{H}}}{V} dV - \int_{\text{compression}} \frac{nRT_{\text{C}}}{V} dV\end{aligned}$$

We can integrate and simplify to get the net work as,

$$W_{\text{net}} = -nR(T_{\text{H}} - T_{\text{C}}) \ln \frac{V_2}{V_1} \tag{9.5}$$

The total heat change in the system (Q_{total}) is the sum of heat exchanged during the power stroke, (Q_{12}) and the compression stroke (Q_{41}) where,

$$Q_{12} = W_{12} = -nRT_{\text{H}} \ln\left(\frac{V_2}{V_1}\right)$$

and,

$$Q_{41} = \Delta U = U_4 - U_1 = \frac{f}{2} nR(T_{\text{H}} - T_{\text{C}})$$

where f is the degrees of freedom of the system for an ideal gas in 3D. Degrees of freedom of molecules is the number of independent ways in which a molecule has translational, rotational, and/or vibratory motion. For example, a monoatomic molecule has a single atom, and hence can have only translational motion along each of the three axes in free 3-D space. Thus a monoatomic molecule has three degrees of freedom. A diatomic molecule, which has two molecules, has six degrees of freedom, three for translational motion along the axes, two for rotational motion along axes perpendicular to the line joining the two atoms and one for vibrational motion along the line joining both molecules. An ideal gas in 3D has three degrees of freedom, i.e., $f = 3$, which gives the internal energy from equation (4.21), as $U = 3/2nRT$. Hence the total heat will be given as,

$$Q_{\text{total}} = Q_{12} + Q_{41} = -nRT_{\text{H}} \ln \frac{V_2}{V_1} + \frac{f}{2} nR(T_{\text{H}} - T_{\text{C}}) \tag{9.6}$$

Putting values in equation (9.4) we get,

$$\eta = \left| \frac{W_{\text{net}}}{Q_{\text{in}}} \right| = \frac{(T_{\text{H}} - T_{\text{C}}) \ln \frac{V_2}{V_1}}{T_{\text{H}} \ln \frac{V_2}{V_1} + \frac{f}{2}(T_{\text{H}} - T_{\text{C}})} \tag{9.7}$$

which is quite complicated, so it is worth noting here that efficiency with the regenerator will be given as,

$$\eta_{\text{with regenerator}} = 1 - \frac{T_{\text{C}}}{T_{\text{H}}} \tag{9.8}$$

Worked example 9.1. A particular ideal Stirling engine uses 0.5 kg of helium as the working fluid. Determine, the amount of heat transfer in the regenerator, and the work output per cycle, supposing it operates between $T_1 = T_2 = 1500$ K and $T_3 = T_4 = 300$ K and pressures of $P_1 = 3000$ kPa and $P_3 = 100$ kPa. Given the gas constant, $R = 2.0769$ kJ kg^{-1} K^{-1} and that the specific heat of helium at room temperature is $c_V = 3.1156$ kJ kg^{-1} K^{-1} and $c_p = 5.1926$ kJ kg^{-1} K^{-1}.

Solution: The amount of heat transfer in the regenerator is given by equation (9.3) as,

$$\begin{aligned} Q_{\text{regen}} &= mc_V(T_1 - T_4) \\ &= (0.5 \text{ kg})(3.11 \text{ kJ kg}^{-1} \text{K}^{-1})(1500 - 300) \text{ K} \\ &= 1866 \text{ kJ}. \end{aligned}$$

and the work output per cycle will be determined as,

$$W_{\text{out}} = \eta_{\text{with regenerator}} Q_{\text{in}} \tag{9.9}$$

where $\eta_{\text{with regenerator}}$ the thermal efficiency of the cycle,

$$\eta_{\text{with regenerator}} = 1 - \frac{T_{\text{C}}}{T_{\text{H}}} = 1 - \frac{300 \text{ K}}{1500 \text{ K}} = 0.80 = 80\%.$$

and Q_{in} is determined as,

$$Q_{\text{in}} = mT_H(S_2 - S_1)$$
$$= mT_H\left(R \ln \frac{V_2}{V_1}\right)$$
$$Q_{\text{in}} = mT_H\left(R \ln \frac{V_3}{V_1}\right)$$

From experimental gas laws from sections 4.1 and 4.3, equation (4.6),

$$\frac{V_3P_3}{T_3} = \frac{V_1P_1}{T_1}$$

Solve this to give,

$$\frac{V_3}{V_1} = \frac{T_3P_1}{T_1P_3}$$

$$Q_{\text{in}} = mT_{\text{H}}\left(R \ln \frac{T_3P_1}{T_1P_3}\right)$$
$$= (0.5 \text{ kg})(1500 \text{ K})(2.0769 \text{ kJ kg}^{-1} \text{ K}^{-1}) \ln \frac{(300 \text{ K})(3000 \text{ kPa})}{(1500 \text{ K})(100 \text{ kPa})}$$
$$= 6377.7 \text{ kJ}$$

Putting values in equation (9.9),

$$W_{\text{out}} = \eta_{\text{with regenerator}} Q_{\text{in}} = 0.80 \times 6377 \text{ kJ} = 5102 \text{ kJ}.$$

9.1.5 Types of Stirling engines

There are three major types of Stirling engines:

1. **α-type:** α-type Stirling engines have two power pistons, one in a hot cylinder, one in a cold cylinder. The working substance is driven between the two by the pistons. α-type is usually in a V-formation with the pistons joined at the same point on a crankshaft, as explained in section 9.1.3.
2. **β-type:** β-type Stirling engine has a single cylinder with a hot end and a cold end, containing a power piston, and a displacer that drives the gas between the hot and cold ends. Figure 9.3 shows a conceptual diagram of a generic Stirling engine. The displacer piston shuttles gas (through the regenerator) between the hot and cold sections. This causes the average temperature and eventually the pressure of the working gas to oscillate. These pressure oscillations drive the power piston which performs mechanical work.
3. **γ-type:** γ-type Stirling engines have two cylinders: one containing a displacer, with a hot and a cold end, and one for the power piston. They are joined to form a single space with the same pressure in both cylinders, where the pistons are in parallel.

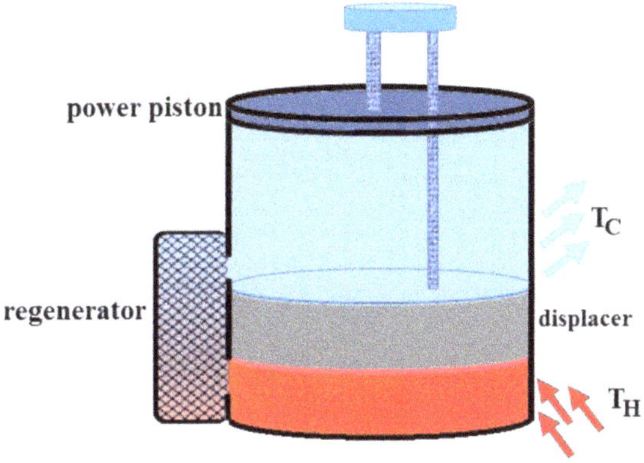

Figure 9.3. Conceptual diagram of a generic Stirling engine.

9.1.6 Advantage of ideal Stirling cycle

Stirling engines are used only in some very specialized applications, like in submarines or auxiliary power generators for yachts, where quiet operation is important. The biggest advantage of the Stirling cycle is that the thermal efficiency of the Stirling cycle with ideal regeneration is equal to the thermal efficiency of the Carnot cycle, which has the maximum possible efficiency of any heat engine system between specified temperature limits. Other advantages include,

1. Real Stirling engines operate very quietly.
2. Real Stirling engines can reach 50% of the maximum theoretical efficiency value, which is incredibly high.

Worked example 9.2. An ideal Stirling engine operates at 300 K and uses air as the working fluid. Suppose it operates with a compression ratio of 10 between specified pressure limits, 400 kPa and 3400 kPa and $p_3 = 100$ kPa. Calculate the heat added to the cycle and the net work produced by the cycle. Assume air behaves as an ideal gas with $c_p = 1.005$ kJ kg^{-1} K^{-1}, $c_v = 0.718$ kJ kg^{-1} K^{-1} and $R = 0.287$ kPa m^3.

Solution: Applying the ideal gas equation to the isothermal process (3–4) gives,

$$p_4 = p_3\frac{V_3}{V_4} = (100\text{ kPa})(10) = 1000\text{ kPa}$$

Since process 4–1 is one of constant volume,

$$T_1 = T_4\frac{p_1}{p_4} = (300\text{ K})\frac{3400\text{ kPa}}{400\text{ kPa}} = 2550\text{ K}$$

Adapting the First law of thermodynamics and work integral to the heat addition process gives,

$$Q_{\text{in}} = W_{12} = RT_1\ln\frac{V_2}{V_1} = (0.287\text{ kJ kg}^{-1}\text{ K}^{-1})(2550\text{ K})\ln(10)$$
$$= 1685\text{ kJ kg}^{-1}.$$

Similarly,

$$Q_{\text{out}} = W_{34} = RT_3 \ln \frac{V_4}{V_3} = (0.287 \text{ kJ kg}^{-1} \text{ K}^{-1})(300 \text{ K}) \ln\left(\frac{1}{10}\right)$$
$$= 198 \text{ kJ kg}^{-1}$$

The net work is then,

$$W_{\text{net}} = Q_{\text{in}} - Q_{\text{out}} = 1487 \text{ kJ kg}^{-1}.$$

Worked example 9.3. Why not use Stirling engines in our cars?

Solution: Stirling engines are very efficient, cause less pollution and can use many different fuels from gasoline to diesel, but we do not use them in our cars. The biggest reason for this is because Stirling engines do not start up immediately. When we get in our cars we do not want to wait around for the cars to start up. In the 1970s the Ford company built a Stirling engine that could start driving 20 seconds after you turned the start key. Even in those days, 20 seconds was considered a very long time to wait.

One of the first prototype cars using an experimental Stirling engine, named P-40, was the 1979 AMC spirit. The P-40 Stirling engine caused less pollution, had 30% better mileage, and the same level of performance as a standard internal combustion engine. But its biggest disadvantage was that cars using Stirling engines do not start up immediately and hence they never caught on.

9.2 Ericsson cycle

The Ericsson cycle, named after the Swedish-American inventor John Ericsson, is much like the Stirling cycle, except that the two constant-volume processes of the Stirling cycle are replaced by two constant-pressure processes. Ericsson's first thermodynamic cycle is now called the 'closed Brayton cycle', which was discussed in section 8.4.

The second cycle invented by Ericsson is nowadays called the Ericsson cycle. In the Ericsson cycle the isothermal expansion is executed in a compressor and the compression processes are executed in a turbine, also called the 'expander', and to improve efficiency, the engine has a regenerator between the compressor and the expander. Hot and cold working fluid streams enter the regenerator from opposite ends.

The Ericsson engine, which is based on the Ericsson cycle, just like the Stirling engine, is known as an *'external combustion engine'*, because it is heated externally, unlike the gasoline or Diesel engines, where the fuel is burned inside the cylinder. The Ericsson engine can be run as either an open-cycle or a closed-cycle, as explained previously in section 8.4.

The Ericsson engine has four processes; two isothermal and two isobaric processes. In the Ericsson cycle, the heat addition (heating) takes place during both the constant pressure process (steps 2–3) and the isothermal process (steps 3–4). Similarly, heat rejection (cooling) takes place both during the constant pressure

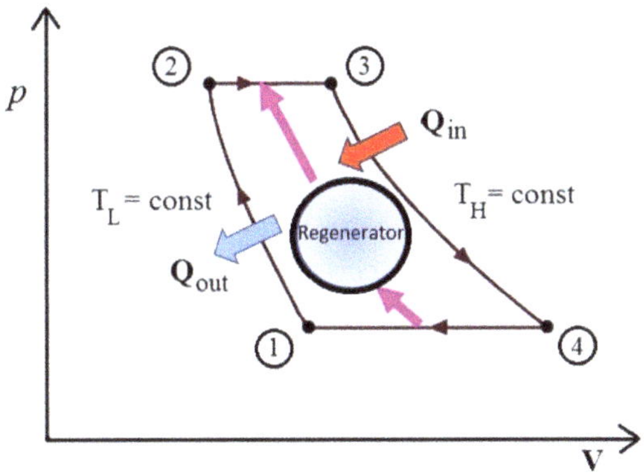

Figure 9.4. pV-diagram for an Ericsson cycle.

(steps 4–1) and the isothermal processes (steps 1–2). The pV-diagram for an Ericsson cycle is shown in figure 9.4. The four processes that occur between the four stages of the ideal Ericsson cycle are as follows:

1. **Step (1) to (2): isothermal (heat removal) compression.** The working fluid (gas) undergoes isothermal compression where heat is rejected from the working fluid isothermally. This compressed air then flows into a storage tank at constant pressure while no heat is lost from the tank walls. Heat rejected during this step is given by,

$$Q_{12} = P_1V_1 \ln\left(\frac{V_1}{V_2}\right)$$

Using the ideal gas law, Q_{12} can be written as,

$$Q_{\text{out}} = nRT_C \ln\left(\frac{V_1}{V_2}\right)$$

The change in entropy during this isothermal step (with $\Delta T = 0$) will be given by equation (5.21).,

$$\Delta S = C_P \ln \frac{T_2}{T_1} - R \ln \frac{P_2}{P_1}$$

since $T_2 = T_1$ the above equation reduces to,

$$\Delta S_{12} = R \ln \frac{P_2}{P_1}$$

The change in entropy from the Second law is also given as,

$$\Delta S_{12} = S_2 - S_1 = \frac{-Q_{\text{out}}}{T_C}$$

2. **Step (2) to (3): isobaric (heat addition) expansion.** From the tank, the compressed air flows through the regenerator where it picks up heat at a

high constant-pressure, say P_2, on its way to the heated power cylinder and its temperature changes from T_C to T_H. The heat (Q_{23}) is given as,

$$Q_{23} = mc_P(T_3 - T_2) = mc_P(T_H - T_C)$$

3. **Step (3) to (4): isothermal (heat addition) expansion.** During this step the power-cylinder expansion-space is heated externally and the gas undergoes isothermal expansion at temperature (T_H). Heat absorbed isothermally during this step is mathematically given as,

$$Q_{34} = P_3V_3 \ln\left(\frac{V_4}{V_3}\right)$$

Using the ideal gas law, Q_{34} can also be written as,

$$Q_{in} = nRT_H \ln\left(\frac{V_4}{V_3}\right)$$

The change in entropy during this step will be given by the formula,

$$\Delta S_{34} = S_4 - S_3 = \frac{Q_{34}}{T_H} = -R \ln \frac{P_4}{P_3}$$

4. **Step (4) to (1): isobaric (heat removal) compression.** Before the compressed air is released as exhaust air, it is passed back through the regenerator, which cools the gas at a low constant pressure and heats the regenerator for the next cycle.

$$Q_{41} = mc_P(T_1 - T_4) = mc_P(T_C - T_H)$$

Thus the cycle keeps on repeating and makes the Ericsson cycle that forms the basis of the Ericsson engine.

The thermal efficiency of the Ericsson cycle can be calculated as,

$$\begin{aligned}\eta &= \frac{W_{net}}{Q_{in}} \\ &= \frac{Q_{in} - Q_{out}}{Q_{in}} \\ &= 1 - \frac{Q_{12}}{Q_{34}} \\ &= 1 - \left(\frac{RT_C \ln \frac{V_1}{V_2}}{RT_H \ln \frac{V_4}{V_3}}\right)\end{aligned}$$

For isothermal processes, steps 1–2 and 3–4 we can write from the ideal gas law ($PV = nRT$) and equation (4.9), the ratios,

$$\frac{\frac{P_1V_1}{T_1}}{\frac{P_2V_2}{T_2}} = \frac{\frac{P_4V_4}{T_4}}{\frac{P_3V_3}{T_3}}$$

which we can simplify, knowing $T_1 = T_2$, $T_3 = T_4$, $P_2 = P_3$, and $P_1 = P_4$,

$$\frac{V_1}{V_2} = \frac{V_4}{V_3}$$

Then we can write the efficiency of the Ericsson cycle as,

$$\eta = 1 - \frac{T_C}{T_H} \tag{9.10}$$

where T_H and T_C are the absolute temperatures in kelvin.

Worked example 9.4. An ideal Ericsson cycle uses air as the working fluid. At the beginning of the isothermal compression process it is at 300 K and 100 kPa and during the isothermal compression 100 kJ kg^{-1} of heat is rejected. Suppose the heat transfer to air happens at 1000 K and the gas constant of air is $R = 0.287$ kJ kg^{-1} K^{-1}. Determine,

(a) the maximum pressure in the cycle,
(b) the net work output per unit mass of air, and
(c) the thermal efficiency of the cycle.

Solution:

(a) The maximum pressure in the cycle will be either P_2 or P_3.

The entropy change during the isothermal compression process (1–2) from the Second law is,

$$\Delta S = S_2 - S_1 = \frac{Q_{out}}{T_C} = \frac{-100 \text{ kJ kg}^{-1}}{300 \text{ K}} = -0.33 \text{ kJ kg}^{-1} \text{ K}^{-1}$$

We can also write,

$$S_2 - S_1 = -R \ln \frac{P_2}{P_1}$$

$$-0.33 \text{ kJ kg}^{-1} \text{ K}^{-1} = (0.287 \text{ kJ kg}^{-1} \text{ K}^{-1}) \ln \left(\frac{P_2}{100 \text{ kPa}}\right)$$

$$P_2 = 328.7 \text{ kPa}$$

(b) The net work output per unit mass of air,

$$W_{out} = Q_{in} - Q_{out}$$

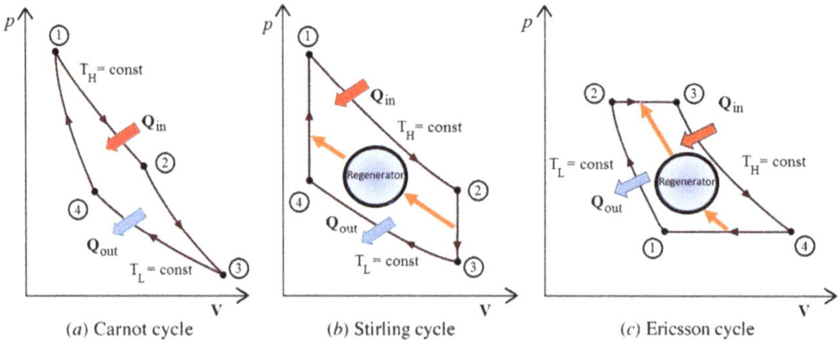

Figure 9.5. pV-diagrams for Carnot cycle, Stirling cycle and Ericsson cycle.

Given that $Q_{\text{out}} = 150 \text{ kJ kg}^{-1}$, we can find Q_{in} for reversible cycles as,

$$\frac{Q_{\text{out}}}{Q_{\text{in}}} = \frac{T_C}{T_H}$$

$$Q_{\text{in}} = Q_{\text{out}}\left(\frac{T_H}{T_C}\right)$$

$$= 100 \text{ kJ kg}^{-1}\left(\frac{1000 \text{ K}}{300 \text{ K}}\right) = 333 \text{ kJ kg}^{-1}$$

Putting values we get, $W_{\text{out}} = 333 - 100 = 233 \text{ kJ kg}^{-1}$.

(c) The thermal efficiency of the cycle,

$$\eta = 1 - \frac{T_C}{T_H} = 1 - \frac{300}{1000} = 70\%$$

9.2.1 Comparing Carnot cycle, Stirling cycle and Ericsson cycle

All three cycles, Carnot cycle, Stirling cycle, and Ericsson cycle, are totally reversible. Hence according to the Carnot principle, all three cycles must have the same thermal efficiency when operating between the same temperature limits T_H and T_C,

$$\eta_{\text{Carnot}} = \eta_{\text{Stirling}} = \eta_{\text{Ericsson}} = 1 - \frac{T_C}{T_H} \tag{9.11}$$

The pV-diagrams for all three cycles is given in figure 9.5.

Concept questions 9.1

1. Compare the thermal efficiencies of these three cycles operating between the same temperature limits,
   (a) The ideal Otto, Stirling, and Carnot cycles.
   (b) The ideal Diesel, Ericsson, and Carnot cycles.

2. What cycle has,
   (a) Two isothermal and two constant-volume processes?
   (b) Two isothermal and two constant-pressure regeneration processes?
3. How does the ideal Ericsson cycle differ from the Carnot cycle?

Exercises 9.1

1. An ideal Stirling engine that is 40% efficient has air as the working fluid and operates between maximum volume 0.5 ft^3 and minimum volume 0.06 ft^3, and P_3 = 68 kPa. Assume air behaves as an ideal gas with c_p = 1.005 kJ kg^{-1} K^{-1}, c_v = 0.718 kJ kg^{-1} K^{-1} and R = 0.287 kPa m^3 and take T_C as room temperature. Determine,
   (a) The temperature of the source reservoir
   (b) The amount of air contained in the engine, and
   (c) The maximum air pressure during the cycle.
2. An ideal Stirling engine at room temperature with air as the working fluid operates with a compression ratio of 12 and between specified pressure limits, 600 kPa and 3600 kPa. Calculate the heat added to the cycle and the net work produced by the cycle. Assume air behaves as an ideal gas with c_p = 1.005 kJ kg^{-1} K^{-1}, c_v = 0.718 kJ kg^{-1} K^{-1} and R = 0.287 kPa m^3.

9.3 Rankine cycle

The Rankine cycle is named after William Rankine, a Glasgow University professor. The Rankine cycle describes processes followed by steam-operated heat engines that are usually found in electricity generating thermal power plants. The Rankine cycle is an ideal cycle for a vapor power cycle, whereas the Carnot cycle is an ideal cycle for a general heat engine. However, the Rankine cycle has a lower efficiency when compared to the Carnot cycle.

In a Rankine cycle the working fluid is continuously evaporated and then condensed, which is why these cycles are also called '***vapor-power cycles***'. In a Rankine cycle, the vapor is superheated before entering the turbine and the condenser completely reduces the steam to a liquid that is convenient for pumping. This is why the selection of the operating fluid depends mainly on the available temperature range for the Rankine cycle.

9.3.1 Steam engine

The Rankine cycle or '**Rankine vapor cycle**' is used by power plants such as coal power plants or nuclear reactors. The history of the steam engine is long and the first engines can be traced back to the first century CE. The first recorded theoretical steam engine was described by the Greek mathematician Hero of Alexandria. A rudimentary steam turbine device was also described by Taqi al-Din in 1551 CE. However, the first commercial steam-powered device was a water pump, developed in 1698 CE, by Thomas Savery, for the purpose of pumping water from underground mines.

Initial Rankine cycles were very inefficient engines and were extremely inefficient and prone to explosions because of the uncontrolled high pressures developed in the boilers that were not strong enough to withstand them. In 1859 CE a Scottish engineer by the name of William J M Rankine developed the Rankine cycle or 'Rankine vapor cycle'. In this engine, a fuel which can either be coal or a nuclear material, is used to produce heat in a boiler, that converts water to steam. The steam then expands in the turbine and produces useful output work.

In the 1990s, Rankine steam cycles generated about 90% of all electric power used throughout the world. The cycle is an arrangement of components that utilizes the phase change of water (boiling water produces steam, condensing exhaust steam produces liquid water) to provide a practical *heat-transfer-to-power* conversion system.

9.3.2 Steps in a Rankine cycle

The Rankine engine consists of a pump, boiler, turbine, and a condenser. The pump delivers liquid water to the boiler. The boiler is heated externally by a fuel, which can be a fossil fuel (coal or natural gas), nuclear material or solar cells. In the boiler liquid water is converted to superheated steam. This superheated steam is used to run the turbine, which powers the generator to produce useful work. Figure 9.6 shows the components that describe the basic Rankine cycle. The pV diagram for the Rankine cycle is shown in figure 9.7. and its four processes are detailed here,

1. **Step (1) to (2): adiabatic pump compression:** Water is pumped from low pressure to high pressure by the pump. The input work during this process is given as,

$$W_{\text{in}} = H_2 - H_1$$

   where the Hs are enthalpies during step 1 and step 2.
2. **Step (2) to (3): vapor (steam) generator (no work):** The high pressure liquid from the pump enters a boiler where it is heated at constant pressure by an

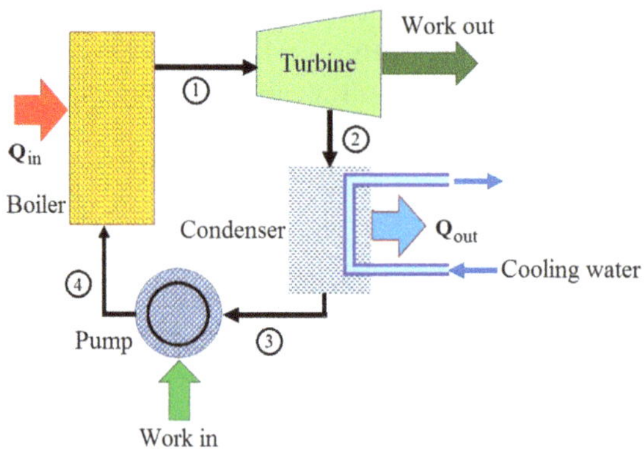

Figure 9.6. Schematic of a basic Rankine cycle.

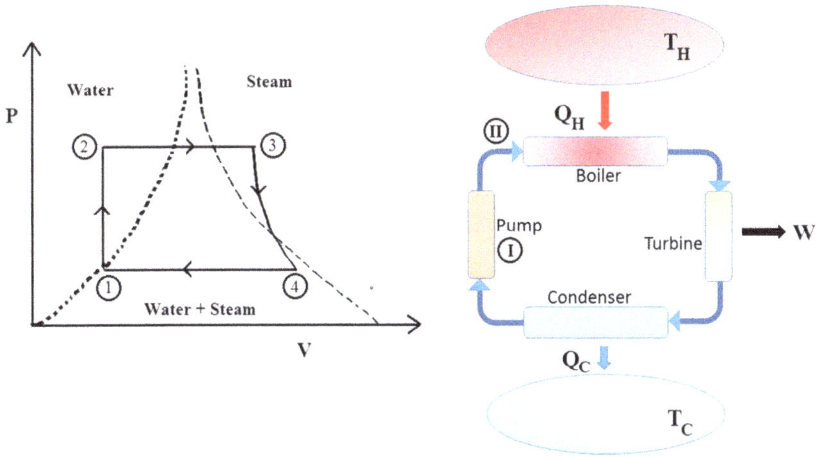

Figure 9.7. pV-diagram and energy flow for a steam engine.

external heat source until it becomes a dry saturated vapor (steam). The heat input during this step is given as,

$$Q_{\text{in}} = Q_{\text{H}} = \text{H}_3 - \text{H}_2$$

3. **Step (3) to (4): adiabatic turbine expansion:** From the boiler, the superheated vapor expands through to the turbine, where it generates power. This results in the decrease of temperature and pressure of the steam and some of the steam condenses back into liquid water. Since this step is adiabatic, which implies $\dot{Q} = 0$, the work done is given mathematically as,

$$W_{\text{out}} = H_3 - H_4$$

4. **Step (4) to (1): condenser (no work):** The partially condensed water from the turbine then enters a condenser, which is just a network of pipes that are in good thermal contact with a low-temperature sink. In the condenser, the vapor is condensed at a constant pressure to a saturated liquid. At this time waste heat is transported to the final external heat sink, for example, the surrounding air or a large body of water, like a nearby lake or river. The rejected heat is mathematically given as,

$$Q_{\text{out}} = Q_{\text{C}} = H_4 - H_1$$

9.3.3 Back work ratio (BWR)

The back work ratio (BWR), as we saw in section 8.4.1, is a measure of performance used in power plants. BWR is the fraction of useful work that the turbine produces to the work consumed by the compressor pump in step (1–2). So, the formula for BWR is given by the following,

$$\text{BWR}_{\text{rankine}} = \frac{W_{\text{in,pump}}}{W_{\text{out,turbine}}} = \frac{H_4 - H_3}{H_1 - H_2} \tag{9.12}$$

The efficiency for a steam engine:

$$\eta = \frac{\text{Desired output}}{\text{Required input}} = 1 - \frac{Q_C}{Q_H} = 1 - \frac{H_4 - H_1}{H_3 - H_1} \tag{9.13}$$

where H are the enthalpies at various steps 1, 3, and 4 and can be easily found by looking them up in 'steam tables'.

Please keep in mind that the efficiency of the Rankine cycle is limited by the heat of vaporization of the fluid. The fluid must be cycled through and reused constantly, therefore, water is the most practical fluid for this cycle. As the water condenses in the condenser, waste heat is given off in the form of water vapor, commonly seen escaping as a cloud flume from the cooling towers. This waste heat is a necessary output in any thermodynamic cycle. Due to this condensation step, the pressure at the turbine outlet is lowered. This means the pump requires less work to compress the water, which results in higher overall efficiencies.

There are three ways to increase the efficiency of a simple ideal Rankine cycle.

1. Increasing the boiler pressure.
2. Superheating the steam to a high temperature
3. Decreasing the condenser pressure.

BWR for gas power cycles is about 40%–80%, whereas BWR for vapor power cycles is about 1%–2%.

9.4 Steam power plant

Steam power plants use water in gaseous form (as steam) as the working fluid. There are many advantages in using water; the biggest advantage is that water is readily available around the world; it is also chemically stable and non-toxic. Additionally, water has a low BWR and its enthalpy change is relatively large.

In steam power plants, water is heated either by burning fossil fuels, for example coal or natural gas, or it is heated by using nuclear fission or solar thermal energy. The modern steam turbine was invented by Sir Charles Parsons in 1884 CE and currently generates about 80% of the electric power in the world. The basic scheme for a steam power plant includes four primary devices, as shown in figure 9.8.

The main parts of a power plant include the following,

1. **Boiler:** The boiler converts liquid water to steam at high pressure and temperature. It adds heat (Q_H) to the system.
2. **Turbine:** Steam from the boiler expands adiabatically into the turbine which turns the turbine and drives a generator to produce useful work.
3. **Condenser:** Exhaust steam from the turbine is pushed into the condenser where it condenses into liquid water and here the heat (Q_C) is removed.
4. **Pump:** The liquid water from the condenser is pumped back into the boiler and the cycle starts again. A small amount of work of the turbine is used to power the pump, which reduces the overall efficiency of the power plant and the pV diagram for the power plant cycle is given in figure 9.9.

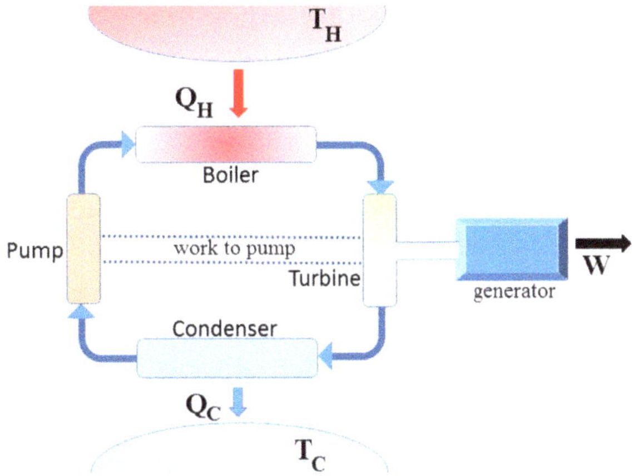

Figure 9.8. Basic components of a power plant.

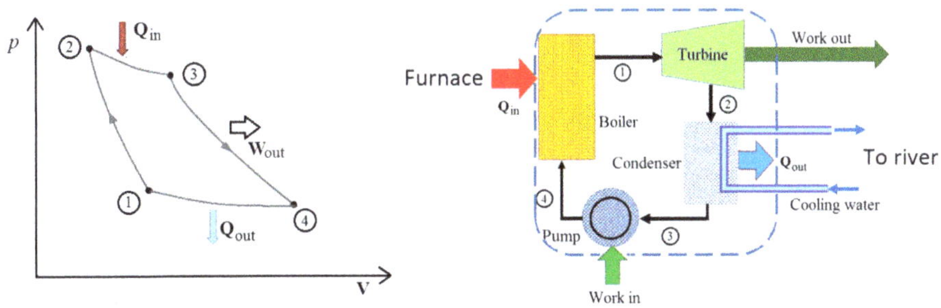

Figure 9.9. *pV*-diagram and a schematic of a power plant cycle.

9.4.1 A typical power plant

A typical power plant is rated at 750 000 kW. The boiler (Q_H) of a typical power plant produces steam at 560 K (550 °F) and 71 atm (1045 psi = 7.2×10^6 Pa) and T_H must be maintained at 550 °F or higher. Thus the surroundings must not have a temperature over 310 K (100 °F) because they acts as a heat sink to absorb heat (Q_C) discarded by the condenser.

For a Carnot cycle operating between these temperatures the thermal efficiency is given as:

$$\eta = 1 - \frac{T_C}{T_H} = 0.45$$

This means that in an ideal system, the efficiency of the power plant is about 44%. That is, at best, ideal power plants can only convert 44% of the heat taken in (Q_H) into useful work. A real power plant is never this efficient. A more realistic efficiency for any real power plants is about 30%–35%.

We can calculate the rate of work, also called the '*rate of production*' of any power plant, by using the formula,

$$\eta = \frac{W}{Q_H}$$

Suppose we have a 750 Mega-watt power plant and we want to calculate the input heat required. At 30% efficiency we find that Q_H is numerically equal to = 2500 MJ and the expelled heat (Q_C) numerically equals (0.70) (Q_H) = 1750 MJ. The difference, which is (Q_H − Q_C = 750 MJ), gives us the rate of work or the rate of production of the power plant.

Worked example 9.5. A particular power plant uses 2×10^5 kg of coal per hour and it produces 800 megawatts of electricity.

(a) Calculate the efficiency of this power plant.

Solution: First we should convert kg per h to watts (J s^{-1}).

$$2 \times 10^5 \text{ kg h}^{-1} = \left(2 \times 10^5 \frac{\text{kg}}{\text{hour}}\right)\left(\frac{2.2 \text{ lb}}{1 \text{ kg}}\right)\left(\frac{13\,200 \text{ BTU}}{1 \text{ lb}}\right)\left(\frac{1055 \text{ J}}{1 \text{ BTU}}\right)\left(\frac{1 \text{ hour}}{3600 \text{ s}}\right)$$
$$= 1.70 \times 10^9 \text{ J s}^{-1}.$$

and 800 MW = 8×10^8 J s^{-1}. Hence the efficiency will be,

$$\text{Efficiency}(\eta) = \frac{\text{Useful energy output}}{\text{Total energy input}} = \frac{8 \times 10^8 \text{ J s}^{-1}}{1.7 \times 10^9 \text{ J s}^{-1}} = 0.47 = 47\%$$

(b) If the efficiency could be improved by 5% over the existing one, how much coal and money could be saved if we suppose each ton of coal costs about $30?

Solution: A 10% improvement means that the new efficiency is (0.47) (1.05) = 0.49, or 49%. In this case, the output of the plant does not change, so the power plant required less input. Suppose the new input is x then it is calculated as follows:

$$\eta = 0.49 = \frac{8 \times 10^8 \text{ W}}{x} \tag{9.14}$$

This implies the new input, $x = 1.63 \times 10^9$ J s^{-1}, which we can convert to kilogram per hour. The new coal input is obtained as follows:

$$\left(1.63 \times 10^9 \frac{\text{J}}{\text{s}}\right)\left(\frac{1 \text{ BTU}}{1055 \text{ J}}\right)\left(\frac{1 \text{ lb}}{13\,200 \text{ BTU}}\right)\left(\frac{1 \text{ kg}}{2.2 \text{ lb}}\right)\left(\frac{3600 \text{ s}}{1 \text{ hour}}\right) = 1.90 \times 10^5 \text{ kg hr}^{-1}.$$

This means we will save 10 000 kg (or about 11 tons) per hour of coal as compared to the less efficient plant given in part (a), which equates to daily cost savings:

$$\text{Cost savings} = \left(\frac{11 \text{ tons}}{1 \text{ hour}}\right)\left(\frac{\$30}{\text{ton}}\right)\left(\frac{24 \text{ h}}{1 \text{ day}}\right) = \$7920 \text{ /day}.$$

Worked example 9.6. Suppose a coal-power plant produces 1 GWatt of electricity at an efficiency of 40%. It is built near a flowing river, where the flow rate of the river is 100 $m^3 s^{-1}$

(a) At what rate does this plant expel waste heat into its environment?

Solution: The power plant is 40% efficient which means 60% of heat is wasted. Then the rate at which the power plant expels waste is:

$$\eta = \frac{W}{Q_H} = \frac{W}{Q_C + W}$$
$$Q_C = \frac{W - \eta W}{\eta}$$

Hence, expelled heat is 1.5×10^9 watts. We know that 1 watt = 1 J s^{-1}, hence in one second 1.5×10^9 J heat has to be dumped.

(b) By how much will the temperature of the river increase?

Solution: In order to find how much the temperature of the river increases we have to first find by how much the heat is spread amongst the molecules of water in the river. We can do that by knowing the density of water (ρ) to be a thousand kilograms per meter cubed. Then,

$$\rho = \frac{\text{Mass flow rate}}{\text{Volume flow rate of river}} = \frac{\dot{m}}{\dot{V}}$$
$$\rho \dot{V} = \dot{m}$$

Using values of $\rho = 1000\frac{kg}{m^3}$ and $\dot{V} = 100$ m^3 s^{-1}, the mass flow rate is calculated to be 10^5 kg s^{-1}. To find the temperature increase we can use equation $Q = \dot{m}c_v\Delta T$ and $c_v = 4186$ J °C^{-1}:

$$\Delta T = \frac{Q}{\dot{m}c_v} = \frac{1.5 \times 10^9}{(10^5)(4186)} \approx 4 \text{ K}.$$

ΔT is calculated to be about 4 °C. This is called '*thermal pollution of the river*'. One way to avoid that is to cool the power plant by evaporation of river water, which is more expensive but more environmentally friendly.

9.5 Comparing engines

In this section, the summary of various heat engines is provided and is shown for easy reference in table 9.1.

1. **Stirling engine versus Carnot engine:**

   Both efficiencies are the same, but Stirling engines are mechanically simpler to build and operate. This is because the walls of the cylinder can

Table 9.1. Summary of various heat engines.

Cycle	Process 1–2 Compression	Process 2–3 Heat addition	Process 3–4 Expansion	Process 4–1 Heat rejection
Carnot	isentropic	isothermal	isentropic	isothermal
Otto (Gas petrol engine)	adiabatic	isochoric	adiabatic	isochoric
Rankine (steam engine)	adiabatic	isobaric	adiabatic	isobaric
Stirling (Stirling engine)	isothermal	isochoric	isothermal	isochoric

be at the same temperature and are always in approximately in thermal equilibrium with the gas that is near them.

2. **Stirling engine versus internal combustion engine:**
   Internal combustion engines are compact since they do not need an external combustion chamber. However, they are limited to certain types of fuel only and are also more polluting since the fuel never burns completely.
3. **Rankine engine versus Carnot engine:**
   The Carnot cycle is an ideal cycle for heat engines, whereas the Rankine cycle is an ideal cycle for a vapor power cycle. Rankine has a lower efficiency when compared to the Carnot cycle.
4. **Internal combustion (Otto) engine versus Diesel engine:**
   The main difference that separates these two cycles is the way they supply heat to the engine to begin the ignition. Otto engines use spark ignition whereas Diesel engines use compression to ignite the fuel. Additionally, in the Otto cycle heat occurs at constant volume, whereas the diesel cycle works on a constant pressure.

Concept questions 9.2

1. Name some differences between the Carnot cycle and Rankine cycles.
2. Is the Rankine cycle reversible?
3. Why is a pump needed in a Rankine cycle?
4. How can we increase the efficiency of a Rankine cycle?
5. Which of these pollutants are produced as a byproduct in a coal-powered plant? Which is not a concern at a natural-gas power plant: nitrogen oxides, sulfur dioxide, carbon dioxide, carbon monoxide?

Exercises 9.2

1. Suppose a power plant produces 1×10^9 Watt of electricity. The turbine of this plant takes in steam at a temperature of 500 °C and the waste heat is expelled into the environment at room temperature (20 °C).
   (a) What is the maximum possible efficiency of this plant?

(b) Suppose there is a leak in the pipe that carries steam from the boiler to the turbine which reduces the temperature of the steam to 300 °C. What will be the new possible maximum efficiency of this plant?

2. Suppose you are the engineer called in to work on the power plant given in Q1 to fix the leak. You successfully fix the leak and are able to restore the efficiency of the power plant to original state $T_H = 500$ °C and $T_C = 20$ °C. While working in the plant you discover by using certain materials you can increase the temperature of the steam to 600 °C.
   (a) What is the new maximum possible efficiency of this plant?
   (b) How much additional power will the power plant produce?
   (c) You negotiate with the owners that you will take the additional profit for two years only. How much money will you make per year if you sell the additional produced electricity at 5 c per kWH?

Further reading

- Cengel Y and Boles M 2006 *Thermodynamics: An Engineering Approach* 5th edn (New York: McGraw-Hill Higher Education)
- Resnick R, Halliday D and Krane K 2002 *Physics* vol 1 5th edn (New York: Wiley)
- Schroeder D 2000 *Thermal Physics* (Reading, MA: Addison-Wesley Longman)

IOP Publishing

Chapter 10

Chemical thermodynamics

Many important thermodynamical processes are not cyclic in nature. A clear example of a type of process that obeys the laws of thermodynamics but is not cyclic is chemical reactions. In chemical reactions the final state of the system is very different from the initial state, so these are clearly not cyclic. In this chapter we apply the laws of thermodynamics to non-cyclic processes, like chemical reactions and phase transformations of matter.

One major difference to keep in mind while studying these systems from the ones we have been discussing so far in this text, is that these systems cannot be thermally and often mechanically isolated from their surroundings. Which implies we cannot consider the energy of these systems to be constant. We can, however, fix other thermodynamic properties like the temperature of the system throughout the process. Similarly, in most systems we cannot keep volume constant but we can fix the pressure of the system.

10.1 Substances and phases of matter

The physical phase of a substance is usually defined by its molecular structure. For example, water is described chemically as H_2O, but it may exist in a number of molecular configurations. At low temperatures, water takes on a rigid crystalline molecular structure called ice. As the temperature of ice is raised the molecular structure of water becomes amorphous and it becomes a liquid. As the temperature of liquid water is further raised, its molecular structure becomes random and it transforms to a vapor. We can easily identify three common structural phases of matter as, solid, liquid, and vapor (or gas).

All substances can exist in two basic states, either as pure substances or as mixtures, as summarized in figure 10.1.

1. **Pure substance:** A pure substance can be defined as anything that contains the same uniform chemical composition in all its physical phases. For example, a mixture of water vapor and liquid water is a pure substance.

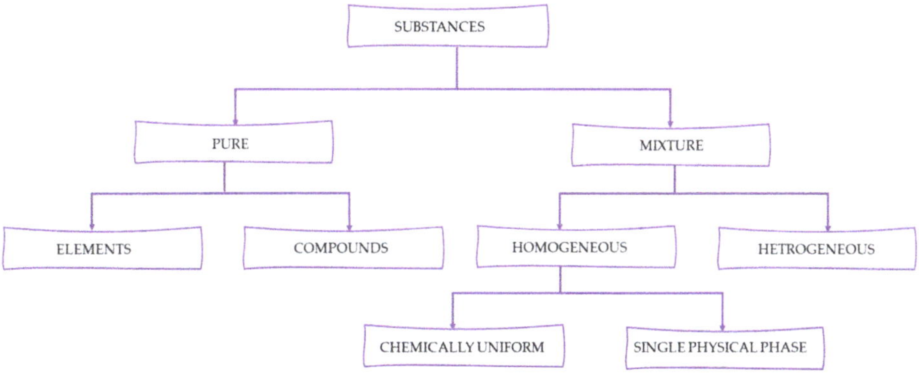

Figure 10.1. Pure substances and mixtures.

On the other hand, air is not really a pure substance, because when it is cooled sufficiently, some of its components condense into their liquid state, thus changing the composition of the remaining gases. Pure substances can be divided into either '*elements*', that contain the same kind of atoms and have well-defined properties or as '*compounds*' that contain atoms of two or more elements, again with well-defined properties.

2. **Mixtures:** Mixtures are defined as substances that contain more than one chemical or phase. Mixtures can be either homogeneous or heterogeneous. Homogeneous mixtures have the same composition throughout. They are either physically uniform, that is, they have a single physical phase or are chemically uniform, that is, they have a single chemical throughout. The components in the mixture are not distinguishable from each other. On the other hand, heterogeneous mixtures do not have the same composition throughout and we are able to distinguish the various compounds in heterogeneous mixtures from each other.

10.1.1 Parts of chemical thermodynamics

The study of chemical thermodynamics is quite broad in its scope so we can divide it based on the various forms of substances as,

1. **Pure component thermodynamics:** Pure component thermodynamics characterizes the behavior of systems that contain a single component.
2. **Solution thermodynamics:** Solution thermodynamics is the part of thermodynamics where the behavior of a mixture with more than one component is studied and characterized, for example, a system that contains more than one type of chemical component is studied under solution thermodynamics.
3. **Chemical reaction equilibrium thermodynamics:** Chemical reaction equilibrium thermodynamics is the part of thermodynamics that characterizes the behavior of a system in which the reactants and products of one or more reversible chemical reactions are at equilibrium.

4. **Phase equilibrium thermodynamics:** Phase equilibrium is the part of chemical thermodynamics where the behavior of multiple phases that exist in equilibrium with each other is studied.

Solution thermodynamics is more complicated than pure component thermodynamics. Luckily we only deal with classical thermodynamics and pure component thermodynamics in this text. Please refer to other texts for details of solution thermodynamics.

10.2 Equilibrium states of systems

We introduced various equilibrium states in chapter 1, section 1.7. Two systems are said to be in '*thermal equilibrium*' with each other when their temperatures are the same and there is no flow of heat energy. We derived a relation for thermal equilibrium in chapter 5, equation (5.33) which is mathematically given as,

$$\left(\frac{\partial S}{\partial U}\right)_{V,N} = \frac{1}{T} \tag{10.1}$$

Similarly, two systems are said to be in mechanical equilibrium when their pressures are the same and there is no exchange of volume. We derived a relation for *mechanical equilibrium* in section 5.4.1, equation (5.40) as,

$$\left(\frac{\partial S}{\partial V}\right)_{U,N} = \frac{p}{T} \tag{10.2}$$

So the next question is what is the relationship when there is no flow of particles between the two systems, i.e., the two systems in contact are in '*diffusive equilibrium*' with each other. In order to answer this question, let's consider two systems A and B that are free to exchange both energy and particles between each other. The systems could be made up of any form of matter, liquid, solid or gas, but for simplicity, consider the two systems under consideration are made entirely of gas molecules. Suppose the internal energy of system A is (U_A), and it has (N_A) number of particles and has entropy (S_A), whereas system B has internal energy (U_B), number of particles (N_B) and entropy (S_B). Also, suppose both systems are at the same temperature (say T). Assume that the total energy and the total particles of the system are fixed. This implies that the total entropy of the system is a function of U_A and N_A and at equilibrium, total entropy will be a maximum. Hence,

$$\begin{aligned}\left(\frac{\partial S_{\text{total}}}{\partial U_A}\right)_{N_A,V_A} &= 0; \qquad \text{condition 1}\\ \left(\frac{\partial S_{\text{total}}}{\partial N_A}\right)_{U_A,V_A} &= 0; \qquad \text{condition 2}\end{aligned} \tag{10.3}$$

Similar to the argument we used in section 5.4.1 we can say that condition 1 implies the two systems are at thermal equilibrium at the same temperature. Condition 2 is new, but we can use the same arguments we made for volume in section 5.4.1, but here applying it to the number of particles as,

$$\frac{\partial S_{\text{total}}}{\partial V_A} = 0$$
$$\frac{\partial S_{\text{A}}}{\partial N_A} + \frac{\partial S_{\text{B}}}{\partial N_A} = 0$$
$$\frac{\partial S_{\text{A}}}{\partial N_A} - \frac{\partial S_{\text{B}}}{\partial N_B} = 0$$

The last step uses the fact that the total number of particles is fixed, hence $dN_A = -dN_B$, therefore we can conclude:

$$\frac{\partial S_{\text{A}}}{\partial N_A} = \frac{\partial S_{\text{B}}}{\partial N_B}; \qquad \text{at equilibrium} \tag{10.4}$$

and if we multiply both sides by $(-T)$ we get:

$$-T\frac{\partial S_{\text{A}}}{\partial N_A} = -T\frac{\partial S_{\text{B}}}{\partial N_B}; \qquad \text{at equilibrium} \tag{10.5}$$

The quantity $(-T\frac{\partial S}{\partial N})$ in equation (10.5) is denoted by μ and is called '*chemical potential*'. Mathematically the chemical potential is given as:

$$\mu \equiv -T\left(\frac{\partial S}{\partial N}\right)_{\text{U,V}} \tag{10.6}$$

Please note that the negative sign in equation (10.6) implies that the system with the larger $(\partial S/\partial N)$ will get more particles. In other words, particles will flow from a high (μ/T) to a low (μ/T). Hence at diffusive equilibrium, the following conditions must hold:

$$T_A = T_B; \quad P_A = P_B; \quad \mu_A = \mu_B \tag{10.7}$$

Chemical potentials are important in many aspects of thermodynamics, including processes like melting, boiling, evaporation, solubility, osmosis, etc.

10.2.1 Physical meaning of chemical potential

If two systems with different chemical potentials are brought into chemical contact with each other, particles move from the system at higher chemical potential to the system at lower chemical potential until both systems reach '*chemical equilibrium*'. Suppose we have two containers of gas that make up systems A and B at different chemical potentials. Suppose system A is at a higher chemical potential with respect to system B. When you put them together and allow them to exchange particles, say, by opening a valve connecting two containers of gas, there will be a net exchange of particles until the chemical potentials between systems A and B become equal.

Figure 10.2. Chemical potential.

Under most conditions in thermodynamics, chemical potential determines the stability of substances and their tendency to chemically react to form new substances. As such, it is the central concept in thermodynamics of materials because all of the thermodynamical properties of substances can be obtained if we know the chemical potentials of the substances.

It is important to remember that high chemical potential is not always caused by how rich a substance is, that is it has a lot of particles; it is caused by how generous a substance is. Similarly low chemical potential is not always because a substance is poor (has a low number of particles) but rather how willing it is to receive particles. A good analogy is a miserly person versus a generous person, as represented in figure 10.2. A person who is rich but miserly, will tend to hoard his money and want to obtain more money, hence they are stingy, i.e., they lack generosity (low chemical potential) but another person who might be poor but generous and willing to give way money will have high chemical potential. So a low chemical potential is assigned to substances that are willing to receive particles, whereas high chemical potential is assigned to substances that are willing to give away particles.

Examples of chemical potentials include,

1. **Humidity**: Humidity is the tendency of an atmosphere to give water molecules, hence it is the chemical potential of water molecules.

$$\ln\left(\frac{P_{\text{water}}}{P_0}\right) = \frac{\mu_{\text{water}} - \mu^0_{\text{water}}}{RT} \tag{10.8}$$

2. **pH**: pH is the tendency of a solution to take H^+, hence it is the chemical potential of a proton.

$$\text{pH} = \frac{\mu^0_{\text{H}} - \mu_{\text{H}}}{RT \ln 10} \tag{10.9}$$

3. **Nernst potential**: Nernst potential is the tendency of system to take electrons, hence it is the chemical potential of electrons.

$$E - E^0 = \left(\frac{RT}{nF}\right) \ln\left(\frac{a_{0x}}{a_{\mathrm{Red}}}\right) \tag{10.10}$$

4. **Ideal gas:** Chemical potential of an ideal gas:

$$\mu(T, p) - \mu_o(T, P_o) = kT \ln \frac{p_{\mathrm{gas}}}{P_o} \tag{10.11}$$

where P_o is the initial pressure, usually taken to be 1 atm.

10.2.2 Physical meanings of equilibria

1. **Thermal equilibrium:** Energy flows from a system with a higher value of temperature to a system with a lower value of temperature. This energy transfer continues until $T_1 = T_2$ and thermal equilibrium is reached. Hence we say the physical meaning of temperature is the '*tendency of a system to give energy*'.
2. **Mechanical equilibrium:** Volume tends to be transferred from a system with a high value of pressure to a system with a low value of pressure. This volume change continues until $P_1 = P_2$ and mechanical equilibrium is reached. Then we can say that the physical meaning of pressure is '*tendency of a system to take volume*'.
3. **Diffusive equilibrium:** Particles tend to flow from a system with a high value of chemical potential (μ_1) to a system with a low value of chemical potential (μ_2). This particle transfer continues until $\mu_1 = \mu_2$ and chemical equilibrium is reached. Then we can say that the physical meaning of chemical potential is '*tendency of a system to give particles*'.

A summary of the various equilibria is provided in table 10.1. for quick reference.

10.3 Thermodynamic potentials

Thermodynamic potentials represent the states of a system. Thermodynamic potentials are also called thermodynamic potential energies. The four main

Table 10.1. Summary of various equilibria.

Equilibrium	Exchanged quantity	Variable	Units	Formula
Thermal	Energy	Temperature	kelvin (K)	$\left(\frac{dS}{dU}\right)_{V,N} = \frac{1}{T}$
Mechanical	Volume	Pressure	Pascal (Pa)	$\left(\frac{dS}{dV}\right)_{\mu,N} = \frac{P}{T}$
Diffusive	Particles	Chemical potential	Gibbs (G = J mol^{-1})	$\left(\frac{dS}{dN}\right)_{U,V} = -\frac{\mu}{T}$

Table 10.2. Summary of thermodynamic potentials.

	Function of	Formula	Thermodynamic identity
Internal energy (U)	$U = U(S,\ V,\ N)$		$dU = TdS - pdV + \mu dN$
Enthalpy (H)	$H = H(S, p, N)$	$H = U + PV$	$dH = TdS + Vdp + \mu dN$
Helmholtz free energy (F)	$F = F(T,\ V,\ N)$	$F = U - TS$	$dF = -SdT - pdV + \mu dN$
Gibbs free energy (G)	$G = G(T, p, N)$	$G = U - TS + PV$	$dG = -SdT + Vdp + \mu dN$

thermodynamic potentials that are useful in chemical thermodynamics of reactions and non-cyclic reactions are internal energy (U), enthalpy (H), Helmholtz free energy (F), and Gibbs free energy (G).

The natural variables are variables that allow us to find other state functions by partial differentiation of the thermodynamic potentials. For example, the natural variables in internal energy (U) are entropy (S), volume (V), number of particles (N), and chemical potential (μ), i.e., $U = U(S,\ V,\ N,\ \mu)$. Natural variables in enthalpy are S, p, N, and μ, i.e., ($H = H(S, p, N, \mu)$), Helmholtz free energy ($F = F(T,\ V,\ N,\ \mu)$) and Gibbs free energy ($G = G(T, p, N, \mu)$).

I have summarized the thermodynamic identities in table 10.2. and the conversions between the four thermodynamic potentials in figure 10.4.

10.3.1 Internal energy (U)

To derive the equation that connects the internal energy to the properties of the system, namely T, p, and μ, we must consider a process in which we change the energy, volume and the number of particles of the system by small amounts and ask by how much will the entropy of the system change under these conditions? To answer this question we must divide the process into three steps. During step one, energy is kept constant ($\Delta U = 0$), during step two, volume is kept constant ($\Delta V = 0$), and during step three, the number of particles is kept constant ($\Delta N = 0$). Then total change in entropy is just the sum of the changes during steps 1, 2, and 3, as,

$$\Delta S_{\text{total}} = \Delta S_1 + \Delta S_2 + \Delta S_3$$

Multiply and divide terms with dU, dV, and dN and rearrange for small changes to get,

$$dS_{\text{total}} = \left(\frac{\partial S}{\partial U}\right)_1 dU + \left(\frac{\partial S}{\partial V}\right)_2 dV + \left(\frac{\partial S}{\partial N}\right)_3 dN$$

Putting the values (from equations (10.1), (10.2), and (10.6)) in the above equation we get,

$$dS = \left(\frac{1}{T}\right)_1 dU + \left(\frac{p}{T}\right)_2 dV + \left(\frac{-\mu}{T}\right)_3 dN$$

Rearranging and simplifying we can write '***the complete thermodynamic identity***' as:

$$dU = TdS - pdV + \mu dN \tag{10.12}$$

where μ is the chemical potential and dN is the change in the number of particles. The combination of parameters on the right side of equation (10.12) is equal to the exact differential of U. We know U is the internal energy and it is conserved for an isolated system. This implies that the natural variables of U are S, V, N, and hence we can write:

$$dU(S, V, N) = \left(\frac{\partial U}{\partial S}\right)_{N,V} dS + \left(\frac{\partial U}{\partial V}\right)_{N,S} dV + \left(\frac{\partial U}{\partial N}\right)_{S,V} dN \tag{10.13}$$

Since these two equations (equations (10.12) and (10.13)) for dU must yield the same result for any dS and dV, the corresponding coefficients must be the same:

$$\left(\frac{\partial U}{\partial S}\right)_{N,V} = T; \quad \left(\frac{\partial U}{\partial V}\right)_{N,S} = -p; \quad \left(\frac{\partial U}{\partial N}\right)_{S,V} = \mu \tag{10.14}$$

This shows that among all macroscopic variables that can be used to characterize a system, for example (p, V, T, μ, N, etc), only three (namely T, p and μ) are independent.

10.3.2 Enthalpy (H)

We defined enthalpy of a system in section 3.5 as the energy needed to create a system plus the work done to make room for it under constant pressure (p) conditions as in equation (3.15):

$$H \equiv U + P_{\text{atm}} V \tag{10.15}$$

In other words, enthalpy (H) is the total energy you need to create a system out of nothing and put it in the environment, since the initial volume (V) of the system is zero, because it does not exist yet, ($\Delta V = 0$). Similarly, for annihilation of any system, H is the energy you recover as a result of the annihilation of the system, i.e., the system's energy plus the work done by the collapsing atmosphere.

Before we move on it is important to remember that H (the enthalpy) is a thermodynamic potential, with natural variables S, p, and N. H is a function of S, p, N, and $[dH = dH(U + PV)]$. We can expand equation (3.17) ($\Delta H = \Delta U + P_{\text{atm}}\Delta V$) to include chemical potential as:

$$\begin{aligned} dH(S, p, N) &= dU + Vdp + \mu dN \\ &= TdS + Vdp + \mu dN \end{aligned} \tag{10.16}$$

The total differential of H in terms of its independent variables is:

$$dH(S, p, N) = \left(\frac{\partial H}{\partial S}\right)_{p,N} dS + \left(\frac{\partial H}{\partial p}\right)_{S,N} dp + \left(\frac{\partial H}{\partial N}\right)_{S,p} dN \tag{10.17}$$

Since these two equations (equations (10.16) and (10.17)) for dH must yield the same result for any dS and dp, the corresponding coefficients must be the same and we can write:

$$\left(\frac{\partial H}{\partial S}\right)_{N,P} = T; \quad \left(\frac{\partial H}{\partial P}\right)_{N,S} = V; \quad \left(\frac{\partial H}{\partial N}\right)_{S,P} = \mu \tag{10.18}$$

Often, however, we are not interested in the total energy needed to create the system or the total energy recovered upon annihilation of the system. If the environment has a constant temperature, the system can extract heat from this environment for free, so all we need to provide to create a system from nothing is the additional work needed. Similarly during annihilation of the system we will not recover all of its energy as work because we have to dispose of its entropy by dumping some of the heat into the environment. At this point I want to talk about two quantities that are very useful, called the Helmholtz free energy (F) and the Gibbs free energy (G). Both of these quantities are related to energy and are similar to the enthalpy (H) of a system.

10.3.3 Helmholtz free energy (*F*)

The Helmholtz free energy (F) was developed by Hermann von Helmholtz, a German physician and physicist, in 1882 CE and is defined as:

$$F \equiv U - TS \tag{10.19}$$

Helmholtz free energy is the total energy needed to create the system out of nothing minus the heat ($T\Delta S = TS$) you get for free from the environment held at temperature (T), where S is the final entropy of the system when the temperature (T) and volume (V) of the system are kept constant. Helmholtz free energy is also called thermodynamic potential at constant volume. Helmholtz free energy (F) is primarily used in the study of explosives since explosive reactions induce pressure changes.

Helmholtz free energy (F) is the energy that must be provided as work if you are creating the system out of nothing. Please note that for creating the system, 'free' is not quite a correct description for Helmholtz free energy (F). The energy that comes for free is (TS) not F, so in the context of creating the system F should in fact be called 'costly' energy. The scientists who named F were thinking of the destruction of a system when they named it 'free', since during destruction you recover F, Helmholtz 'free' energy, as work[1]. Helmholtz free energy is equal to the maximum amount of useful work that can be extracted from the system. Hence, Helmholtz free energy (F) measures the '*useful*'[2] work obtainable from a closed thermodynamic system at a constant temperature and volume. The Helmholtz free energy (F) is used for systems held at constant volume where no work is performed on the environment and is used to define fundamental equations of state of pure substances.

[1] Please note, here the word work means 'all' work, including any work that is done automatically by the surroundings.

[2] '*Useful*' work here is the same as the boundary work introduced in chapter 2.

Helmholtz free energy is a thermodynamic potential at constant T and V, with natural variables T, V, and N, and $dF = dF(U - TS)$ and can be mathematically written as,

$$dF(T, V, N) = -SdT - pdV + \mu dN \tag{10.20}$$

because ($dT = 0$) and ($dV = 0$). The total differential of Helmholtz free energy (F) in terms of its independent variables is given as:

$$dF(T, V, N) = \left(\frac{\partial F}{\partial T}\right)_{V,N} dT + \left(\frac{\partial F}{\partial V}\right)_{T,N} dV + \left(\frac{\partial F}{\partial N}\right)_{T,V} dN \tag{10.21}$$

Since these two equations for dF (equations (10.20) and (10.21)) must yield the same result for any dT and dV, the corresponding coefficients must be the same:

$$\left(\frac{\partial F}{\partial T}\right)_{V,N} = -S; \quad \left(\frac{\partial F}{\partial V}\right)_{T,N} = -p; \quad \left(\frac{\partial F}{\partial N}\right)_{S,p} = \mu \tag{10.22}$$

Another interesting thing to note is that the second condition in equation (10.22) can also be written as the following, by using the definition of Helmholtz free energy given in equation (10.19) ($F = U - TS$):

$$\begin{aligned} \left(\frac{\partial F}{\partial V}\right)_{T,N} &= -p \\ \left(\frac{\partial U}{\partial V}\right)_{T,N} + T\left(\frac{\partial S}{\partial V}\right)_{T,N} &= -p \end{aligned} \tag{10.23}$$

The first term ($\partial U/\partial V$) in equation (10.23) is called the '***energy pressure***' and is dominant in most solids. The second term ($\partial S/\partial V$) is called the '***entropy pressure***' and is dominant in gases. Also notable is the fact that for an ideal gas, U does not depend on V, and only the second term survives.

10.3.4 Gibbs free energy (*G*)

Gibbs free energy (G) was developed in the 1870s by the American scientist Josiah Willard Gibbs. The total Gibbs free energy is a thermodynamic potential that can be used to calculate the maximum or reversible work that may be performed by a thermodynamic system at a constant temperature and pressure. It is the work you need to create the system or the work you can recover when you destroy it. The change in Gibbs free energy (G) tells us about the spontaneity of a process. In other words, Gibbs free energy tells us if a process will happen on its own or not. In order to explain Gibbs free energy we have to discuss a few other things first.

To create a rabbit out of nothing and place it on the table the magician needs enthalpy, ($H = U + PV$). But she can get some help since some energy, equal to ($T\Delta S$), can flow in spontaneously as heat from the environment held at temperature (T). So the magician does not need the total enthalpy (H), she in fact only needs (($U + PV$) $-$ TS) which is less than the total enthalpy (H), as we see in figure 10.3. Less energy needed to create the system is always good because that

Figure 10.3. To create a rabbit and put in on the table the magician does not need the entire enthalpy (H). Some energy (= TS) enters spontaneously as heat, so she only needs ($H - TS$) energy, called Gibbs free energy (G).

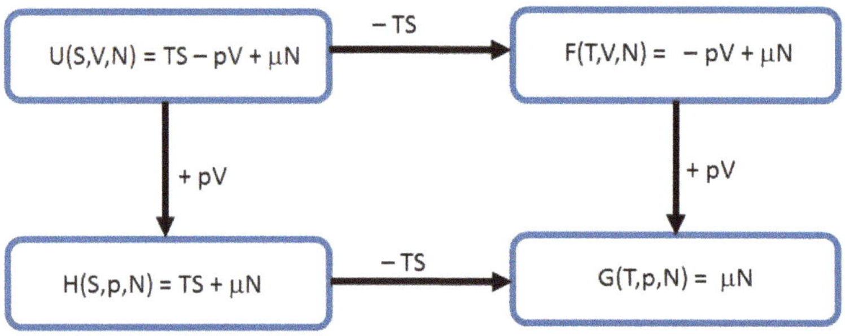

Figure 10.4. The four thermodynamic potentials considered above: U, F, H, G form a square, keeping N constant.

implies we need less input work. We define this 'less' energy as the Gibbs free energy (G), mathematically given as:

$$G \equiv H - TS = U - TS + pV \tag{10.24}$$

Taking the total differential, we have,

$$dG = dU - TdS - SdT + pdV + Vdp \tag{10.25}$$

Replace dU with the result from the complete thermodynamic identity, equation (10.12) ($dU = TdS - pdV + \mu N$), we get:

$$dG = -SdT + Vdp + \mu N \tag{10.26}$$

When the system is under constant pressure (p) and temperature (T), the work needed to create the system or vice versa the work recovered when the system is destroyed, is given by Gibbs free energy (G).

Another way to think about the Gibbs free energy is that it is a thermodynamic potential which is minimized when a system reaches chemical equilibrium at

Table 10.3. Summary of formulas.

Quantity	Function of	Equals	Thermodynamic identity	
Internal energy (U)	$U = U(S, V, N)$	$\left(\frac{\partial U}{\partial S}\right)_{N,V} = T$	$\left(\frac{\partial U}{\partial V}\right)_{N,S} = -p$	$\left(\frac{\partial U}{\partial N}\right)_{S,V} = \mu$
Entropy (S)	$S = S(U, V, N)$	$\left(\frac{dS}{dU}\right)_{V,N} = \frac{1}{T}$	$\left(\frac{dS}{dV}\right)_{\mu,N} = \frac{P}{T}$	$\left(\frac{dS}{dN}\right)_{\mathrm{U,V}} = -\frac{\mu}{T}$
Enthalpy (H)	$H = H(S, p, N)$	$\left(\frac{\partial H}{\partial S}\right)_{N,P} = T$	$\left(\frac{\partial H}{\partial P}\right)_{N,S} = V$	$\left(\frac{\partial H}{\partial N}\right)_{S,P} = \mu$
Helmholtz energy (F)	$F = F(T, V, N)$	$\left(\frac{\partial F}{\partial T}\right)_{V,N} = -S$	$\left(\frac{\partial F}{\partial V}\right)_{T,N} = -p$	$\left(\frac{\partial F}{\partial N}\right)_{S,p} = \mu$
Gibbs energy (G)	$G = G(T, P, N)$	$\left(\frac{\partial G}{\partial T}\right)_{P,N} = -S$	$\left(\frac{\partial G}{\partial P}\right)_{T,N} = V$	$\left(\frac{\partial G}{\partial N}\right)_{T,P} = \mu$

constant pressure and temperature. The total differential of G in terms of its independent variables (T, p, and N) is given as:

$$dG(T, p, N) = \left(\frac{\partial G}{\partial T}\right)_{p,N} dT + \left(\frac{\partial G}{\partial p}\right)_{T,N} dp + \left(\frac{\partial G}{\partial N}\right)_{T,p} dN \tag{10.27}$$

Since these two equations (equations (10.26) and (10.27)) for dG must yield the same result for any dT and dp, the corresponding coefficients must be the same, that is,

$$\left(\frac{\partial G}{\partial T}\right)_{p,N} = -S; \quad \left(\frac{\partial G}{\partial p}\right)_{T,N} = V; \quad \left(\frac{\partial G}{\partial N}\right)_{T,p} = \mu \tag{10.28}$$

Alternative names for the Helmholtz energy are Helmholtz function, Helmholtz free energy, or work function. Alternative names for the Gibbs energy are Gibbs function and Gibbs free energy. Both the Helmholtz energy and Gibbs energy have been called simply free energy, and the symbol F has historically been used for both, which leads to all kinds of confusion if we are not very careful.

All multiple formulas we derived for the thermodynamic potentials are summarized in table 10.3. for easy reference.

10.3.5 Difference between Helmholtz energy (*F*) and Gibbs free energy (*G*)

Both are thermodynamic functions and are called thermodynamic potentials. Helmholtz energy is defined when volume (V) and temperature (T) are kept constant, whereas Gibbs free energy is defined when pressure (p) and temperature (T) of a system are kept constant. Helmholtz energy remains constant during isothermal and isochoric processes, and Gibbs free energy remains constant during isothermal and isobaric processes. Helmholtz energy is called thermodynamic potential at constant volume and Gibbs free energy is called thermodynamic potential at constant pressure.

For isothermal-and-isobaric processes, Helmholtz energy tends to minimize as the system tends to equilibrium. For isothermal-and-isobaric processes Gibbs free energy tends to minimize as the system tends to equilibrium. Summarizing, we can say Helmholtz energy (F) is the maximum amount of useful (non-PV) work that can be extracted from a thermodynamically-closed system at constant V and T. Gibbs free energy (G) is the energy available to do useful (non-PV) work in a thermodynamically-closed system at constant V and T.

Worked example 10.1. Derive non-trivial Maxwell relations for U, H, F, and G.

Solution:

1. **Maxwell relation for** $U(S, V)$:

   Before we solve for Maxwell relations, we have to understand that the functions and physics are well-behaved in that the mixed partial derivatives of the functions do not depend on the order of differentiation. This implies, when 'N' is kept fixed, we can write,

   $$\frac{\partial}{\partial V}\left(\frac{\partial U}{\partial S}\right) = \frac{\partial}{\partial S}\left(\frac{\partial U}{\partial V}\right) \tag{10.29}$$

   where $\frac{\partial}{\partial V}$ is taken when S is fixed and $\frac{\partial}{\partial S}$ is taken when V is fixed. From the thermodynamic identity for U, $dU = TdS - pdV$, we know that,

   $$\left[\left(\frac{\partial U}{\partial S}\right)_V = T\right] \quad \text{and} \quad \left[\left(\frac{\partial U}{\partial V}\right)_S = -p\right]$$

   Putting this value in equation (10.29) we see that,

   $$\left(\frac{\partial T}{\partial V}\right)_S = -\left(\frac{\partial p}{\partial S}\right)_V \tag{10.30}$$

   Which is called the **First Maxwell relation for internal energy**, $U(S, V)$.

2. **Maxwell relation for** $H(S, p)$:

   Here we can assume the function is well behaved so we can write,

   $$\frac{\partial}{\partial p}\left(\frac{\partial H}{\partial S}\right) = \frac{\partial}{\partial S}\left(\frac{\partial H}{\partial p}\right) \tag{10.31}$$

   From the thermodynamic identity for H, $dH = TdS + pdV$ we know that,

   $$\left[\left(\frac{\partial H}{\partial S}\right)_p = T\right] \quad \text{and} \quad \left[\left(\frac{\partial H}{\partial p}\right)_S = V\right]$$

   Putting this value in equation (10.31) we see that,

   $$\left(\frac{\partial T}{\partial p}\right)_S = -\left(\frac{\partial V}{\partial S}\right)_p \tag{10.32}$$

   which is called the **Second Maxwell relation for enthalpy**, $H(S, P)$.

3. **Maxwell relation for** $F(T, V)$:

   Here we can assume the function is well enough behaved that we can write,

$$\frac{\partial}{\partial V}\left(\frac{\partial F}{\partial T}\right) = \frac{\partial}{\partial T}\left(\frac{\partial F}{\partial V}\right) \tag{10.33}$$

   From the thermodynamic identity for F, $dF = -SdT - pdV$ we know that,

$$\left[\left(\frac{\partial F}{\partial T}\right)_V = -S\right] \quad \text{and} \quad \left[\left(\frac{\partial F}{\partial V}\right)_T = -p\right]$$

   Putting this value in equation (10.33) we see that,

$$\left(\frac{\partial S}{\partial V}\right)_T = -\left(\frac{\partial p}{\partial T}\right)_V \tag{10.34}$$

   Which is called the **Third Maxwell relation for Helmholtz free energy**, $F(T, V)$.

4. **Maxwell relation for** $G(T, P)$:

   Here we can assume the function is well enough behaved that we can write,

$$\frac{\partial}{\partial p}\left(\frac{\partial G}{\partial T}\right) = \frac{\partial}{\partial T}\left(\frac{\partial G}{\partial p}\right) \tag{10.35}$$

   From the thermodynamic identity for G, $dG = -SdT + VdP$ we know that,

$$\left[\left(\frac{\partial G}{\partial T}\right)_p = -S\right] \quad \text{and} \quad \left[\left(\frac{\partial G}{\partial p}\right)_T = V\right]$$

   Putting this value in equation (10.35) we see that,

$$-\left(\frac{\partial S}{\partial p}\right)_T = \left(\frac{\partial V}{\partial T}\right)_p \tag{10.36}$$

   Which is called the **Fourth Maxwell relation for Gibbs free energy**, $G(T, P)$.

Concept questions 10.1

1. Why is Helmholtz free energy called 'free'?
2. Does the chemical potential depend on temperature? Does it depend on pressure? Explain in your own words.
3. Why is Gibbs free energy negative?

Exercises 10.1

1. It is a common practice to use a thermodynamic potential called Grand free energy or grand potential when considering open complex systems including both fluid and solid phases. Grand free energy or grand potential is defined as,

$$\Omega(T, V, \mu) = F - \mu N = U - TS - \mu N \tag{10.37}$$

   (a) Show the thermodynamic identity for Ω is,

$$d\Omega = -pdV - SdT - Nd\mu$$

   (b) Find related formulas for partial derivatives with respect to T, V, and μ.

2. Suppose we have a mole of water at room temperature (22 °C) and 1 atm. Hint: entropy of a mole of water under standard conditions is 69.91 J K^{-1}.
   (a) Compare the Gibbs free energy if we raise the temperature to 28 °C keeping pressure constant.
   (b) To compensate for the change in Gibbs free energy should we increase or decrease the pressure on the water?
   (c) How much pressure will be required?

10.4 Change in Helmholtz energy (ΔF) and Gibbs free energy (ΔG)

Usually we deal with processes that are much less dramatic than the complete creation or destruction of entire systems. In situations like these, instead of Helmholtz free energy (F) and Gibbs free energy (G), we focus on the changes in these quantities, that is, ΔF and ΔG.

1. **Changes in Helmholtz free energy** (ΔF):

$$\Delta F = \Delta U - T\Delta S - S\Delta T$$

   Since temperature is constant ($\Delta T = 0$) we can write, using the First law of thermodynamics, the changes in Helmholtz free energy as,

$$\Delta F = (Q + W) - T\Delta S$$

   where Q is the heat added to the system and W is the useful work done on the system.

   (a) **Case 1:** When no new entropy is created during the process $\Delta S = Q/T$ implies we can write the Helmholtz free energy as: $\Delta F = (T\Delta S + W) - T\Delta S$,

$$\Longrightarrow \Delta F = W$$

   (b) **Case 2:** When there is new entropy created during the process, $T\Delta S > 0$, and we write the Helmholtz free energy as,

$$\Delta F < W$$

From these two conditions we can generalize that at constant temperature, the change in Helmholtz free energy is,

$$\Delta F \leqslant W \tag{10.38}$$

where W includes all work in or out of the system including any work done automatically by expanding and/or collapsing environments.

2. **Changes in Gibbs free energy** (ΔG):

When the environment is at constant pressure and constant temperature, and we are not interested in keeping track of the work that the environment does automatically on the system, we have to consider Gibbs free energy (G) instead of Helmholtz free energy (F). For small changes, the change in Gibbs free energy is,

$$\Delta G = \Delta U - T\Delta S - S\Delta T + p\Delta V + V\Delta p$$

Since temperature and pressure are constant ($\Delta T = 0$ and $\Delta p = 0$) we can write, using the First law of thermodynamics, the changes to Gibbs free energy (G) as,

$$\Delta G = (Q + W) - T\Delta S + p\Delta V$$

or,

$$\Delta G = (Q - T\Delta S) + W + p\Delta V$$

where W includes the work done by the environment on the system, ($W = -p\Delta V + W_{\text{other}}$) and W_{other} can include any other work done on the system, for example electrical work or the spark needed to convert hydrogen (H_2) and oxygen (O_2) into water (H_2O). Also note that the term ($Q - T\Delta S$) will always be either zero or negative. This leads to two conditions,

(a) **Case 1:** When the term ($Q - T\Delta S$) is zero: $\Delta G = W + p\Delta V$ and $W = -p\Delta V + W_{\text{other}}$ implies,

$$\Delta G = W_{\text{other}}$$

(b) **Case 2:** When the term ($Q - T\Delta S$) is negative,

$$\Delta G < W_{\text{other}}$$

From these conditions we can generalize that at constant temperature and constant pressure,

$$\Delta G \leqslant W_{\text{other}} \tag{10.39}$$

Gibbs free energy (G) is a very useful quantity, hence values of ΔG for many chemical reactions and other processes have been measured and tabulated in chemistry textbooks. The easiest way to measure ΔG is to first measure ΔH for the reaction under consideration by measuring the heat absorbed when the reaction takes place at constant pressure and no other work is done. Then calculate ΔS from the entropies of initial and final states of the system, determined using heat capacity data tables found in most chemistry textbooks and putting both of these values together as,

$$\Delta G = \Delta H - T\Delta S \tag{10.40}$$

10.5 Thermodynamic potentials and chemical reactions

The thermodynamic quantities, ΔU, ΔS, ΔF, ΔH are '*extensive properties*' of matter, which means they will change as the amount of matter in the sample changes. For example, doubling the amount of matter will double the internal energy of the system. To make these quantities '*intensive properties*' of matter, we must define precisely the temperature, pressure, mass and the physical state of the substance.

We defined a standard state in chapter 6, section 6.5.4, as a state of the system when the pressure is 1 atm, the temperature is 25° and 1 mole of the substance is present. When the thermodynamic quantities are determined at the standard state, they are called intensive properties and a subscript '°' is added to their symbol to signify the state as: $\Delta U°$, $\Delta S°$, $\Delta F°$, $\Delta H°$. For chemical reactions, the standard state involves the number of moles designated by the stoichiometric coefficients in the simplest balanced chemical equation with the smallest possible whole number coefficients.

10.5.1 Standard enthalpy changes and the standard heat of reaction ($\Delta H°$)

The heat energy or enthalpy change, ΔH, produced by a chemical reaction is an extensive property. Reacting a larger amount of chemicals produces a larger amount of heat. For example, when propane is burnt as,

$$CH_3CH_2CH_3(g) + 5O_2(g) \rightarrow 3CO_2(g) + 4H_2O(g) \tag{10.41}$$

more heat is generated as more propane is burnt. To make the heat produced by a reaction an intensive property, we must specify the amount of chemical that reacts. The standard heat of reaction, $\Delta H°$, is defined as the heat produced when the number of moles specified in the balanced chemical equation reacts. In the above reaction, $\Delta H° = -2044$ kJ. The negative sign indicates that a large amount of heat is released, which makes propane an excellent fuel for cooking and heating.

10.5.2 Hess's law

Hess's law states that mathematical operations performed on a chemical reaction are the same when applied to the heat of reaction. Hess's law allows chemists to measure ΔH°_{reac} for several reactions and then by combining these equations and their heats to calculate ΔH°_{reac} for other, sometimes completely different reactions:

1. If the coefficient of a chemical equation is multiplied by a constant, $\Delta H°$ must also be multiplied by the same constant.
2. If two or more equations are added to obtain an overall reaction, the heats of these equations are also added to give the overall heat of the reaction.

10.5.3 Formation reaction

A formation reaction is defined as one in which the reactants are elements in the standard state (1 atm and 25°) and there is only one mole of product. The symbol for standard heat of formation is ΔH_f°. It is important to note that fractional coefficients may be used in formation reactions. Since there is always one mole of product, the standard heats of formation are usually written as the heat produced per mole of product.

Heats of formation of the elements are always zero, regardless of whether they are molecules or atoms; this is because the formation reaction for an element such as oxygen is defined as,

$$O_2(g) \rightarrow O_2(g)$$

Since oxygen is at standard conditions both as the product and as the reactant, the initial and final states of oxygen are the same. This means the difference must be zero.

$$\Delta H_f^\circ = 0$$

Heats of formations can be calculated from heats of combustion. For the combustion of propane, from equation (10.41). ($CH_3CH_2CH_3(g) + 5O_2(g) \rightarrow 3CO_2(g) + 4H_2O(g)$), ΔH_f° can be calculated using the formation reaction and the tabulated heats of formation. This means we will need a formation reaction for each of the reactants and products. Please keep in mind, heats of formation of elements (in this case O_2) are always zero.

$$3C(s) + 4H_2(g) \rightarrow CH_3CH_2CH_3(g);\ \Delta H_f^\circ = -\,103.8\ \text{kJ mol}^{-1}$$
$$O_2(g) \rightarrow O_2(g);\ \Delta H_f^\circ = 0$$
$$C(s) + O_2(g) \rightarrow CO_2(g);\ \Delta H_f^\circ = -\,393.5\ \text{kJ mol}^{-1}$$
$$H_2(g) + \frac{1}{2}O_2(g) \rightarrow H_2O(g);\ \Delta H_f^\circ = -\,241.8\ \text{kJ mol}^{-1}$$

The following operations must be performed on these reactions so they can be combined to use the combustion reaction,

1. The propane formation reaction is reversed:

$$CH_3CH_2CH_3(g) \rightarrow 3C(s) + 4H_2(g);\ \Delta H_f^\circ = (-1\ \text{mol}) \times -103.8\ \text{kJ mol}^{-1}$$

2. the carbon dioxide formation reaction is multiplied by 3:

$$3C(s) + 3O_2(g) \rightarrow 3CH_3(g);\ \Delta H_f^\circ = (3\ \text{mol}) \times -393.5\ \text{kJ mol}^{-1}$$

3. the water formation reaction is multiplied by 4:

$$4H_2(g) + \frac{4}{2}O_2(g) \rightarrow 4H_2O(g);\ \Delta H_f^\circ = (4\ \text{mol}) \times -241.8\ \text{kJ mol}^{-1}$$

From these operations we can calculate the corresponding sum of heats to give the heat of the reaction as:

$$\Delta H^\circ = [(-1 \text{ mol}) \times -103.8 \text{ kJ mol}^{-1}] + [(3 \text{ mol}) \times -393.5 \text{ kJ mol}^{-1}] + [(4 \text{ mol}) \times -241.8 \text{ kJ mol}^{-1}] = -2043.9 \text{ kJ} \quad (10.42)$$

After more examples using heats of formation, scientists see a pattern. The heat of any reaction will be the sum of ($\text{mol}_{\text{product}} \times \Delta H_f^\circ$) for all the products minus the sum of ($\text{mol}_{\text{reactant}} \times \Delta H_f^\circ$) for all the reactants, where $\text{mol}_{\text{product}}$ and $\text{mol}_{\text{reactant}}$ are the stoichiometric coefficients and in the balanced equation for the reaction.

$$\Delta H_f^\circ = \Sigma(\text{coeff} \times \Delta H_f^\circ)_{\text{products}} - \Sigma(\text{coeff} \times \Delta H_f^\circ)_{\text{reactants}} \quad (10.43)$$

Worked example 10.2. Calculate the heat of combustion of CH_4,

$$CH_4 + 2O_2 \rightarrow CO_2 + 2H_2O$$

where all of the reactants and the products are in their gaseous form. Given the heats of formation: $\Delta H_f^\circ(CH_4(g)) = -74.8 \text{ kJ mol}^{-1}$, $\Delta H_f^\circ(CO_2(g)) = -110.5 \text{ kJ mol}^{-1}$, and $\Delta H_f^\circ(H_2O\ (g)) = -241.8 \text{ kJ mol}^{-1}$.

Solution: The heat of reaction for the combustion of CH_4 is calculated by using equation (10.43). The heat of formation of O_2 by definition is zero.

$$\begin{aligned}\Delta H_f^\circ &= \left[(\text{coeff} \times \Delta H_f^\circ)_{CO_2} + (\text{coeff} \times \Delta H_f^\circ)_{H_2O}\right] \\ &\quad - \left[(\text{coeff} \times \Delta H_f^\circ)_{CH_4} + (\text{coeff} \times \Delta H_f^\circ)_{O_2}\right] \\ &= [(1 \text{ mol})(-110.5 \text{ kJ mol}^{-1}) + (2 \text{ mol})(-241.8 \text{ kJ mol}^{-1})] \\ &\quad - [(1 \text{ mol})(-74.8 \text{ kJ mol}^{-1}) + (2 \text{ mol})(0 \text{ kJ mol}^{-1})] \\ &= -519.3 \text{ kJ}.\end{aligned}$$

The heat of reaction for the combustion of CH_4 is -519.3 kJ.

10.5.4 Equilibrium law

In a chemical reaction the actual concentrations of the reactants and products at equilibrium, depend on the initial concentration of the reacting mixture. Experiments with different initial concentrations result in equilibrium mixtures with different concentrations. Although these mixtures may be different, they all obey the '*equilibrium law*', which states that the ratio of the concentrations of all of the products multiplied together to the concentrations of all reactants multiplied together equals a number called the equilibrium constant, K_G. The value of K_G depends on the specific reaction and the temperature of the reaction mixture when equilibrium is reached.

For the following reaction,

$$C_2H_4 + 3O_3 \rightarrow 2CO_2 + 2H_2O$$

The equilibrium law is written as,

$$K_G = \frac{[CO_2]^2[H_2O]^2}{[C_2H_4][O_2]^3}$$

Worked example 10.3. Calculate the equilibrium constant K_G for the reaction when sulfur dioxide reacts with oxygen to give sulfur trioxide,

$$O_2 + 2SO_2 \rightarrow 2SO_3$$

where all of the reactants and the products are in their gaseous form and the concentrations at equilibrium are found to be $[O_2] = 2 \times 10^{-8}\ M$, $[SO_2] = 3.4 \times 10^{-9} M$ and $[SO_3] = 9.7 \times 10^{-1}\ M$

Solution: The equilibrium law for the reaction will be written as,

$$K_G = \frac{[SO_3]^2}{[SO_2]^2[O_2]}$$

Putting the values of concentrations at equilibrium in the equilibrium law we find the equilibrium constant K_G for the reaction as,

$$K_G = \frac{[9.7 \times 10^{-1}\ \mathrm{M}]^2}{[3.4 \times 10^{-9}\ \mathrm{M}]^2[2 \times 10^{-8}\ \mathrm{M}]} = 4.2 \times 10^{24}.$$

10.5.5 Gibbs free energy at standard state

The value of the change in Gibbs free energy (ΔG) given by equation (10.40) tells us whether a reaction will continue and in which direction it will go.

1. When ΔG is negative, the reaction proceeds in the forward direction (to the right) in order to reach equilibrium.
2. When ΔG is positive, the reaction proceeds in the backward direction (to the left) in order to reach equilibrium.
3. When ΔG is zero, the reaction is at equilibrium and no further reaction occurs. In this case,

$$\Delta G^\circ = -RT \ln K_G$$

where K_G is called the equilibrium constant. The numerical value of the equilibrium constant is determined by measuring the concentration of each reactant and product in the mixture. In general, we say that equilibrium constants are dimensionless quantities, however, sometimes we have to assign units for expediency. When we have to assign units to equilibrium constants, we use either $(\mathrm{atm})^{\Delta n}$ or $(\mathrm{mol\ L^{-1}})^{\Delta n}$, where $\Delta n \equiv \Sigma$(moles of products) $-\Sigma$(moles of reactants) in a balanced chemical reaction. All reactions that have equilibrium constants greater than 1 are spontaneous reactions.

10.5.6 Gibbs free energy at standard state

A spontaneous reaction does not require additional input energy after the reactants are mixed and the reaction is initiated. The change in Gibbs free energy (ΔG) given by equation (10.40) does not tell us about the spontaneity of any process under consideration. ΔG tells us the direction of a reaction as it approaches equilibrium. So it is customary to use ΔG° at the standard state to tell us if the reactions are spontaneous or not.

ΔG° is the symbol used for Gibbs free energy change at the standard state. ΔG° is the maximum amount of energy available from any chemical reaction. Two forces drive chemical reactions: the first is the enthalpy, ΔH°, which represents the change in internal potential energy of the atoms; the second is that the entropy of the system always increases. The combination of these two driving forces is given as,

$$\Delta G^\circ = \Delta H^\circ - T\Delta S^\circ$$

where S° has units of J K^{-1} mol^{-1}. It should be noted that the standard entropy values for elements are not zero as they are for standard heat formation, ΔH_f° and standard Gibbs free energy, $\Delta G_f{}^\circ$. Generally, if ΔG° is negative, the process is spontaneous and when ΔG° is positive, the process is not spontaneous. So we can use this equation to see how a spontaneous process could be enthalpically favored or entropically favored or both, but not neither. Let's look at the conditions:

1. **Case 1:** Suppose, $\Delta S^\circ > 0$ and $\Delta H^\circ < 0$: In this case ΔH° is negative which implies this is an exothermic[3] reaction and is energetically favored. ΔS° is a positive number which implies that there is an increase in entropy, which is also favorable. Hence ΔG° will be negative and the process will be spontaneous.
2. **Case 2:** Suppose, $\Delta S^\circ < 0$ and $\Delta H^\circ > 0$: In this case ΔH° is positive, which implies this is an endothermic[4] reaction and is energetically unfavored. ΔS° is a negative number, which implies that there is an decrease in entropy, which is also not favored. Hence, ΔG° comes out to be positive and the process will not be spontaneous.
3. **Case 3:** For the situation where one of the two, either ΔH° or ΔS° is unfavored we have to do math to figure out whether the process is spontaneous or not,
   (a) When both ΔS° and ΔH° are positive, we have a situation where this process is energetically unfavorable, however, it is entropically favorable. A process like this is spontaneous at high temperatures.
   (b) When both ΔS° and ΔH° are negative, we have a situation where the process is energetically favorable but entropically unfavorable. A process like this is spontaneous at low temperatures.

Relations between ΔS°, ΔH°, and ΔG° are summarized in table 10.4. for reference. The condensation of a gas and the crystallization of a liquid are two physical

[3] An exothermic reaction indicates that the energy is being lost from the system to the surroundings. The prefix *exo* indicates a negative sign for the numerical thermodynamic quantities. For an exothermic reaction, the heat of reaction (ΔH°) is a negative number.

[4] An endothermic reaction is where the energy is gained from the surroundings. An endothermic reaction absorbs heat energy and appears to cool as it progresses. For an endothermic reaction the heat of reaction (ΔH°) is a positive number.

Table 10.4. Relation between $\Delta S°$, $\Delta H°$, and $\Delta G°$ at the standard state.

$\Delta S°$	$\Delta H°$	$\Delta G°$	Reaction type
Positive	Negative	Always negative	Always spontaneous at all temperatures
Negative	Positive	Always positive	Never spontaneous at all temperatures
Positive	Positive	Becomes negative as T increases	Becomes spontaneous as T increases
Negative	negative	Becomes positive as T increases	Becomes non-spontaneous as T increases

processes that are spontaneous at low temperatures and non-spontaneous at higher temperatures.

Worked example 10.4. Given enthalpy of reaction, also called the heat of reaction, $\Delta H° = -1235.5$ kJ and entropy change $\Delta S° = 217.39$ J K^{-1}, calculate the $\Delta G°$ for the reaction of liquid ethanol at room temperature, and state if this process is spontaneous or not.

C_2H_5OH (liquid ethanol) + $3O_2$(gas) → $2CO_2$ (gas) + $3H_2O$ (gas).

Solution: We are given enthalpy of reaction $\Delta H° = -1235.5$ kJ. Since this is a negative value, this is an exothermic reaction and is energetically favored. We are also given entropy change $\Delta S° = 217.39$ J K^{-1}, which means the change in entropy is positive, which is also favored. This means this process will be spontaneous.

To calculate the numerical value of ΔG we use equation (10.40) and the given values,

$$\Delta G° = \Delta H° - T\Delta S° = -1235.5 \times 10^3 \text{ J} - (298 \text{ K})(217.39 \text{ J K}^{-1})$$
$$\Delta G° = -1300\,282 \text{ J} = -1300.3 \text{ kJ}$$

The value of $\Delta G°$ is a negative number, which again shows this is a spontaneous process.

Worked example 10.5. Calculate the change in entropy ($\Delta S°$) for the combustion of propane, discussed in section 10.5.1.

$$CH_3CH_2CH_3(g) + 5O_2(g) \rightarrow 3CO_2(g) + 4H_2O(g) \quad (10.44)$$

Given that the entropy changes at standard conditions (1 atm and 25 °C) as: $\Delta S°(CH_3CH_2CH_3\ (g)) = 270.2$ J mol^{-1} K^{-1}, $\Delta S°(O_2(g)) = 205$ J mol^{-1} K^{-1}, $\Delta S°(CO_2(g)) = 213.6$ J mol^{-1} K^{-1}, and $\Delta S°(H_2O(g)) = 188.7$ J mol^{-1} K^{-1}.

Solution: We can also use the same method as we used when we calculated the heat of reaction for the combustion of CH_4 by using equation (10.43).

$$\Delta S° = \Sigma(\text{coeff} \times \Delta S°)_{\text{products}} - \Sigma(\text{coeff} \times \Delta S°)_{\text{reactants}} \quad (10.45)$$

Then for this reaction we can write,

$$\begin{aligned}\Delta S^\circ &= [(\text{coeff} \times \Delta S)_{CO_2} + (\text{coeff} \times \Delta S^\circ)_{H_2O}] \\ &\quad - [(\text{coeff} \times \Delta S^\circ)_{CH_3CH_2CH_3} + (\text{coeff} \times \Delta G^\circ)_{O_2}] \\ &= [(3\text{ mol})(213.6\text{ J mol}^{-1}\text{ K}^{-1}) + (4\text{ mol})(188.7\text{ J mol}^{-1}\text{ K}^{-1})] \\ &\quad - [(1\text{ mol})(270.2\text{ J mol}^{-1}\text{ K}^{-1}) + (5\text{ mol})(205\text{ J mol}^{-1}\text{ K}^{-1})] \\ &= +\,100.4\text{ J K}^{-1}.\end{aligned}$$

The change in entropy is a positive number, which means this reaction results in an increase in entropy, which is as expected since there are seven moles of products and only six moles of reactants. The increase in the number of moles of gas in this reaction is 1, which implies that the entropy change is expected to be positive.

Worked example 10.6. Is the reaction for the combustion of propane, discussed in section 10.5.1.

$$CH_3CH_2CH_3(g) + 5O_2(g) \rightarrow 3CO_2(g) + 4H_2O(g) \tag{10.46}$$

spontaneous or non-spontaneous? Given $\Delta S^\circ = 100.4$ J, from example 10.5, and the free energy at standard conditions (1 atm and 25°) as: $\Delta G^\circ_{298}(CH_3CH_2CH_3\ (g)) = -23.5$ kJ mol^{-1}, $\Delta G^\circ_{298}(CO_2(g)) = -394.4$ kJ mol^{-1}, and $\Delta G^\circ_{298}(H_2O\ (g)) = -228.6$ kJ mol^{-1}.

Solution: This question can be done two ways:

Method 1: Use equation (10.40) and the given values from equation (10.42),

$$\begin{aligned}\Delta G^\circ &= \Delta H^\circ - T\Delta S^\circ \\ &= -\,2044\text{ kJ} - (298\text{ K})(0.100\,4\text{ kJ K}^{-1}) \\ &= -\,2074\text{ kJ}\end{aligned}$$

The value of ΔG° is a negative number, which shows this is a spontaneous process.

Method 2: We can also use the same method as we used when we calculated the heat of reaction for the combustion of CH_4 using equation (10.43).

$$\Delta G^\circ = \Sigma(\text{coeff} \times \Delta G^\circ)_{\text{products}} - \Sigma(\text{coeff} \times \Delta G^\circ)_{\text{reactants}} \tag{10.47}$$

Then for this reaction we can write,

$$\begin{aligned}\Delta G^\circ &= [(\text{coeff} \times \Delta G)_{CO_2} + (\text{coeff} \times \Delta G^\circ)_{H_2O}] \\ &\quad - \left[(\text{coeff} \times \Delta G^\circ_{298})_{CH_3CH_2CH_3} + (\text{coeff} \times \Delta G^\circ_{298})_{O_2}\right] \\ &= [(3\text{ mol})(-394.4\text{ kJ mol}^{-1}) + (4\text{ mol})(-228.6\text{ kJ mol}^{-1})] \\ &\quad - [(1\text{ mol})(-23.5\text{ kJ mol}^{-1}) + (5\text{ mol})(0\text{ kJ mol}^{-1})] \\ &= -\,2074.1\text{ kJ}.\end{aligned}$$

The value of ΔG° is again a negative number, which shows this is a spontaneous process.

10.5.7 Gibbs free energy at other temperatures

Using the standard free energy of formations we can calculate the Gibbs free energy change at 298 K. To calculate the standard Gibbs free energy changes and other temperatures we need to know the standard entropy change, $\Delta S^{\circ}_{\text{react}}$, and the standard heat of reaction, $\Delta H^{\circ}_{\text{react}}$, for the reaction. These values can be assumed to not change significantly with temperature and hence we can use them to find the values of Gibbs free energy at any temperature other than 298 K.

Worked example 10.7. For a certain reaction, $\Delta S^{\circ}_{\text{react}} = +12.3$ J K^{-1} and $\Delta H^{\circ}_{\text{react}} = +3.0$ kJ. Calculate the G° at 200 K, 298 K, and 400 K.

Solution: Convert the values into kJ, $\Delta S^{\circ}_{\text{react}} = +0.0123$ kJ K^{-1} and $\Delta H^{\circ}_{\text{react}} = +3.0$ kJ. Use equation (10.40) and the given values,

1. At 200 K:

$$\begin{aligned}\Delta G^{\circ}_{200K} &= \Delta H^{\circ} - T\Delta S^{\circ}\\ &= +3.0\text{ kJ} - (200\text{ K})(0.0123\text{ kJ K}^{-1})\\ &= +0.54\text{ kJ}\end{aligned}$$

This reaction is not spontaneous at 200 K.

2. At 298 K:

$$\begin{aligned}\Delta G^{\circ}_{298} &= \Delta H^{\circ} - T\Delta S^{\circ}\\ &= +3.0\text{ kJ} - (298\text{ K})(0.0123\text{ kJ K}^{-1})\\ &= -0.67\text{ kJ}\end{aligned}$$

This reaction is spontaneous at 298 K.

3. At 400 K:

$$\begin{aligned}\Delta G^{\circ}_{400K} &= \Delta H^{\circ} - T\Delta S^{\circ}\\ &= +3.0\text{ kJ} - (400\text{ K})(0.0123\text{ kJ K}^{-1})\\ &= -1.92\text{ kJ}\end{aligned}$$

This reaction is spontaneous at 400 K.

This example shows that the reaction is not spontaneous at 200 K but increasing the temperature to 298 K makes the reaction spontaneous.

10.5.8 Gibbs free energy and equilibrium

When a system is not a standard state, the Gibbs free energy change is represented by ΔG and not by ΔG°. The relationship between ΔG and ΔG° is,

$$\Delta G = \Delta G^{\circ} + RT \ln Q_G \tag{10.48}$$

where Q_G is the reaction quotient. The reaction quotient is a measure of the relative amounts of products and reactants present in a reaction at a given time.

When Q_G does not change with time, the reaction is in a state of equilibrium. When the value of Q_G is not equal to the equilibrium constant (K_G), the reaction continues and further reactions take place until the system reaches equilibrium. When $Q_G = 1$, $\ln 1 = 0$, and $\Delta G = \Delta G°$. When $K_G = Q_G$, the reaction is in a state of equilibrium. If $K_G > Q_G$, the reaction proceeds in the forward direction and when $K_G < Q_G$, the reaction proceeds in the backward direction. Please note I am using the subscripts G with the symbols in this text to not confuse them with the symbols of Boltzmann constant (k) and heat energy (Q). Other texts do not use subscripts with the symbols.

Worked example 10.8. For a certain reaction, the value of the equilibrium constant (K_G) at 298 K = 45 and the reaction quotient (Q_G) = 35. Show that the value of G indicates the same direction for the reaction as Q_G predicts.

Solution: When Q_G is less than K_G, the reaction will proceed in the forward direction. Next we calculate the value of $\Delta G°$ as,

$$\Delta G° = -RT \ln K_G = -(8.314 \text{ J mol}^{-1} \text{ K}^{-1})(298 \text{ K}) \ln(45) = -9.43 \text{ kJ mol}^{-1}.$$

The value of $\Delta G°$ is calculated as,

$$\begin{aligned}\Delta G &= \Delta G° + RT \ln Q_G \\ &= 9.43 \text{ kJ mol}^{-1} + (8.314 \text{ J mol}^{-1} \text{ K}^{-1})(298 \text{ K}) \ln(35) \\ &= -\ 0.62 \text{ kJ mol}^{-1}.\end{aligned} \tag{10.49}$$

The value of ΔG (negative) shows that the reaction will proceed in the forward direction. The fact that Q_G is less than K_G also indicates that the reaction will proceed in the forward direction.

10.5.9 Gibbs free energy and chemical potential

We know that Gibbs free energy (G) is an extensive property and is a function of T, p, and N. This means if the number of particles is increased by an arbitrary constant b, then we can write,

$$bG(T, p, N) = G(T, p, bN) \tag{10.50}$$

This implies we can represent Gibbs free energy (G) as:

$$G(T, p, N) = N\mu \tag{10.51}$$

which gives us a new interpretation of the chemical potential, at least for systems with only one type of particles; the chemical potential is just the Gibbs free energy per particle. Hence we can redefine chemical potential for the systems with only one type of particle as,

$$\mu \equiv \left(\frac{\partial G}{\partial N}\right)_{T,p} \tag{10.52}$$

This means both equations for μ, equations (10.51) and (10.52) are equivalent and we can define chemical potential as,

$$\mu \equiv \left(\frac{\partial G}{\partial N}\right)_{T,p} = \frac{G}{N} \tag{10.53}$$

When we add one particle to a system, holding T and p constant, the Gibbs free energy of the system will increase by μ. By adding more particles, we do not change the value of μ since we do not change the density. Note that U, H, and F, whose differentials also have the term μdN, depend on N non-linearly, because in processes with the independent variables (S, V, N), (S, p, N), and (V, T, N), $\mu = \mu(N)$ might vary with the number of particles (N).

1. **System having more than one type of particle:** For a system containing more than one type of particle, equation (10.51) generalizes to:

$$G(T, P, N) = N_1\mu_1 + N_2\mu_2 + \ldots\ldots$$
$$= \sum_i N_i\mu_i$$

2. **Chemical potential of an ideal gas:** Suppose we have a certain (N) amount of an ideal gas at a fixed temperature (T) but under variable pressure. Then from equation (10.28) $\left(\left(\frac{\partial G}{\partial P}\right)_{T,N} = V\right)$ and (10.51) we can write:

$$\frac{\partial \mu}{\partial p} = \frac{1}{N}\frac{\partial G}{\partial p} = \frac{V}{N}$$

and from the ideal gas Law ($pV = NkT$):

$$\frac{\partial \mu}{\partial p} = \frac{kT}{p}$$

Integrating both sides we get the chemical potential of an ideal gas ($\mu(T, p)$) as:

$$\mu(T, p) - \mu_o(T, P_o) = kT \ln \frac{p_{\text{gas}}}{P_o} \tag{10.54}$$

where P_o is the initial pressure, usually taken to be 1 atm. The values of $\mu(T, P_o)$ are usually given in tables of Gibbs free energies in chemistry and physics textbooks.

Please note, if we have a mixture of gases, equation (10.54) applies to each gas individually if you take p as the partial pressure of that particular gas. This works because one, gases are mostly empty space and two, the way one ideal gas exchanges particles with the environment does not get affected by the presence of others in the mixture.

Concept questions 10.2

1. What is the difference between U, ΔU and ΔU^o?
2. Does the chemical potential depend on temperature? Does it depend on pressure? Explain in your own words.
3. Which of the following can be precisely determined for a chemical substance; entropy (S), enthalpy (H), Gibbs free energy (G), internal energy (U)?

Exercises 10.2

1. At what temperature (in °C) will $K_G = 1.0$, given $\Delta S° = 22.6$ J K^{-1} and $\Delta H° = 15.3$ kJ for a chemical reaction.
2. Water boils at 100 °C. What is the entropy change when water condenses,

$$H_2(g) \rightarrow H_2O(l)$$

   Hint: use molar heat of vaporization = +43.9 kJ and during phase change $\Delta G° = 0$.
3. Suppose we have a mole of water at room temperature (22 °C) and 1 atm. Hint: entropy of a mole of water under standard conditions as 69.91 J K^{-1}.
   (a) Compare the Gibbs free energy if we raise the temperature to 28 °C keeping pressure constant.
   (b) To compensate for the change in Gibbs free energy should we increase or decrease the pressure on the water?
   (c) How much pressure will be required?

10.6 Practical application of chemical thermodynamics

There are many applications of chemical thermodynamics. In this text we discuss a few common ones that include, phase transformations, electrolysis and fuel cells.

10.6.1 Phase transformations

Gibbs free energy is essential in understanding phase transformations of substances. A phase transformation is the process of changing the state of a substance, for example when ice transforms into liquid water or when liquid water transforms to a gas. Phase changes provide an important example of a system in which there is a large change in internal energy with volume at constant temperature. The variation of internal energy (U) with volume for a system of water plus steam at constant temperature (T) is mathematically given by the equation,

$$\left(\frac{\partial U}{\partial V}\right)_T = \frac{L}{\Delta V} = -p \tag{10.55}$$

where L is the latent heat and p is the pressure.

10.6.1.1 Phase diagram of water (H_2O)

A graph called a '*phase diagram*', as shown in figure 10.5, shows equilibrium phases as a function of temperature and pressure.

The phase diagram is divided into three main regions indicating conditions under which ice (shown here in blue), water (shown by the gray region) or the vapor state (represented by green) are the most stable phases. Blue lines on the phase diagram represent the points at which two phases can coexist in equilibrium. The point on a phase diagram where all three phases, solid, liquid and gas, coexist is called the '***triple point***'. It is represented with a green dot in figure 10.5. For water the triple point exists at a temperature of 0.01 °C and pressure of 0.006 atm (608 Pa). At low pressures ($P < 0.006$ atm) liquid water cannot exist in equilibrium and ice sublimates directly into vapor. When a gas is able to coexist with its liquid or solid form, the pressure of the system is called the '*vapor pressure*', whose value varies with temperature. The vapor pressure of water at room temperature is approximately 0.023 atm (2.34 kPa).

Please note, the boundary between the liquid and gas phases always has a positive slope. So when you have a system where liquid and gas are in thermal equilibrium and you increase the temperature, you must increase the pressure on the system to keep the liquid from vaporizing. As pressure increases, the gas becomes dense, such that the difference between a liquid and a gas lessens. Eventually, a point is reached where there is no discontinuous change from liquid to a gas. This point is called a '***critical point***', shown by the yellow dot in figure 10.5. Critical points for some substances are given in table 10.5.

10.6.1.2 Phase change in carbon—diamond versus graphite

Elemental carbon exists in two phases, both of which are solids. It can exist as diamond or as graphite. Both of these phases are solids, but they are made of different crystalline structures. At ordinary pressures graphite is the more stable phase, which

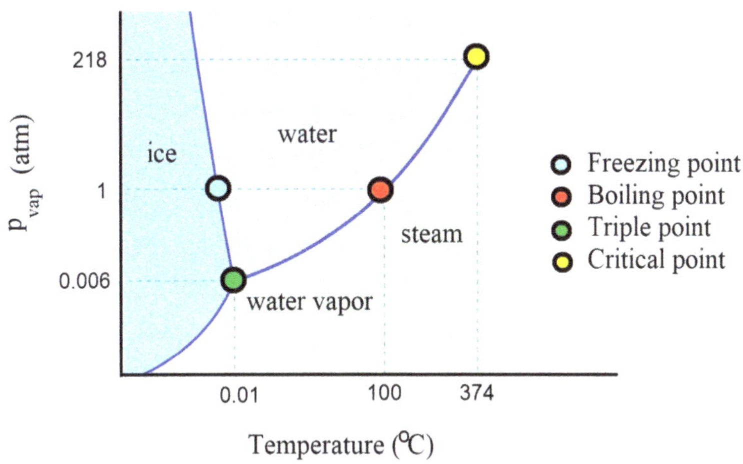

Figure 10.5. Phase diagram for water.

Table 10.5. Temperatures and pressures for critical points for some substances.

	Temperature	Pressure
For water	$T = 374$ °C	$P = 221$ bars = 221 kPa
For CO_2	$T = 31$ °C	$P = 74$ bars = 74 kPa
For N_2	$T = -147.15$ °C	$P = 34$ bars = 34 kPa

means diamonds spontaneously convert to graphite. Luckily, at room temperatures, this process occurs very, very slowly so you do not need to worry about the diamond ring you bought five years ago becoming graphite overnight. At high temperatures ($\approx$ 1500 °C), however, this conversion happens very quickly, so I do suggest to never put your diamond ring in the fireplace, for the longevity of the ring!

At any given temperature (T) and pressure (p) the more stable phase has a lower value of Gibbs free energy (G). The G value of one mole of graphite at standard conditions of 1 atm and 298 K, is 2900 J less than the G of one mole of diamond at 1 atm and 298 K. We can divide the dependence of phases into two parts:

1. **Temperature dependence of *G*:** The temperature dependence of G is mathematically derived in equation (10.28). and is given as,

$$\left(\frac{\partial G}{\partial T}\right)_{p,N} = -S \tag{10.56}$$

The negative implies that as temperature increase G decreases. This decrease is more rapid for graphite because it has a higher entropy than a diamond. As temperature increases, graphite becomes more and more stable relative to the diamond state of carbon and hence diamonds convert to graphite at high temperatures.

2. **Pressure dependence of *G*:** We know that the pressure dependence of G, as shown in equation (10.28). is,

$$\left(\frac{\partial G}{\partial p}\right)_{T,N} = V \tag{10.57}$$

This equation shows that the rate of change of G with respect to pressure (p) is equal to the volume (V). One mole of graphite has a greater volume than one mole of diamond, which implies a mole of graphite will grow more rapidly as p is increased. If we treat their volumes as constant (neglecting the compressibility of both substances), each curve will be a straight line, with slopes:

For graphite $V = 5.31 \times 10^{-6}$ m^3

For diamond $V = 3.42 \times 10^{-6}$ m^3

These lines intersect at $p \approx 15$ katm as shown in figure 10.6. At any pressure greater than this, diamond is the more stable form rather than graphite. This implies that diamond formation requires a very high pressure

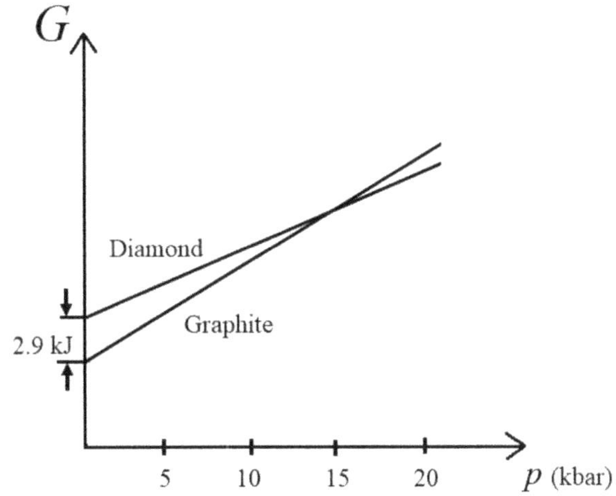

Figure 10.6. Molar Gibbs free energies of diamond and graphite as a function of pressure at room temperature.

and so natural diamonds are only formed deep within the Earth's crust. To find out how deep, we can start with an educated guess, that is, an average rock is three to four times as dense as water so we can estimate that for every 10 meters of depth the pressure increases by 3 atm. Then a pressure of ≈ 15 katm would require a depth of well over 50 km. So diamonds are formed at depths of about 100 km underground and have an average life of greater than 1 billion years! This analysis is very important to geochemists who study rocks, their ages and the and formation conditions of various kinds of rocks.

10.6.2 Clausius–Clapeyron relation

We know that entropy determines the temperature dependence of Gibbs free energy (G) (equation (10.56)), and volume determines the pressure dependence of G (equation (10.57)). This implies that the shape of any phase boundary line on a pT diagram is related in a very simple way to the S and V of the two phases. At a phase boundary, Gibbs free energy (G) is the same between the two phases. That means at a phase boundary,

$$\begin{gathered} G_{\mathrm{L}} = G_{\mathrm{g}} \\ G_{\mathrm{ice}} = G_{\mathrm{L}} \end{gathered} \tag{10.58}$$

where G_{L} is the Gibbs free energy of water in its liquid phase, G_{g} is the Gibbs free energy of water in its gas phase, and G_{ice} is the Gibbs free energy of water in its solid phase.

If we change the temperature of water by (dT) and the pressure by (dp) in such a way that the two phases remain equally stable, then, since the Gibbs free energies of the two phases must be equal to each other to remain on the phase boundary and the changes in Gibbs free energy (G) must also be equal, we have

$$\begin{aligned} dG_{\mathrm{L}} &= dG_{\mathrm{g}} \\ dG_{\mathrm{ice}} &= dG_{\mathrm{L}} \end{aligned} \tag{10.59}$$

We know that Gibbs free energy is defined as,

$$G \equiv U + PV - TS$$

and from equation (10.26).

$$dG = -SdT + VdP + \mu dN.$$

If we keep the number of particles (N) constant then the change in the number of particles, $dN = 0$, and the expression for dG reduces to,

$$dG = -SdT + VdP \tag{10.60}$$

We can plug this into the expression showing the equality of the change in Gibbs free energy G (equation (10.59)) on the phase boundary between liquid and gas, ($dG_{\mathrm{L}} = dG_{\mathrm{g}}$), to remain on the boundary, to get,

$$\begin{aligned} dG_L &= dG_g \\ -S_{\mathrm{L}}dT + V_{\mathrm{L}}dp &= -S_{\mathrm{g}}dT + V_{\mathrm{g}}dp \\ (-S_{\mathrm{L}} + S_{\mathrm{g}})dT &= (V_{\mathrm{g}} - V_{\mathrm{L}})dp \\ \left(\frac{S_{\mathrm{g}} - S_{\mathrm{L}}}{V_{\mathrm{g}} - V_{\mathrm{L}}}\right) &= \left(\frac{dp}{dT}\right) \end{aligned}$$

where $\left(\frac{dp}{dT}\right)$ is the slope. This shows that the slope is determined by the volume and the entropies of the two phases.

$$\frac{\Delta S}{\Delta V} = \frac{dp}{dT} = \text{slope} \tag{10.61}$$

Usually, it is more convenient to write the difference ($S_{\mathrm{g}} - S_{\mathrm{L}} = \Delta S$) as,

$$(S_{\mathrm{g}} - S_{\mathrm{L}}) = \Delta S = \left(\frac{L}{T}\right) \tag{10.62}$$

where L is the total latent heat for converting the material, in this case from a liquid to a gas. Then we can rewrite the slope in equation (10.61) as,

$$\frac{L/T}{\Delta V} = \left(\frac{L}{T\Delta V}\right) = \left(\frac{dp}{dT}\right) \tag{10.63}$$

This expression is known as the '***Clausius–Clapeyron relation***'. Please note that both L and ΔV are extensive properties, and we know from appendix D, their ratio is intensive, which is independent of the amount of material. It is also important to note that this relation applies to the slope of any phase boundary line on a pT diagram, not just the line separating liquid from gas (figure 10.7).

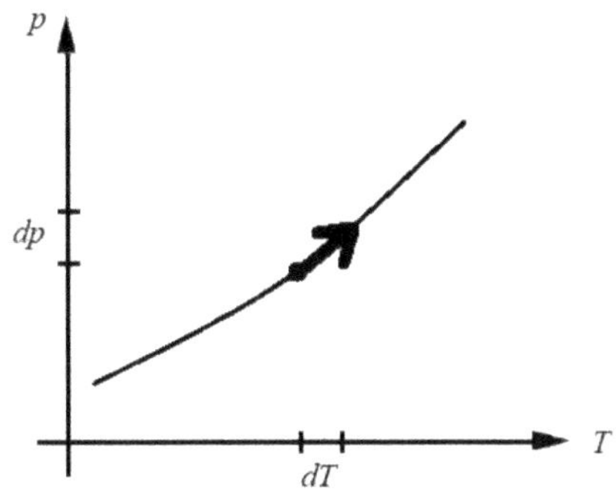

Figure 10.7. Infinitesimal changes in pressure and temperature.

1. **Large difference in entropy**: A small change in temperature (dT) may be enough to shift the equilibrium from one phase to another. This is shown by a very steep phase boundary curve.
2. **Large difference in volume**: A small change in pressure (p) may be enough to shift equilibrium from one phase to another. This is shown by a very shallow phase boundary curve.

Worked example 10.9. As one mole of diamond converts to graphite at room temperature, its entropy (S) increases by +3.4 J K^{-1}, and its volume increases by $+1.9 \times 10^{-6}$ m^3. Find the rate of change of pressure.

Solution: To find the rate of change of pressure use equation (10.61).

$$\frac{dP}{dT} = \frac{\Delta S}{\Delta V}$$

$$\frac{dP}{dT} = \frac{3.4 \text{ J K}^{-1}}{1.9 \times 10^{-6} \text{ m}^3} = 1.8 \times 10^{6} \text{ Pa K}^{-1}$$

1.8×10^6 Pa K^{-1} = 18 atm K^{-1} at room temperature (298 K).

We know from the last section that diamonds are stable at pressures larger than 15 katm. So if in this case we increase temperature by 100 K, $\left(\frac{dp}{dT}\right)$ = 1800 atm K^{-1}. This means we will need 1.8 katm K^{-1} of pressure to make diamonds stable.

10.6.2.1 Clausius–Clapeyron equation in meteorology

There are two version of the Clausius–Clapeyron equation: one that relates the vapor pressure to the dew-point and the other that relates the saturation vapor pressure to the temperature. The dew-point is the atmospheric temperature below which water droplets begin to condense and dew (liquid water) can form. The dew-point in terms of relative humidity is stated as the temperature to which air needs to be cooled, at constant pressure, to achieve a relative humidity (RH)[5] of 100%.

[5] Relative humidity is the ratio of the pressure of water vapor in a pocket of air to the saturation pressure of water vapor in the same pocket of air at a particular temperature. 100% relative humidity is when dew-point and temperature are the same.

Vapor pressure is the pressure exerted by water vapor in millibars. As the amount of moisture increases, the vapor pressure increases.

At a particular pressure, the dew-point temperature is said to be reached when the water vapor in air is in equilibrium with the liquid water. This means the water vapor condenses at the same rate at which the liquid water evaporates. Most people are comfortable with a dew-point temperature of about 60 °F or lower. Below the dew-point, liquid water condenses on solid surfaces, for example on flowers or on solid particles in the atmosphere, such as dust, which forms either clouds at higher altitudes or fog at lower altitudes. Above the dew-point, >70 °F, most people feel hot or sticky because the amount of water vapor in the air slows the evaporation of perspiration and keeps their body from cooling.

To determine the amount of moisture in outside air we use the version of Clausius–Clapeyron equation that relates vapor pressure to dew-point. To determine the maximum amount of moisture in the air for a given temperature we use the version that relates saturation vapor pressure to the temperature.

1. The Clausius–Clapeyron equation that relates vapor pressure (e) to dew-point (T_d) is given as,

$$\ln\left(\frac{e}{6.11}\right) = \frac{L}{R}\left(\frac{1}{273} - \frac{1}{T_d}\right) \tag{10.64}$$

   where e is the vapor pressure (in mbar), L is latent heat of vaporization (2.453×10^6 J kg^{-1}), R is the gas constant for moist air (461 J kg^{-1} K^{-1}), and T_d is the dew-point in K.

2. The Clausius–Clapeyron equation that relates saturation vapor pressure (E_s) to the temperature is given as,

$$\ln\left(\frac{e_s}{6.11}\right) = \frac{L}{R}\left(\frac{1}{273} - \frac{1}{T}\right) \tag{10.65}$$

   where e_s is the saturation vapor pressure (in mbar), L is latent heat of vaporization (2.453×10^6 J kg^{-1}, R is the gas constant for moist air (461 J kg^{-1} K^{-1}), and T is the temperature in K.

Please note both versions of the equation are equivalent when the relative humidity is 100%. If the air is unsaturated, then use the dew-point to determine the amount of moisture in the air.

Worked example 10.10. What is the vapor pressure when the dew-point is 50 °F (283 K)?

Solution: We wish to find the vapor pressure (e), so we use equation (10.64), and solve,

$$\begin{aligned} \ln\left(\frac{e}{6.11}\right) &= \frac{L}{R}\left(\frac{1}{273} - \frac{1}{T_d}\right) \\ \ln\left(\frac{e}{6.11}\right) &= \frac{2.453 \times 10^6}{461}\left(\frac{1}{273} - \frac{1}{283.15}\right) \end{aligned} \tag{10.66}$$

and simplify to get $e = 12.288$ mbars.

10.6.3 Electrolysis

A process in which electrical energy is used to produce a chemical change in a substance (usually a compound) is called '*electrolysis*'. Electrolysis is a slow process that occurs at constant temperature (isothermal) and at constant pressure (isobaric). A device in which electrical energy is used to bring about chemical changes is called an '*electrolytic cell*'. The equipment used for electrolysis is made up of three basic parts:

1. A source of direct current, usually a direct current (DC) battery.
2. Two electrodes, commonly two strips of platinum metal, called anode and cathode.
   (a) Anode, from the Greek word *anodos* which translates to 'way up', from *ana* ('up') and *hodos* ('way'). Defined as 'the positively charged electrode by which the electrons leave a device'.
   (b) Cathode, from the Greek word *kathodos* which translates to 'way down', from *kata* ('down') and *hodos* ('way'). Defined as 'the negatively charged electrode by which electrons enter an electrical device'.
3. Electrolyte, usually a few drops of sulfuric acid added to water. Electrolyte is a substance that conducts an electric current when dissolved in water.

10.6.3.1 Electrolysis of water

The most familiar example of electrolysis is the electrolysis of water (H_2O). Water can be decomposed into diatomic hydrogen (H_2) and diatomic oxygen (O_2) gases by providing energy from a DC battery. The reaction requires a potential difference to split water into H_2 and O_2. Electrolysis of water begins around a minimum of 1.2 volts and increases in rate as the voltage is increased. Typically, the electrolysis is carried out around 6 volts. Electrolysis is used to make hydrogen gas (that can be used as fuel) and to make breathable oxygen.

The electrolytic cell consists of a pair of platinum electrodes immersed in water to which a small amount of an electrolyte such as sulfuric acid (H_2SO_4) is added to increase the concentration of hydrogen ions, because pure water will not carry enough charge due to the lack of ions. It is important to keep in mind that H_2SO_4 is not consumed during the reaction so it does not take part in the reaction. At the anode, water is oxidized to oxygen gas and hydrogen ions. At the cathode, water is reduced to hydrogen gas and hydroxide ions, as seen in figure 10.8.

Electrolysis process:

When the battery is turned on, current flows from anode to the cathode. Positive hydrogen ions (H^+) in the electrolyte pick up electrons from that anode and become neutral hydrogen molecules (H_2):

$$2H^+ + 2e^- \rightarrow H_2$$

The hydrogen gas escapes and is seen as tiny bubbles escaping from the electrolyte at the cathode, as seen in figure 10.8. On the other electrode, the anode, we also observe escaping bubbles. At this electrode, electrons are being taken out of the electrolyte

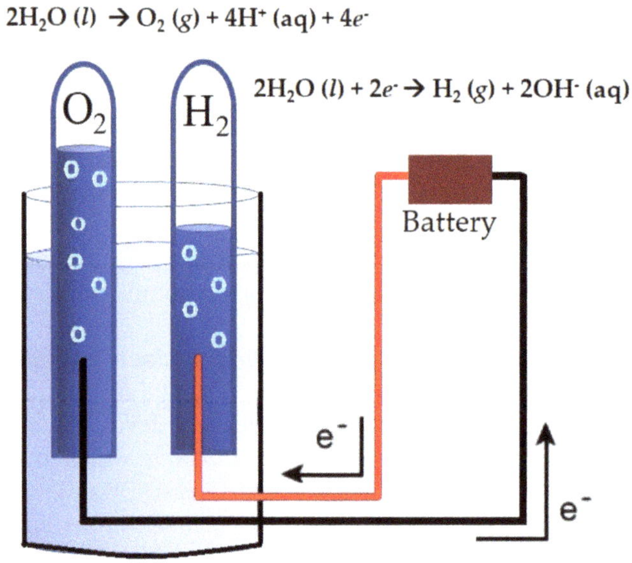

Figure 10.8. Electrolysis of water. Water decomposes into oxygen and hydrogen gas during electrolysis. The volume of hydrogen gas collected is twice the volume of oxygen gas collected, due to the stoichiometry of the reaction.

and fed back into the battery. The electrons come from negatively charged hydroxide ions (OH^-), which have an excess of electrons.

$$4OH^- - 4e^- \rightarrow O_2 + 2H_2O$$

This equation says that electrons are taken away from hydroxide ions and oxygen gas is produced in the reaction. The oxygen gas bubbles are given off at the anode, while the extra water formed as a result of the reaction remains behind in the electrolyte. The overall chemical reaction that occurs during the electrolysis of water is:

Anode (oxidation): $2H_2O$ (liquid) $\rightarrow O_2$ (gas) $+ 4H^+$ (aq) $+ 4e^-$

Cathode (reduction): $2H_2O$ (liquid) $+ 2e^- \rightarrow H_2$ (gas) $+ 2OH^-$ (aq)

Overall reaction: $2H_2O$ (liquid) $\rightarrow 2H_2$ (gas) $+ O_2$ (gas)

This reaction says that one mole of liquid water produces one mole of hydrogen and half a mole of oxygen. That is, the number of hydrogen molecules produced is two times the number of oxygen molecules produced. Assuming equal temperature and equal pressure for both gases, the produced hydrogen gas has twice the volume of the produced oxygen gas. The number of electrons pushed through the water is two times the number of hydrogen molecules generated and four times the number of oxygen molecules generated.

In a nutshell, we can say that electrons from the battery are added to the hydrogen ions in the electrolyte, which changes them into hydrogen gas, whereas, electrons are taken from hydroxide ions in the electrolyte and transferred to the battery. Over time, water molecules are broken down to form hydrogen and oxygen molecules.

According to standard reference tables, the change in enthalpy (ΔH) for one mole of water is 286 kJ. This is the amount of heat needed to form 1 mole of water from hydrogen and oxygen molecules. Recall, from our definition of enthalpy in chapter 3 section 3.5, some part of 286 kJ will be needed to push the atmosphere away to make room for the gases produced. The calculated value for that energy is ($p\Delta V$) = 4 kJ, whereas the rest of the energy (282 kJ) remains within the system.

The question that now arises is do we have to supply '*all*' of the energy as work or is there another way to add it to the system? The change in entropy of the system helps us here. Suppose you begin with 2 moles of water, and know from standard reference tables, $S_{\text{water}} = 70\ \text{J K}^{-1}$, $S_{H_2} = 131\ \text{J K}^{-1}$ and $S_{O_2} = 205\ \text{J K}^{-1}$. Then in the reaction, we get,

$$2H_2O\ (\text{liquid}) \rightarrow 2H_2\ (\text{gas}) + O_2\ (\text{gas})$$

Using this reaction and values of entropies,

$$2(S_{\text{water}}) \rightarrow 2(S_{H_2}) + (S_{O_2})$$
$$2(70\ \text{J K}^{-1}) \rightarrow 2(131\ \text{J K}^{-1}) + (205\ \text{J K}^{-1})$$

This means the change in entropy of the system (ΔS) is, [2(131) + 205] − 2(70) = +327 J K^{-1} and the entropy of the system '*increases*'. The maximum amount of heat that can enter the system at room temperature is [$T\Delta S$ = (298 K)(327 J K^{-1}) = 97.5 kJ]. The amount of energy that must enter as electrical work must be (2(286 J K^{-1}) - 97.5) = 474.5 kJ. This value (474.5 kJ) is the Gibbs free energy, i.e., it is the minimum '*other work*' required to make the reaction possible,

$$\Delta G = \Delta H - (T)(\Delta S)$$
$$474.5\ \text{kJ} = 572\ \text{kJ} - (298\ \text{K})(327\ \text{J K}^{-1})$$

In general, the values of ΔG are included in data tables in most chemistry textbooks, which saves us the hassle of solving for them ourselves every time!

10.6.3.2 Other applications of electrolysis

There are many other practical applications of electrolysis that include,

1. **Copper refinement:** Electrolysis can be used for refining copper. Purity of 99.999% is usually required in electrical components and electrolysis is the easiest way to produce this amount of purity in copper. An electrolytic cell for refining copper contains very pure copper at the cathode and impure copper at the anode, and uses copper sulfate as the electrolyte. When the anode and cathode are connected to a battery, electrons flow into the cathode, where they combine with copper ions (Cu^{++}) in the electrolyte, as,

$$Cu^{++} + 2e^- \rightarrow Cu^0$$

Pure copper metal (Cu^0) is formed at the cathode. Copper atoms (Cu^0) at the anode lose electrons and become copper ions (Cu^{++}) in the electrolyte, as,

$$Cu^0 - 2e^- \rightarrow Cu^{++}$$

Any impurities in the anode are just left behind, and nearly 100% pure copper builds up on the cathode and can be extracted easily.

2. **Isolating elements from compounds:** Electrolysis is used to break down very stable compounds. For example, aluminum oxide is a very stable compound and is the main source of aluminum, which is in common usage, from pots in our kitchen to space ships.

   Extraction of aluminum from aluminum oxide is very difficult. However, in 1886 CE a 21 year-old American student, Charles Hall at Oberlin College in Ohio, discovered that electrolytic processes can be used to extract aluminum from aluminum oxide. Hall found that melting aluminum oxide in a certain way forms ions of aluminum and oxygen. He found pure aluminum metal deposited at the cathode, which can be extracted from there. Gas bubbles of oxygen were found at the anode as they escaped from the anode.
3. **Electroplating:** Electroplating is a process of putting a very thin coating of expensive metals, such as silver and gold, on the surfaces of cheaper metals. Jewelers do this in the attempt to keep jewelry prices affordable.

   In silver plating, a solution of silver cyanide (AgCN) is used as the electrolyte. The object to be plated, say a diamond ring, is used as the cathode and a bar of silver metal is used as the anode. Connecting a battery allows the electrons to flow into the cathode where they combine with silver ions (Ag^+) in the electrolyte to form silver atoms (Ag^0). These silver atoms plate out as a thin coating on the cathode, whereas, at the anode, silver atoms give up electrons and become silver ions in the electrolyte. Silver is hence cycled from the anode to the electrolyte to the cathode, where it is plated out.

10.6.4 Fuel cells

A fuel cell is a device that converts chemical energy into electricity through a chemical reaction. The first fuel cells were invented in 1838 CE, but the first commercial fuel cell was developed for NASA space programs over a hundred and fifty years later. Nowadays, fuel cells have many applications, from cars to submarines and as backup power for commercial and residential buildings.

There are many different types of fuel cells but all of them have basically the same structure. They all consist of an anode, a cathode, and an electrolyte between the anode and cathode as shown in figure 10.9. Usually, very fine platinum powder is used as an anode, which breaks down the fuel into electrons and ions. Nickel is used as a cathode and it turns the ions into waste chemicals like water or carbon dioxide. A stand-alone fuel cell produces a small voltage of about 0.6 V. However, if we need

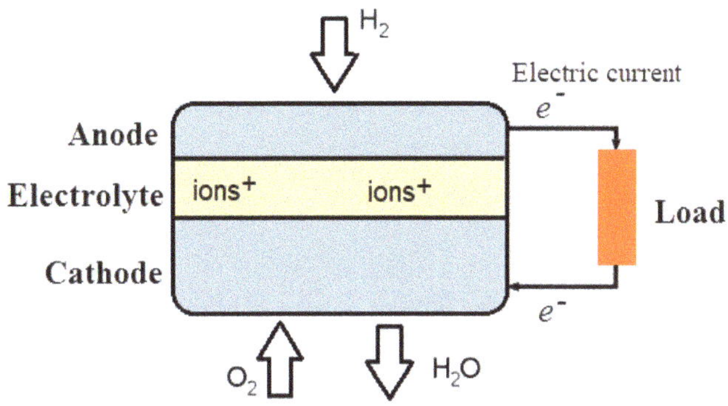

Figure 10.9. Fuel cell.

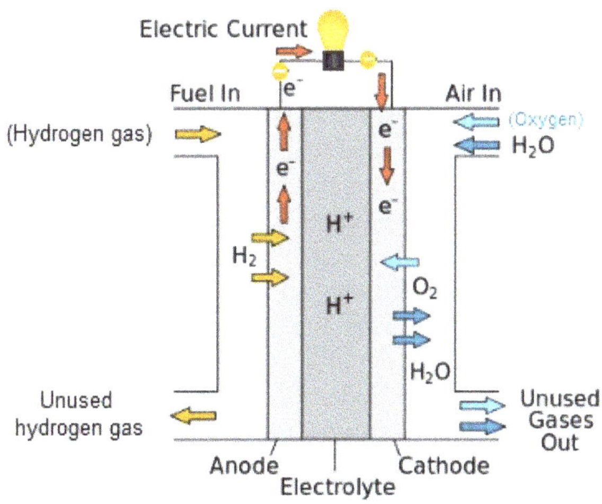

Figure 10.10. Hydrogen fuel cell.

higher output voltage, fuel cells can be combined in series, and in parallel to get a higher current output. These designs are called '*fuel cell stacks*'.

The design of a fuel cell depends on the type of fuel cell and is very complex. However, some basic components are found in many fuel cell systems: like a fuel cell stack, fuel processor, power conditioners, air compressors and humidifiers. Common types of fuel types include polymer electrolyte membrane (PEM) fuel cells, direct methanol fuel cells, alkaline fuel cells, molten carbonate fuel cells, etc.

Hydrogen and oxygen can be combined in a fuel cell to produce electrical energy. Fuel cells differ from a battery in that the fuel (H_2 and O_2) is continuously supplied as seen in figure 10.10.

We can run the process of electrolysis in reverse, as a controlled reaction between H_2 and O_2 to produce water (H_2O). Electrolysis extracts about 237 kJ of electrical work for each mole of H_2 consumed. The entropy of the gases decreases by 49 kJ mol^{-1} since there are a fewer number of water molecules than the initially combining numbers of

H_2 and O_2 molecules. The Second law of thermodynamics states that the total entropy cannot decrease in the reaction, hence the excess entropy must be expelled to the environment as heat. The calculated efficiency of an ideal fuel cell is therefore

$$\eta = \frac{W_{\text{net}}}{Q_{\text{in}}} = \frac{237\ \text{kJ}}{286\ \text{kJ}} \times 100\% = 83\%! \tag{10.67}$$

This efficiency is far greater than the ideal efficiency of a heat engine, which sounds great. However, in practice, the waste heat is much higher and the efficiency is much lower, but a typical hydrogen fuel cell still beats almost any other kind of engine.

10.6.4.1 Hydrogen fuel cell vehicles

Vehicles that use hydrogen fuel cells have zero emissions and run on compressed hydrogen via a '*fuel cell stack*'. The fuel cell stack produces electricity to power the vehicle. The fuel cells used in cars are polymer electrolyte membrane (PEM) fuel cells, which are also called '*proton exchange membrane fuel cells*'. Each fuel cell is a combination of a cathode, anode, and a proton exchange membrane (*pem*). The PEM membrane allows only the positively charged ions to pass through it on their way to the cathode. The negatively charged electrons must travel along an external circuit to the cathode, which will create an electrical current that can be used to power the engine. Compressed hydrogen, from an external tank enters the fuel cell at the anode side.

Oxygen from the surrounding air enters the cathode side of the fuel cell. As a hydrogen molecule encounters the membrane, a catalyst splits it into a proton and electron. The proton then goes through the fuel cell stack and the electron goes to an external circuit, which delivers current to the electric motor for the motion of a vehicle. The proton gets to the cathode side via the fuel cell, where the proton and electron join again. These then combine with oxygen to form water, which is the only emission of the car.

Concept questions 10.3

1. Why are there no degrees of freedom (f = 0) at the triple point of a pure substance? Explain in your own words.
2. Why does the freezing point of a pure liquid decrease when a solute is added to it? Your explanation should include the effects that solutes have on the chemical potential of a solvent.
3. We know that diamond becomes more stable than graphite at about 15 000 atm. How is diamond more stable than graphite when it has less entropy?

Exercises 10.3

1. How much pressure will you have to apply to one gram of ice to make it melt at −1 °C? Hint: Density of ice is 917 kg m^{-3}, density of water is 1000 kg m^{-3}, latent heat of 1 g of ice is 333 J.

2. Geochemists often use (kJ kbars^{-1}) instead of m^3 as units for volume. Show that these are equivalent units.
3. What is the dew-point when the vapor pressure is 12.29 mbar?

Further reading

- Kittle C and Kroemer H 1980 *Thermal Physics* 2nd edn (San Fransisco, CA: W.H. Freeman and Company)
- Reif F 1985 *Fundamentals of Statistical and Thermal Physics* (Singapore: McGraw-Hill International Edition)
- Schroeder D 2000 *Thermal Physics* (Reading, MA: Addison-Wesley Longman)
- Sears F and Salinger G 1975 *Thermodynamics, Kinetic Theory and Statistical Thermodynamics* 3rd edn (Reading, MA: Addison-Wesley Publishing Company)
- Van Ness H C 1969 *Understanding Thermodynamics* (New York: McGraw-Hill Book Company)

Section II

Statistical thermodynamics

Chapters 11–13

Similar to section I, I will subdivide the study of statistical thermodynamics into the three main subdivisions for each concept:

1. Historical development of concepts of statistical thermodynamics.
2. Formalism developed for statistical mechanics.
3. Applications of statistical mechanics

Statistical thermodynamics in a nutshell

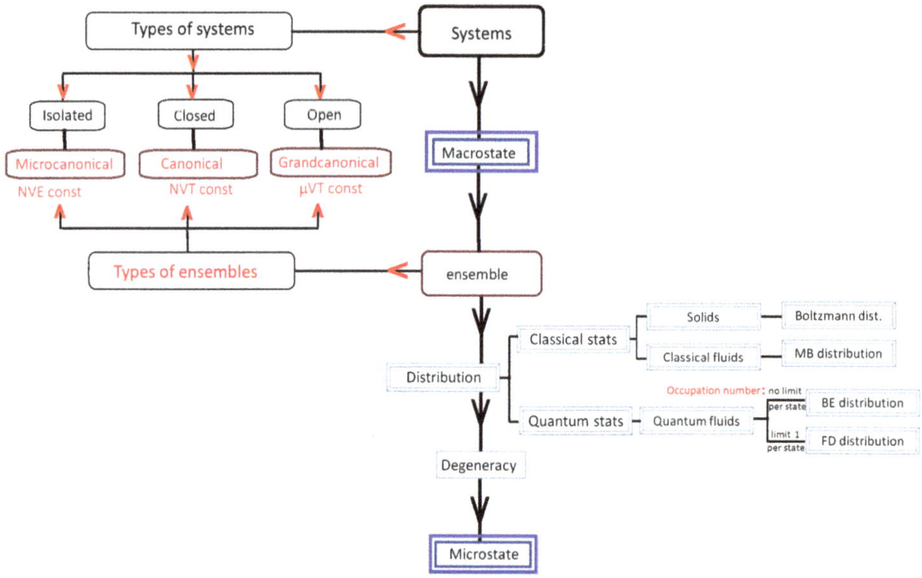

Chapter 11

Statistical mechanics and thermodynamics

11.1 Types of mechanics

Classical thermodynamics deals only with the conclusions drawn from experimental laws, and it makes no hypothesis about the actual nature of matter. This means classical thermodynamics is a purely empirical science. Thermodynamical principles predict relationships between properties of matter, such as the differences between specific heat capacities, and it predicts the variation of these quantities with pressure. However, it is not possible to derive the absolute magnitude of heat capacities using classical thermodynamical considerations alone.

To overcome this limitation of classical thermodynamics, the 'molecular theory of gases' was developed around the mid-19th century, which states that matter is not continuous but is rather made up of particles called '*molecules*'. To predict the properties of matter, starting with molecular theory, two distinct theories were developed:

1. The first theory was called the '*kinetic*' or '*dynamic*' theory of matter. The kinetic theory of matter[1] says that all matter consists of many tiny particles (that could be either molecules, atoms or ions) and that these tiny particles are physically spaced away from each other with gaps between them. In fluids, which include materials in the form of gas or liquid, these tiny particles are in a continuous and random state of motion, where the speeds of particles are determined by the amount of energy they have and their relationship to other particles around them.

   The kinetic theory applies the laws of mechanics to individual particles in the system under consideration. From these laws, the kinetic theory derives mathematical expressions for the properties of state, such as the internal energy, specific heat capacity and the pressure for fluids, etc. The kinetic

[1] Kinetic theory of matter is also called particle theory.

doi:10.1088/978-0-7503-3083-1ch11

theory of matter also explains why matter exists in different phases (solid, liquid or gases) and how matter changes from one phase to the other.

2. The second theory that was developed was called '*statistical thermodynamics*'. The statistical thermodynamic theory ignores the detailed considerations of molecules as individuals and applies probability considerations to the large number of molecules that make up the substance.

Both kinetic theory and statistical thermodynamics were first developed under the assumption that laws of classical mechanics could be applied to all particles without requiring any modification. However, it soon became apparent through experimentation that this was not the case. Classical laws that apply to large-scale systems failed for small scale systems, which led to the development of quantum theory and hence quantum mechanics.

The modern field of mechanics is divided into two basic types: classical mechanics and quantum mechanics. Both types of mechanics use similar approaches to problem-solving as we see here,

1. In both types of mechanics, we express a complete state of the system as a function of time. In classical mechanics a complete state can be treated as if it is located at the center of mass of the system, whereas in quantum mechanics a complete state is just the pure quantum state vector[2].
2. For both classical and quantum mechanics we can write an equation of motion that expresses the complete state as it moves forward in time. For classical mechanics that describes the state of a system as a function of time there is Newton's force equation given as ($\Sigma\vec{F} = m\vec{a}$), where m is the mass and $\vec{a}$ is the corresponding acceleration, whereas in quantum mechanics Schrödinger's equation ($i\hbar\frac{\partial\psi}{\partial t} = -\frac{\hbar^2}{2m}\nabla^2\psi + V(r)\Psi$)[3] is used to describe the state as a function of time.

However, both classical and quantum mechanics, in practice, do not tell us at a microscopic level the simultaneous position and velocities of each molecule while carrying out processes at the human scale (for example, when performing a chemical reaction)[4]. Statistical mechanics fills this gap between the laws of mechanics and the practical experience of incomplete knowledge, by adding some uncertainty about the state of the system, as we shall see in the following sections.

11.2 Development of statistical thermodynamics

The methods of statistical thermodynamics were first developed by Josiah Gibbs (1839–1903 CE) in the United States and Ludwig Boltzmann (1844–1906 CE), an

[2] A state vector is not a property of a physical system, but is an experimental procedure for preparing or testing one or more physical systems.

[3] I will not describe the details of Schrödinger's equation. Please refer to a quantum mechanics textbook for the details.

[4] Please note, that in principle they do tell us these things, however, as a practical matter we cannot do it because of the exhaustive calculation and computations requirements to do that.

Austrian physicist and philosopher, working in Germany in the latter half of the 18th century. Bose (Satyendranath Bose, 1894–1974 CE), Einstein (Albert Einstein, 1879–1955 CE), Fermi (Enrico Fermi, 1901–1954 CE), and Dirac (Paul Dirac, 1902–1984 CE) all modified Boltzmann's original ideas of statistical thermodynamics to clear up some features of Boltzmann statistics that needed some modification. This work lead to the development of the fields of quantum mechanics and statistical mechanics.

Statistical mechanics, also called statistical thermodynamics, does not deal with the detailed considerations of systems. It rather takes advantage of the fact that in the case of a system composed of a large number of particles the average properties of like-molecules can be calculated without knowing any information about individual molecules.

The primary goal of statistical thermodynamics is to derive the classical thermodynamics of materials in terms of the properties and interactions of the constituent particles. Statistical thermodynamics provides a connection between the macroscopic properties of materials in thermodynamic equilibrium, and the microscopic behaviors and motions occurring inside the material. A good way to think about this is to imagine yourself drinking coffee and reading a book inside a moving train. Newtonian (classical) mechanics will treat the train as a particle and will not be concerned with you inside the train reading a book. Statistical mechanics cares about each individual component of the train, you, the train, the book and the coffee that you are drinking. However, statistical mechanics will not care about the number of words in the book and the number of molecules in your coffee!

As introduced above, statistical thermodynamics takes advantage of the fact that in a system composed of a large number of particles, the average properties of like-particles are calculated without knowing detailed information about individual molecules themselves. A good way of imagining this is that life insurance companies can predict to a high degree of accuracy life expectancies of people in any given year without necessarily knowing the individual health of each person. Statistical mechanics can be applied to many different systems, from molecules, to elastic waves in a solid, to wave functions in quantum mechanics. In this text, I shall use the term 'particle' interchangeably for all possible systems.

11.2.1 Introduction to statistical mechanics

Statistical mechanics can be called the most fundamental field of physics because it provides us with techniques that can be applied to physical systems. It is important to note that statistical mechanics does not give us any natural laws, like the Newton's laws of motion in classical mechanics or the Schrödinger equation in quantum mechanics. Statistical mechanics provides us with a set of techniques that we can use to understand the microscopic nature of our systems.

There are two main concerns that arise when studying microscopic systems: one is that macroscopic systems are generally made up of an enormous number of particles; and two, in microscopic systems the particles are in complete chaotic motion that is pretty much unpredictable.

We will see in detail later that dealing with individual molecules quickly becomes tedious and keeping track of each molecule in a sample of just 1 mole of substance (of the order of 10^{23} molecules) gets to be impossible. This is where statistical mechanics comes to our rescue when we still want to use our knowledge of the molecular (microscopic) nature of matter to understand the macroscopic nature of matter. This understanding then leads to the development of statistical mechanics and then further to the development of statistical thermodynamics.

The most obvious method to develop statistical mechanics is to apply statistics to the properties of each molecule in our system. However, this treatment is not general because the molecules interact with each other and some of their properties change as a result. For example, the kinetic energies are unique to the individual particles (molecules) but the potential energies are due to the forces acting between molecules and hence are shared among all the molecules in the system.

In chapter 4, we developed the statistical average of kinetic energy and found that it equals the internal energy of the system. However, we also saw that we cannot calculate the internal energy by simply averaging the kinetic energy for real gases. So the concern is, if we cannot deal with individual molecules and we do not have the computational power to deal with the entire system, then what is the alternative? The solution is to instead deal with a large collections of macroscopically identical systems and call the collection an '***ensemble***'. Statistical mechanics uses *statistical ensembles* to calculate the state of a system at any time. An ensemble represents various possible states a single system could be in. More broadly, a statistical ensemble can be defined as a probability distribution over all possible states of the system. We will discuss statistical ensembles in section 12.3.1.

11.2.2 Equilibrium statistical mechanics

Statistical thermodynamics is also called '*equilibrium statistical mechanics*' and it provides a connection between the macroscopic properties of a material and its microscopic properties at thermodynamic equilibrium. Thermal equilibrium is the point where the ensemble as a whole stops evolving and does not change anymore. This state is also called the '*steady state*'. Please note statistical equilibrium does not mean the particles have come to rest but rather it means that the ensemble no longer evolves with respect to time.

11.3 Statistical description of physical systems

In classical physics, most quantities are continuous as they can take on any value in a continuous range. Consider a box resting on a ramp, as shown in figure 11.1, the gravitational potential energy (GPE) of this box is continuous since the box can be placed anywhere on the ramp. The GPE can have any value from the maximum GPE at the top of the ramp to the minimum GPE at the bottom of the ramp. However, the same box resting on a staircase can only be placed on top of every stair and never in between the stairs. In this case GPE can have only certain allowed values. Quantities are called '*quantized*' when their allowed values are limited to a discrete set of values.

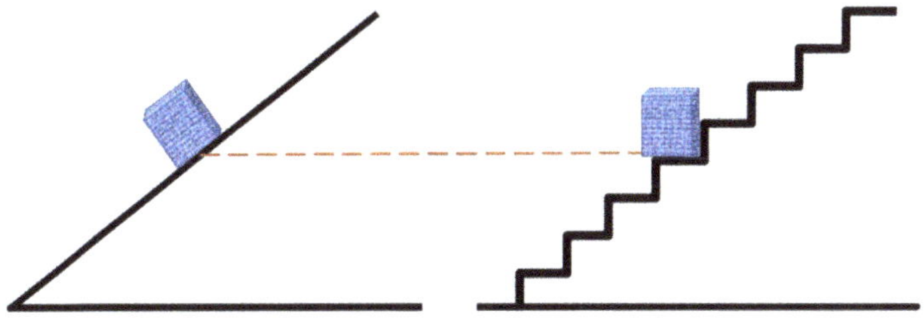

Figure 11.1. (a) A box resting on a ramp, with continuous gravitational potential energy. (b) The box resting on a staircase with quantized gravitational potential energy.

In classical mechanics, while the box is being moved from one step to another its GPE passes through all intermediate values. By contrast, in quantum mechanics, the box will not pass through intermediate values; its GPE will suddenly change from one GPE value to another as we go up or down the steps. This means rules of classical mechanics do not apply on a microscopic scale and hence the rules of classical mechanics must be replaced by rules of quantum mechanics at the microscopic scale.

11.3.1 Energy states and energy levels in quantum mechanics

In quantum mechanics the energy of a particle cannot take on any arbitrary value and can only exist in a particular state with only a specific energy. The energy is said to be '*quantized*'. Since this is a statistical mechanics textbook, I will not go into details of quantum mechanics and quantum mechanical rules but will state them as necessary. I will leave it to you to study a quantum mechanics textbook for the details. Generally, for this text, it is enough to know that energy is quantized in quantum mechanics.

So let's start with a quick discussion of quantum mechanics. Quantum mechanics, also called '*wave mechanics*', uses Schrödinger's time independent equation ($H\psi = E\psi$, where ψ is an eigenstate, E represents the eigenvalue and H is the Hamiltonian[5]). Schrödinger's equation was introduced by Erwin Schrödinger in the mid-1920s, to solve problems at the quantum scale. This equation, for most problems, is equivalent to the wave equation set up in a string of length (L) fixed at both ends, called a '*standing wave*'.

A standing wave is a particular type of a wave that is set up in a string such that it has a node (N), a point of zero amplitude, at each end and an anti-node (A), a point of maximum amplitude at the center, as shown in figure 11.2. From classical mechanics we know that a standing wave can only vibrate in a steady state when it has a certain allowed frequency. The allowed frequencies in a standing wave are integral multiples of the fundamental frequency given as ($f_n = nf_0$), where f_0 is the fundamental frequency.

[5] A Hamiltonian ($H = KE + PE$) is an operator corresponding to the sum of the kinetic plus the potential energies for all the particles in the system, and it represents the total energy of the system in most situations.

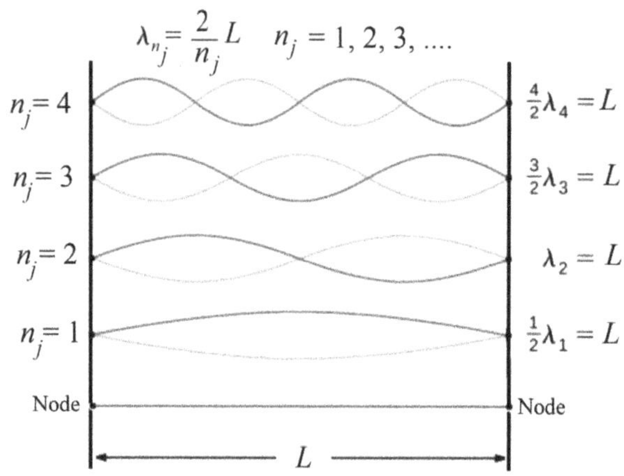

Figure 11.2. Standing waves created on a string of length (L). The standing wave patterns possible on the string are known as the normal modes. A node, with zero amplitude, occurs at each end of the string.

For a steady-state condition, when the system does not evolve with time, there must be an integral number of antinodes and the distance between the antinodes must be one-half of the wavelength. The wavelengths (λ) of the possible standing waves are given as $\lambda_1 = 2L$, $\lambda_2 = L$, $\lambda_3 = \frac{2}{3}L$ and so on, which can be generalized as,

$$\lambda_j = \frac{2}{n_j}(L) \tag{11.1}$$

where n_j (= 1, 2, 3, ...) is an integer and is equal to the number of anti-nodes.

A standing wave is equivalent to two waves propagating in opposite directions. Each wave is reflected at the ends of the string as they traverse through the string and the superposition of these two waves creates a standing wave. This is analogous to the motion of a particle moving back and forth, as we saw in chapter 4, section 4.3, along a straight line of length (L) in a box between two walls suffering elastic collisions with the walls and being reflected back. According to quantum mechanics, a standing Schrödinger wave is in fact equivalent to this particle, and the wavelength (λ) of this standing wave is related to the momentum (p) of the particle by the equation,

$$p\lambda = h \tag{11.2}$$

where h is called Planck's constant, with numerical value, $6.626\,2 \times 10^{-34}$ Js.

The momentum of the particle (p) in quantum mechanics is therefore permitted to have only the particular set of values given by the formula,

$$p_j = n_j \frac{h}{2L}. \tag{11.3}$$

In this case, n_j (= 1, 2, 3, ...) are integers called quantum numbers. In a cubical box of side L in rectangular coordinate systems x, y, z, the components of its momentum are permitted to have only the values,

$$p_{xj} = n_{xj}\frac{h}{2L};\ p_{yj} = n_{yj}\frac{h}{2L};\ p_{zj} = n_{zj}\frac{h}{2L}; \tag{11.4}$$

where n_{xj}, n_{yj} and n_{zj} are quantum numbers that can have values, (= 1, 2, 3, …). Each set of these quantum numbers therefore corresponds to a certain '*direction*' of the momentum. The resultant momentum (p^2_{total}) corresponding to quantum numbers, n_{xj}, n_{yj} and n_{zj}, is mathematically given as,

$$\begin{aligned} p^2_{\text{total}} &= p^2_{xj} + p^2_{yj} + p^2_{zj} = \left(n^2_{xj} + n^2_{yj} + n^2_{zj}\right)\frac{h^2}{4L^2} \\ p^2_{\text{total}} &= n^2_j\frac{h^2}{4L^2} \end{aligned} \tag{11.5}$$

Note, the values of n^2_j define the state of the system. The kinetic energy (KE) of a particle of mass (m), speed (v) and momentum (p) can hence be written as,

$$KE = \frac{1}{2}mv^2 = \Sigma_j\frac{p^2_{\text{total}}}{2m} = \Sigma_j n^2_j\frac{h^2}{8mL^2} \tag{11.6}$$

The values of n_j in 3D space (n_x, n_y, n_z) define the state of the particle, and the energies corresponding to the different possible values of n^2_j are the possible energy levels. It is worth noting here that energy levels depend only on the values of n^2_j and not on the individual values of n_x, n_y, n_z. In other words, the quantum mechanical energy depends only on the magnitude of the momentum (p_j) and not on its direction. Please note that here I used the representation in Cartesian coordinates. If we use other coordinate representation, we will need to adjust as necessary. For example, it will be (r, θ, ϕ) in spherical coordinates and (r, ϕ, z) in cylindrical coordinates.

In general a number of different states that corresponds to different directions of momentum may have the same energy. The energy level is then said to be '***degenerate***' and in general the symbol (g_i) is used to designate the degeneracy of level (i), which is the number of states having the same energy (E_i).

11.3.1.1 Schematic interpretation

A good way to think about energy levels is to think about apartments in a high-rise building with multiple floors. In this example, energy levels correspond to the floors at each elevation while the energy states correspond to the apartments themselves. The degeneracy (g_j) of the level (j) is the number of apartments on each floor. The number of people living in the building are distributed among the various apartments. Here the number of people in any one apartment correspond to the number of particles in a particular *quantum-state*.

It is important to note that the people in the apartments on any floor may be in different apartments (states), but they all have to same energy. That means the particles in the same energy level have the same energy. The total number of people in the apartments at any level (j) is called the '***occupation number***' (N_j) of that level.

Occupation number in quantum mechanics and quantum statistics is a number that indicates the extent to which quantum states are filled by the particles of a quantum-mechanical system consisting of many identical particles. The sum of the occupation numbers (N_j) over all levels tells us the total number of particles (N) in our system. Mathematically given as,

$$\Sigma_j N_j = N$$

and since all particles that are in states at energy level (j) have the same energy, (E_j), the total energy of the particles in level (j) is,

$$E = N_j \times E_j.$$

This can be simplified by the example shown in figure 11.3, where the four energy levels correspond to the four floors in an apartment building. The number of apartments on each floor give the degeneracy of the state and the number of people on each floor gives us the occupation state of each energy level.

11.3.2 A free particle in a cubical box

For a free particle in a cubical box of length (L) in Cartesian coordinates (x, y, z), we can write its energy from equation (11.6), as,

$$E_i = n_i^2 \frac{h^2}{8mV^{\frac{2}{3}}} \tag{11.7}$$

where V is the volume of the cube and $n_i = n_x, n_y, n_z$ along the x, y, and z axes, respectively. Please note here that this result can be applied to any container of any shape as long as the dimensions of the container are large as compared to the wavelength of the Schrödinger wave. Equation (11.7) states that the energy of the ith level is directly proportional to the quantum number (n_j^2) and inversely proportional to the volume (V) of the container. Therefore, decreasing the size of the container will increase the energy of the ith level.

To clarify these concepts let's discuss here the first two energy levels for a free particle in a cubical box of length (L), level 1 and level 2.

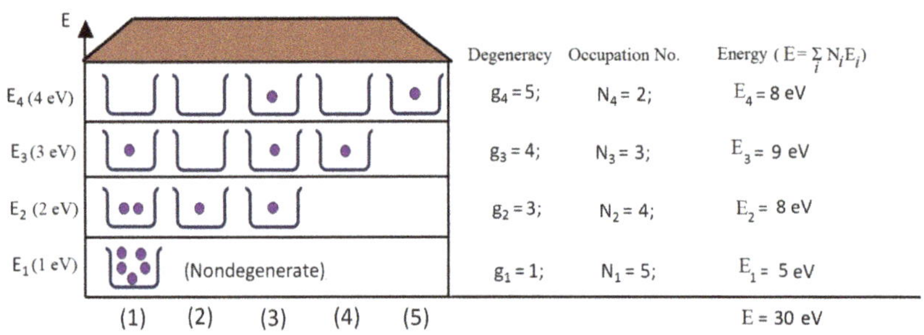

Figure 11.3. Schematic representation of energy levels, their degeneracies and their occupation numbers.

1. **Level 1:** Level 1 corresponds to the lowest energy level. The lowest energy level will have quantum numbers $n_x = 1$, $n_y = 1$, $n_z = 1$, in 3D. The possible states in level 1 are given in table 11.1.

   Then the total quantum number for level 1 will be, $n_j^2 = n_x^2 + n_y^2 + n_z^2 = 1^2 + 1^2 + 1^2 = 3$ and equation (11.7) can be written as,

$$E_1 = \frac{3h^2}{8mV^{\frac{2}{3}}} \tag{11.8}$$

   Since only one set of quantum numbers have this particular energy there can only be one possible state. This implies that the lowest energy level is '***non-degenerate***' with $g_1 = 1$ and x, y, and z components of the corresponding momentum (p_1) all equal to $h/2L$, given as,

$$p_{1x} = \frac{h}{2L}; \; p_{1y} = \frac{h}{2L}; \; p_{1z} = \frac{h}{2L}; \tag{11.9}$$

2. **Level 2:** Level 2 is the next level above level 1. In level 2, any one of the following states, shown in table 11.2 are possible. The quantum number for level 2 is given mathematically as, $n_2^2 = n_x^2 + n_y^2 + n_z^2$. Therefore, mathematically $n_2^2 = 6$, which means we can write equation (11.7) as,

$$E_2 = \frac{6h^2}{8mV^{\frac{2}{3}}}. \tag{11.10}$$

   Since three different states have the same energy, the degeneracy of level 2 is $g_2 = 3$ and x, y and z components of the corresponding momentum (p_2) of the first of the three states are,

Table 11.1. Possible states in level 1.

State	n_x	n_y	n_z
1	1	1	1

Table 11.2. Possible states in level 2.

State	n_x	n_y	n_z	n_j^2
1	2	1	1	6
2	1	2	1	6
3	1	1	2	6

$$\text{State 1:} \quad p_{2x} = 2\left(\frac{h}{2L}\right); \quad p_{2y} = \frac{h}{2L}; \quad p_{2z} = \frac{h}{2L}.$$

$$\text{State 2:} \quad p_{2x} = \frac{h}{2L}; \quad p_{2y} = 2\left(\frac{h}{2L}\right); \quad p_{2z} = \frac{h}{2L}.$$

$$\text{State 3:} \quad p_{2x} = \frac{h}{2L}; \quad p_{2y} = \frac{h}{2L}; \quad p_{2z} = 2\left(\frac{h}{2L}\right).$$

11.4 Description of a system in statistical mechanics

To start the discussion of the description of a system in statistical mechanics, let me start with asking a couple of questions. Given a physical system, how do you apply statistics to describe it? What does that even mean? As we talked about before, the main goal of statistical physics is to make a bridge between the over-elaborate detail of kinetic mechanics and the obscure generalities of classical mechanics.

Systems in statistical mechanics may be made up of many different particles or they may be made up of identical particles. When the system is made of identical particles, it is generally called an '***assembly***' or an '***ensemble***'. The description of the identical particle ensemble depends on the type of particles that make up the system. Identical particles can either be '*distinguishable*' or '*indistinguishable*' from each other. For example, in a solid molecules can be labeled using spatial recognition (actual position of the particle in the crystal lattice). Therefore, molecules in a solid can be distinguished based on their actual physical locations in the lattice structure and the ensemble made of solid particles is considered to be made up of '*distinguishable particles*'. On the other hand, in an ensemble of gaseous particles, molecules are not fixed in place but are rather in a constant motion so it is not possible to label molecules to distinguish them from one another; here the particles are considered to be completely '*indistinguishable*' from each other.

When the particles are indistinguishable, the total number of particles in each energy state defines the '*microstate*' of the ensemble. When the particles are distinguishable, '*microstate*' of the ensemble is defined by specifying the energy state of each particle. In this case we must specify not only how many particles are in each state but which particular particles they are. However, whether the particles are distinguishable or not, the specification of the number of particles in each energy level defines a '*macrostate*' of the ensemble.

1. **Microstate:**
   The system can exist in a discrete but possibly infinite sets of microstates. A microstate defines the values of all possible microscopic variables. For classical systems, a microstate defines the position and momentum of every particle in the system. For a typical statistical system, the number of microstates is very large and they give more detail than we usually worry about in statistical mechanics. For example, in our piston and cylinder

system from chapter 6, we cannot keep track of the exact position and momentum of every gas particle at any given time. Additionally, even if we could somehow keep track of every particle, we may not care about that much detail in our system.

Instead, we are usually interested in a small number of macroscopic variables that we can keep track of and can measure. These variables usually are ones that have practical importance. For example, in our piston and cylinder system we are usually interested in the total energy of the system, or the total number of gas molecules, or the volume of space it occupies, etc, rather than keeping track of what each gas molecule is up to.

2. **Macrostate:**
The basic task of statistical physics is to take a system in a well-defined thermodynamic state and compute its various thermodynamic properties from an assumed microscopic model. A macrostate is another way of defining a thermodynamic system which contains just enough information for its thermodynamic state to be defined and no more information than that.

A macrostate is defined by specifying the value of every macroscopic variable, for example temperature, pressure, volume and density, etc. There may be a huge number of microstates that all correspond to the same macrostate. For example, if you are interested in measuring the total energy of the cylinder and piston system, there are a large number of arrangements of individual gas particles that result in the same total energy. This means that in this case all you can guarantee is that your system is in one of these arrangements. However, you may not have any idea of which particular microstate (or configuration) the system is in.

Worked example 11.1. Before continuing let me summarize to clarify the difference between the types of specifications of a system, let's consider the example of a school (figure 11.4).

Macrostate: The entire school is an example of a macrostate.

Ensembles: System of identical particles, in this case the number of students in various classes, say the students in PHYS-404 versus PHYS-305 classes.

Distribution: Typical distribution would give the number of students achieving grades in a certain range, say A in range 90%–100%; B in range 75%–90%, C in range 60%–75%, and so on. In a distribution, the identity of each student will be hidden but we will retain statistical detail of the macrostate.

Degeneracy: Each distribution will be divided further as:

1. A (in range 95%–100%) and A- (in range 90%–95%); with degeneracy = 2
2. B+ (in range 85%–90%), B (in range 80%–85%), B- (in range 75%–80%); with degeneracy = 3.
3. C+ (in range 70%–75%), C (in range 65%–70%), C- (in range 60%–65%); with degeneracy = 3.

Microstate: Tells us about the grade of each student. There is nowhere to hide.

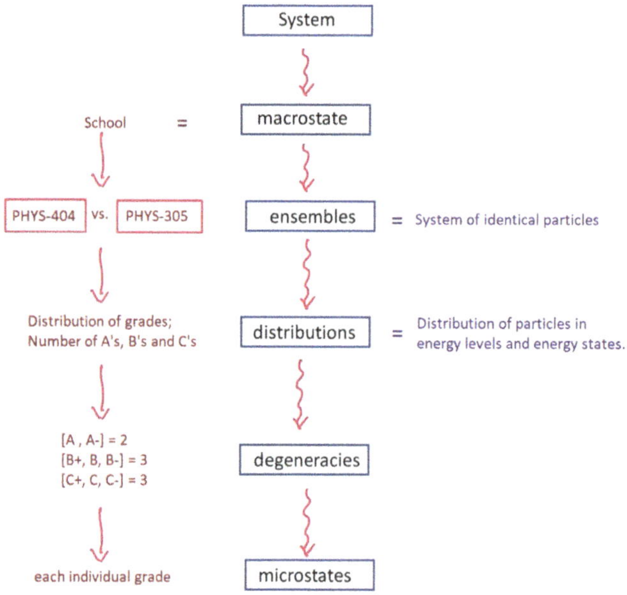

Figure 11.4. Vocabulary/description of a statistical system.

Table 11.3. Penny toss, where H = heads, T = tail.

States	Penny 1	Penny 2	Penny 3
1	H	H	H
2	H	H	T
3	H	T	H
4	H	T	T
5	T	T	T
6	T	H	T
7	T	H	H
8	T	T	H

Worked example 11.2. Suppose we have three distinguishable pennies and we toss them. How many possible outcomes are there? Assume that all coins are fair and the probability of each outcome is equally probable.

Solution: For this situation, we see in table 11.3 that there are ($2^3 = 8$) eight possible outcomes. Each of the eight outcomes is called a *microstate*. In general, to specify the microstate of a system we must specify the state of each individual particle, in this case the particular state of each coin. If we specify the state more generally by saying how many heads or tails there are, we call that a *macrostate*.

Of course, if you know the microstate of the system (say HTT) then you know its macrostate (in this case two tails). However, the reverse is not true. Knowing that there are two tails does not tell you the particular microstate, since there are three microstates (#4, #6, #8) corresponding to this macrostate. The number of microstates corresponding to a given macrostate is called **multiplicity of that microstate** (Ω).

Probability of any particular macrostate: In this example, the multiplicity is: Ω(3 heads) = 1, Ω(2 heads) = 3, Ω(1 head) = 3, Ω(0 heads) = 1.

The total multiplicity = total number of microstates = 1 + 3 + 3 + 1 = 8.

Then the probability of any particular macrostate is given as:

$$\text{Probability of } n \text{ heads} = \frac{\Omega(n)}{\Omega(\text{all})} \tag{11.11}$$

Here, the probability of getting the state with two heads and one tail = 3/8.

Generalizing for *N* coins If we have 'N' distinguishable coins, the multiplicity of the macrostate with 'n' heads will be given as:

$$\Omega(N, n) = \frac{N!}{n!(N-n)!} = \binom{N}{n} \tag{11.12}$$

which is the binomial coefficient, the number of choosing 'n' objects out of total 'N' objects.

Worked example 11.3. Suppose we have 20 equally weighted fair coins.

1. What are the possible outcomes (microstates) when tossing the coins? Hint: you do not have to list them all, just tell the total number of microstates.

   Solution: Each coin has two possibilities, head and tail. Coins are independent, hence the total number of microstates as,

   $$2^n = 2^{20} = 1048\,576$$

   which is over a million possibilities, hence there are over a million microstates!
2. What is the probability of getting the sequence [THTHTHTHTTHTHTH THTTT] once the coins are tossed?

   Solution: Probability of getting one particular microstate is,

   $$= \frac{1}{2^{20}} = \frac{1}{1048\,576}$$
3. What is the probability of finding 5 heads (in any order) out of 20 tossed coins?

   Solution: We can use equation (11.12),

   $$\Omega(20, 5) = \binom{20}{5} = \frac{20!}{5!(20-5)!} = 15\,504.$$

Probability of $\Omega(20, 5)$:

$$= \frac{15\,504}{1048\,576} = 1.5\%$$

4. What is the probability of finding 8 heads (in any order) out of 20 tossed coins?

   Solution: We can use equation (11.12),

$$\Omega(20, 8) = \binom{20}{8} = \frac{20!}{8!(20-8)!} = 125\,970.$$

Probability of $\Omega(20, 8)$:

$$= \frac{125\,970}{1048\,576} = 12\%$$

Worked example 11.4. You wish to choose three marbles at random from a box containing four marbles labeled A, B, C, D. Determine the number of permutations and combinations for this scenario.

Solution: In a simple sense, in mathematics, when the order of things do not matter, it is called a '**combination**', whereas, when the order does matter it is called a '**permutation**'. In other words, a permutation is an ordered combination. In this example, there are four marbles ($N = 4$) and we have to choose three ($M = 3$), then the number of permutations will be,

$$P(N, M) = \frac{N!}{(N-M)!} = \frac{4!}{(4-3)!} = 24.$$

The number of combinations,

$$C(N, M) = \frac{N!}{M!(N-M)!} = \frac{4!}{3!(4-3)!} = 4.$$

11.4.1 Density of states

The number of microstates corresponding to a macrostate is called the '*density of states*'. It is written mathematically as $\Omega = \Omega(E, V, \ldots)$. In statistical mechanics, microstates are always assumed to be discrete, that means Ω is also a discrete function. Each variable can only take on specific values, and it is only defined for those values. In practice, however, we usually treat Ω as a continuous function. As long as the allowed values are sufficiently close together, the total number of states in any given interval will be huge and we can simply think of Ω as the number of microstates per unit volume in the space defined by the macroscopic variables. That is why we call it the '*density*' of states.

Concept questions 11.1

1. In your own words explain the difference between a microstate, macrostate, distribution and ensemble. Can you think of an example to explain these concepts?
2. How do you calculate microstates?
3. How do you calculate macrostates?

Exercises 11.1

1. You have four fair coins that you can flip.
   (a) List all the possible outcomes (microstates) in a table.
   (b) List all the different 'macrostates' and their probabilities.
   (c) Compute the multiplicity using equation (11.12).
2. How many permutations of just 4 different letters chosen from the twenty six letters of the alphabet are there?
3. In a typical lottery, a player chooses 6 different numbers out of 49 possible numbers. You feel lucky this week and decide to enter the lottery by buying a ticket and selecting your six favorite numbers. What are your chances of winning all 6 numbers this weekend and becoming a multimillionaire?

11.5 Statistical entropy

It is thought (somewhat incorrectly) that thermodynamics developed before the development of atomic theory. In reality, both theories developed simultaneously. The first book on applied thermodynamics was published in 1859 CE by Rankine. The same year Maxwell published his first paper on the dynamical theory of gases. So we can say that thermodynamics, statistical mechanics, and kinetic theory of molecules developed together.

However, the similarities and interdependencies between thermodynamics and atomic theory were not clearly understood until much later. This led to Maxwell's famous problem called '*Limitations of the Second Law*' posed in his book titled *Theory of Heat* that was printed in 1872 CE. Maxwell introduced a thought experiment to illustrate the difficulties that arise when trying to apply the Second law of thermodynamics to molecules.

Maxwell and his thought experiment:
In his thought experiment Maxwell made up a fictional character which was later called '*Maxwell's demon*' by Lord Kelvin in 1874 CE. Maxwell assumed that this demon was capable of sorting molecules based on the temperature difference of the molecules themselves into hot and cold molecules. The Second law of thermodynamics (chapter 5) says that in an isolated system, the entropy of a system either remains the same or increases but never decreases. In creating a difference in temperature the demon apparently violates the Second law of thermodynamics by decreasing the entropy of the system.

According to the Zeroth law of thermodynamics (chapter 1), we know that when two objects of different temperatures are brought into thermal contact with each other, after a certain relaxation time, they both attain the same temperature and are said to be in thermal equilibrium with each other.

Maxwell assumed in this thought experiment that one container is divided into two sides, and called the right side, side A and the left side, side B. There is a trapdoor between the two containers which is guarded by our '*demon*' as seen in figure 11.5(a). He said the container is initially filled with a gas at temperature (T). The demon has the authority to open or close the trap-door at will, but is bound to follow the following rules:

1. When a faster-than-average molecule from side A hits the trapdoor, the demon opens it and allows it to go to side B.
2. When a slower-than-average molecule from side B knocks at the trapdoor, the demon lets it pass from side B to side A.

The demon follows these rules precisely and after a little while we observe that the average speed of the molecules in side B increases while the average speed of molecules in side A decreases. We know that the average molecular speed corresponds directly to temperature; this means the temperature on side A decreases while the temperature on side B increases, as seen in figure 11.5(b). This implies that our demon is actually a '*molecular-sorter*'. It allows molecules to pass through the trap door such that it produces a temperature difference. The gas on side A becomes colder while the gas on side B becomes hotter as time goes by. This is contrary to the Zeroth law of thermodynamics, which states that two objects with different temperatures will reach thermal equilibrium when they are brought into contact with each other.

It is worth noting that at this point, we could start a heat engine that operates between the two temperatures which could in principle deliver work continuously as long as the demon continues to work as shown in figure 11.6. In this system, we will get a heat engine that operates in a cycle and converts all the input heat into useful work. Maxwell imagined that his demon itself did no work, he imagined it as a reversible agent that releases energy when the frictionless trap door opened and recovers expended energy when the door closes. Please note that in this case the Second law of thermodynamics is still violated.

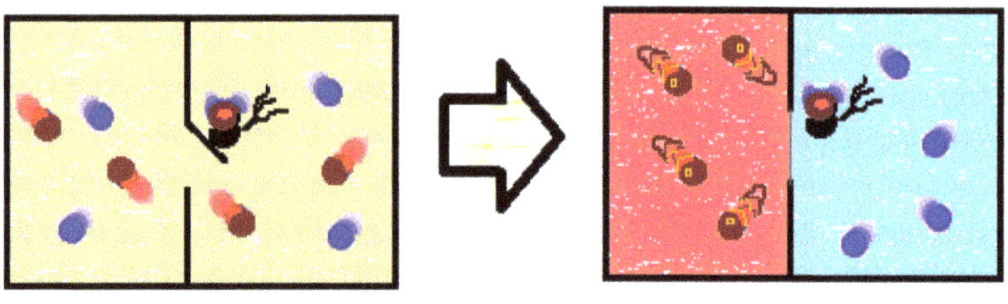

Figure 11.5. Maxwell's demon in action; red molecules are at higher temperature (and hence travel at a higher speed) and blue molecules are at lower temperature (and hence travel at slower speeds).

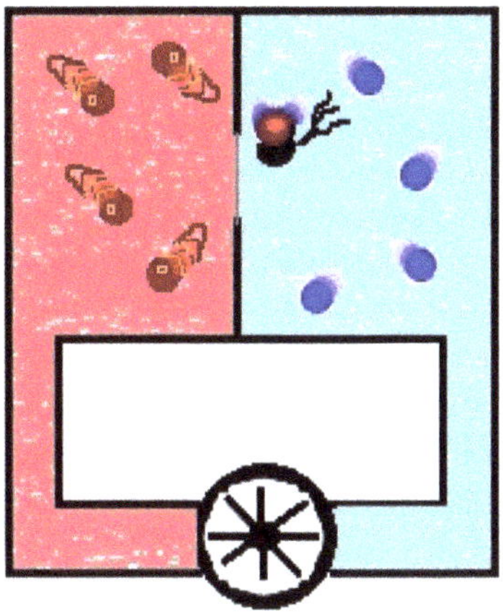

Figure 11.6. Perpetual engine with a Maxwell's demon.

Figure 11.7. Ninja at work.

There are many suggestions on how to build a device that violates the Second law, however, no one has ever been able to build one successfully. The question to think about is how should we deal with the '*molecule sorter*' called the Maxwell demon? The easiest way would be to say that there is '***no***' such being as a Maxwell's demon. But it is easier said than done, after all, the microscopic world is more mysterious than we know.

Maxwell postulated that his demon was intelligent and could deal with individual molecules. This might be our key to success. To do this, let's start by considering another scenario: let's start with our original container still divided into two parts, side A and side B, and a trapdoor in the middle. However, this time fill one side with gas and let the other side be a complete vacuum. This time suppose the trapdoor is guarded by a 'ninja' whose job is to keep an eye on the trapdoor and gas molecules, as seen in figure 11.7.

Experience tells us that if we open the trapdoor the gas expands and fills the total volume of the container. The question is, how long does the 'ninja' have to wait until the gas, on its own, returns to its initial location which had all particles in side A and a complete vacuum in side B. Since the gas molecules are in continuous random motion, we can conclude that there is a finite, albeit a very, very tiny, but not zero '**probability**' that at some time the original configuration is reproduced. Now if that happens the 'ninja' closes the trap door immediately and we get a state of lower entropy than the system had originally. Whether you believe this will actually happen or not, it represents a process seemingly at odds with the Second law of thermodynamics. It is also important to note that in this case the 'ninja' has to continuously monitor the system and close the trap door at just the right moment for this to work.

For both hypothetical processes presented above, with a 'demon' or the 'ninja', there are some common characteristics. Both processes use the idea of *molecular-ordering* and an *agent* to monitor the process. In the case of Maxwell's demon the ordering was done on the basis of the speed of the molecules and in the case of the ninja, ordering was done in the sense that molecules were collected from a larger region in a more restricted region of space.

These are only two ways out of many ways that ordering could be achieved. Ordering could be done on the basis of the size of molecules or mass of molecules when the system is made of a mixture of gases. In this case, the demon or ninja could open or close the trap-door in such a way as to separate the molecules on the basis of a particular '*type*' of molecules. In short, '**ordering**' is very important in statistical thermodynamics. Molecules left to their own devices will not become ordered and we need an 'external agent' (a 'demon' or a 'ninja') to do the ordering for us, which increases order in the system and reduces the entropy of the system. This leads us to the obvious conclusion:

'*Increasing order corresponds to decreasing entropy.*'

All we need is a method of expressing order or disorder in a quantitative way. Maxwell demons sort molecules on the basis of some particular '*information*', that is, the demon will not act until he knows some particular property of the molecules, for example, the speed or the size of the molecules. Even a ninja, as in the second case, just sits around and waits and while waiting must always be informed continuously about the locations of molecules. Otherwise, he would miss the opportunity to close the trapdoor at the correct time. There are two ideas that come out of these observations:

1. There has to be some connection between entropy and information.
2. The information gathering activities of the 'demon' and the 'ninja', may be the key to whether or not the 'demon' and the 'ninja' can do their work.

This link between information and entropy has been exploited and developed and it falls under the heading of '*information theory*', which is very important in the design and applications of communication systems. The fundamental equation of information theory is identical to the equation of entropy in statistical mechanics. This might lead to confusion but the quantity we calculate that tells us the information content of messages is also called '*entropy*'!

In statistical mechanics, we calculate the properties of matter by applying statistics to a large number of molecules. In information theory, we find the information-carrying capacity of communication systems by applying statistics to a large number of messages.

11.5.1 Does Maxwell's demon really violate the Second law?

In order to answer this question we have to consider a few things:

1. Is Maxwell's Demon alive?

   Answer: not necessarily. The demon is just a device that responds to the information signal, which means it can be automated for the tasks in a more efficient method than a living being. Additionally the demon has to be placed inside the system, mostly for long intervals of time. So the best Maxwell's demons are probably 'automated devices' and not alive.
2. How is a Maxwell demon activated?

   Answer: a sorting device that sorts the molecules cannot be bigger than the molecules themselves, otherwise sorting will not happen as the device will interfere with the motion of the molecules. Hence the Maxwell's demon cannot be bigger than the size of the molecules. However, a molecular sized Maxwell's demon undergoes the same thermal motion as the molecules themselves, and so Maxwell's demon moves as a result. Hmm, this is not productive, which means that we are very limited in our choices of Maxwell's demons. A device that transmits electromagnetic radiation and uses light as the sorting device is most likely an example of a Maxwell's demon.
3. Does the Maxwell's demon violate the Second law?

   Answer: no. For Maxwell's demon to work it has to be enclosed within the container and the only way for it to be as un-obstructionist as possible is that it has to be:
   1. Sufficiently sized;
   2. Mechanically reversible; and
   3. Be able to sense molecules without disturbing their motion.

   As in the answer to question 2 above, the best device will be something that radiates energy and senses the location of molecules through their reflection or radiation. This energy is absorbed throughout the system and can be shown to cause a greater entropy increase than any decrease caused by proper working of the device. Hence the total entropy of system increases, which is in line with the Second law of thermodynamics. Hence we conclude that Maxwell's demon does not violate the Second law.

An interesting fact to note is that this question whether the Second law holds or not has been around for more than 150 years and is still an open question. The reason for such a device that violates the Second law holds a lot of interest for the scientific community. This is because if we were to actually find a system where the Second law was violated, we would get a system for energy generation that could in principle run forever and never produce any waste products. So far we have not been able to find such a system, but the hope lives on!

You may be wondering why I went through this discussion. My purpose here was to illustrate the many problems we encounter when we deal with individual molecules. We still need to use our knowledge of the molecular (microscopic) nature of matter to help us understand the macro nature of matter. This need to understand the macroscopic nature of matter is what made the development of statistical thermodynamics necessary and even an obvious choice.

11.5.2 Number of ensembles that are truly different

Clearly the easiest way to develop statistical thermodynamics is to apply statistics to the properties of the molecules themselves. However, in general, this simple solution does not work because molecules interact with each other. Hence, the statistical arrangement of private properties of molecules does not provide any meaningful description of the macroscopic system. For example, molecules in a system have their own individual kinetic energies but not their own potential energies, because potential energy is of the system as a whole.

So in an isolated ensemble with 'n_i' members, the total number of members (n) is just a sum of individual particles written as:

$$n = \Sigma_i n_i = \text{Constant} \tag{11.13}$$

and the total internal energy (E) of the members that are each in their own quantum state with energies (E_i) from a set of energies at any given time, is just the sum of energies (E_i) times the number of members (n_i) in the particular state. Thus, internal energy (E) for an isolated system, is given as:

$$E = \Sigma_i n_i E_i = \text{Constant} \tag{11.14}$$

These two conditions give us the constraints on the ensemble.

Next, suppose we have a system of 20 members where each member of the ensemble is defined to be in a particular quantum state characterized by a particular value of E_i at any given time. Assume there are four possible quantum states 1, 2, 3, and 4 with energies E_1, E_2, E_3, and E_4, respectively, as seen in table 11.4. Each of the 20 members of the ensemble must be in one of these four quantum states. Let's assign a quantum state to each member. Each arrangement such as given in table 11.4 represents a particular distribution of energy states.

Table 11.4. Distribution of energy states.

Q. State	Energy (E_i)	Members in state (i)	No. of Members (E_i)
1	E_1	a, b, c, d, e, f, g, h	$n_1 = 8$
2	E_2	i, j, k, l, m, n	$n_2 = 6$
3	E_3	o, p, q, r	$n_3 = 4$
4	E_4	s, t	$n_4 = 2$

Figure 11.8. Ensemble lattice for quantum state 4.

Assuming state 4 has two members (*s*) and (*t*), we have two possible orders: order 1, (*s*) – (*t*) or order 2, (*t*) – (*s*). This means to arrange these members on a lattice there are (2!) ways, as shown in figure 11.8. However, since all members are identical and indistinguishable the actual order will not make any difference.

In state 3, there are four identical and indistinguishable members, namely (*o*, *p*, *q*, *r*), so there are (4!) ways to arrange them on the lattice. Similar arguments can be made for state 2 and state 1 with six and eight members respectively. Hence, for 20 members separated into four states, the number of ensemble or lattice arrangements that are really different are given mathematically as:

$$\Omega = \frac{20!}{8!6!4!2!} = 1.75 \times 10^9. \tag{11.15}$$

which we can generalize to say that we must divide the total number of members ($n!$) by the product of ($n_i!$) which are the number of members in each quantum state, mathematically written as,

$$\Omega = \frac{n!}{\prod_i n_i!} \tag{11.16}$$

where $\prod_i$ is the running product of all the factorials and Ω is called the '**multiplicity**' Note, if we had assigned any other members to the quantum states or if we had changed the number of quantum states, then we would get a different answer than our example above.

11.5.3 Entropy in statistical thermodynamics

As we see from equation (11.15) the number of ensembles arrangements (Ω) for an ensemble made of just 20 members is 1.75×10^9. In statistical mechanics, we usually deal with systems that have a large number of members, for example, just 1 mol of a gas will contain $N_A = (6.022 \times 10^{23})$ members! This leads to a very, very large number of ensemble arrangements or multiplicities (Ω). Here mathematics comes to our aid.

In mathematics, we can make large numbers tangible by taking the logarithm, so in order to make multiplicity tangible, we do not deal with (Ω) but rather deal with the logarithm of our multiplicity (Ω) and say that the entropy of the ensemble is some function of the logarithm of (Ω) and develop the functional relationship that leads to results that agree with experiments. The Austrian physicist named Ludwig

Boltzmann first pointed out the relationship between entropy and probability in 1877 CE that leads to success. It is now called the '*Boltzmann's entropy equation*', mathematically given as:

$$S = k\ \ln(\Omega) \tag{11.17}$$

where S is the entropy of the entire ensemble; k is the constant of proportionality called the Boltzmann Constant, numerically equal to 1.38×10^{-23} J K^{-1}, and (Ω) is the multiplicity associated with the configuration whose entropy (S) we wish to calculate.

Another important factor to remember is that equation (11.17) is a fundamental postulate of statistical mechanics and as such there is no way to prove it. We guess it to be true and then test the consequences with experiment. The rest of the steps are mathematical. In this scenario the ensemble of members have to be isolated from their surroundings, which implies at equilibrium the total entropy of the system must be a maximum with respect to all possible variations within the ensemble. From these arguments we can reach the only possible assumption, which is:

> '*The preferred distribution of ensemble members is the one that maximizes the entropy of the entire ensemble*'.

11.5.4 Redefining the Second law of thermodynamics

In nearly all systems where the system has enough particles and units of energies so that we can apply the statistics of very large numbers, the particles and the energy of the system tend to rearrange themselves until the multiplicity is at or very near, its maximum value. Hence we can redefine the Second law of thermodynamics as:

> '*In any large system the multiplicity tends to increase and at equilibrium the system will be found in the macrostate with the greatest multiplicity*'.

Since the multiplicities tend to be very large numbers, we work with logarithms instead and the entropy will be given by equation (11.17). In general, the number of particles in a system is directly proportional to the energy of the system. This implies systems with a large number of particles will have larger multiplicities and hence larger entropy values.

11.5.5 Why I do not like to think of entropy as disorder

Some people find it useful to think of entropy as 'disorder' in the system. But what is disorder? We all agree that a shuffled deck of cards is more disorderly than a sorted one. This means that shuffling increases the entropy because it increases the number of arrangements of the system. This is simple enough.

On the other hand, if we take an equal amount of water and crushed ice, a glass of crushed ice will appear more disorderly. However, water has more multiplicity and hence a larger entropy, because there are so many more ways of arranging the molecules and so many more ways of arranging the larger amount of energy between them. So I like to think about entropy as the possible number of arranging microstates instead of thinking of entropy as disorder.

11.5.6 Nice property of entropy

The total entropy of a composite system is equal to the sum of the entropies of its parts. Suppose a system is made of two parts A and B. Then the total entropy of the system will be the sum of entropies of part A and part B, assuming macrostates of A and B systems have been specified separately.

$$\begin{aligned} S_{\text{total}} &= k\ \ln(\Omega_{\text{total}}) \\ &= k\ \ln(\Omega_{\text{A}}) + k\ \ln(\Omega_{\text{B}}) \\ &= S_A + S_B \end{aligned} \tag{11.18}$$

If systems A and B interact with each other, then the macrostates A and B can fluctuate over time and to calculate the entropy over long time scales, we have to compute (Ω_{total}) by summing over '*all*' the macrostates of the two systems.

Entropy, like multiplicity, is a function of the number of accessible microstates and this number depends on the time-scale under consideration. However in practice, this distinction rarely matters. Just assume the composite system is in its most likely macrostate and get essentially the same entropy as if we were summing over all macrostates.

11.5.7 Steps to follow to solve problems

Steps to follow to solve problems in statistical mechanics include:

1. Use equation (11.16) to get multiplicity.
2. Find ln (Ω) and simplify as much as possible. Usually the multiplicity (Ω) is a very large number and when you try to find the factorial you may get an overflow problem with your calculator. Fortunately, there is a good approximation called Stirling's approximation for ln (n!),

$$\ln(n!) \approx n\ \ln(n) - n \tag{11.19}$$

   to achieve this step[6].
3. Substitute this result in equation (11.17) to find entropy.
4. Maximize the entropy to obtain the preferred distribution of quantum states.

Worked example 11.5. For an ensemble of 24 members defined in section 11.5.2, suppose the four quantum states have the following member assignments $n_1 = n_2 = n_3 = 0$; $n_4 =$ all. Then find the multiplicity (Ω) and entropy (S).

Solution: There are four quantum states in our assignment, namely $n_1 = n_2 = n_3 = 0$; $n_4 =$ all. This means,

[6] It is interesting to note that the Stirling of the Stirling approximation and the one that invented the Stirling engine are not the same person.

$$\Omega = \frac{24!}{24!0!0!0!} = 1.$$

Hence there is only one way of arranging our system. Then from equation (11.17) we can find the entropy as,

$$S = k\ \ln(\Omega) = k\ \ln(1) = 0.$$

Worked example 11.6. Next suppose for our ensemble of example 11.5, we change our member assignments to $n_1 = n_2 = n_3 = n_4 = 6$. Then find multiplicity (Ω) and entropy (S).

Solution: There are four quantum states in our assignment, namely $n_1 = n_2 = n_3 = n_4 = 6$. This means the number of ensemble arrangements or multiplicity is,

$$\Omega = \frac{24!}{6!6!6!6!} \approx 1.9 \times 10^{19}$$

That is, there are 1.9×20^{19} ways of arranging our system. Which implies from equation (11.17),

$$S = k\ \ln(\Omega) = (1.38 \times 10^{-23})(19.3) = 2.66 \times 10^{-22}\ \text{J K}^{-1}$$

Worked example 11.7. Suppose you have a molecule (XY) composed of two elements X and Y, that can have two possible orientations $[X - Y]$ and $[Y - X]$. Assuming these orientations are completely random, calculate the residual entropy of one mole of a system made of molecule (XY).

Solution: There are Avogadro's number (N_A) of molecules in the system. Each of the N_A (XY)-molecules has two equally likely orientations, which implies the multiplicity (Ω) of the system is $2N_A$, hence the entropy of this system will be,

$$S = k\ \ln(\Omega) = k\ \ln\left(2^{N}_{A}\right) = N_A K\ \ln 2$$

Then the entropy in one mole of the system is,

$$S = (6.022 \times 10^{23})(1.38 \times 10^{-23}\ \text{J K}^{-1})\ln 2 = 5.76\ \text{J K}^{-1}$$

Worked example 11.8. Computer memory is a physical object whose basic unit is called a '*bit*'. A bit can exist in two states 0 or 1. A byte of computer memory is a collection of 8 bits.

A kilobyte = 2^{10} bytes = 2^{13} bits.
A megabyte = $2^{10} \cdot 2^{10}$ bytes = $2^{10} \cdot 2^{10} \cdot 2^{3}$ bits = 2^{23} bits.
A gegabyte = $2^{10} \cdot 2^{10} \cdot 2^{10}$ bytes = 2^{33} bits.

1. Suppose one day your roommate erases and overwrites one gegabyte of memory on your computer, which included your thermodynamics take-home final exam. Oh no! You no longer have any record of doing your final exam. Explain and mathematically calculate how much entropy this process creates.
2. If this entropy is dumped into a room how much heat must be added to the room. Is this a significant amount?

Solution:

1. Before the memory was erased it was possible that the system was in any one of the ($2^{2^{33}}$) states. This means that the multiplicity (Ω) is $2^{2^{33}}$.

$$S = k\ \ln(\Omega) = k\ \ln(2^{2^{33}}) = k(2^{33})\ln(2) = 8.2 \times 10^{-14}\ \text{J K}^{-1}$$

2. At room temperature (T = 298 K),

$$\Delta S = \frac{Q}{T}$$

$$\Longrightarrow Q = T\Delta S = (298\text{K})(8.2 \times 10^{-14}\ \text{J K}^{-1}) = 2.5 \times 10^{-11}\ \text{J}$$

Concept questions 11.2

1. Is the entropy of the Universe increasing or decreasing? Explain in your own words.
2. Which has more entropy: a shuffled deck of playing cards or an unshuffled deck of card?
3. What would happen if there was no entropy in the Universe?

Exercises 11.2

1. Consider the initial ninja situation of section 11.5, with the container divided into two equal parts. Suppose the container has N gas molecules that are initially on the right side of the container only, whereas the left side of the container is completely empty.
   (a) Find the multiplicity of this initial state.
   (b) After the trap door is opened the gas fills the entire container uniformly, and eventually $N/2$ molecules result on each side of the trap door. Find the multiplicity of the final state.
   (c) Show that the change in entropy (ΔS) is given as,

   $$\Delta S = kN\ \ln 2$$

   where k is the Boltzmann's constant.
2. Suppose you have a container with 300 molecules.
   (a) In how many independent ways can you divide the 300 molecules equally into two halves of the container?
   (b) How many microstates are there that correspond to 200 molecules in one half of the container and 100 molecules in the other half?
   (c) In how many independent ways can you divide 300 molecules equally into three parts of the container?
3. Suppose there are two identical balls in a box marked A and six identical balls in another box marked B. Next, suppose the two boxes are brought together and balls are mixed in such a way that there are four balls in each box. Show that the change in entropy during this process is,

   $$\Delta S = k\ \ln(2.5)$$

Further reading

- Hill T 2013 *An Introduction to Statistical Thermodynamics* (North Chelmsford, MA: Courier Corporation)
- Kittle C and Kroemer H 1980 *Thermal Physics* 2nd edn (San Fransisco, CA: W.H. Freeman and Company)
- Mandl F 1988 *Statistical Physics* 2nd edn (The Manchester Physics Series) (New York: Wiley)
- Schroeder D 2000 *Thermal Physics* (Reading, MA: Addison-Wesley Longman)
- Sears F and Salinger G 1975 *Thermodynamics, Kinetic Theory and Statistical Thermodynamics* 3rd edn (Reading, MA: Addison-Wesley Publishing Company)
- Reif F 1985 *Fundamentals of Statistical and Thermal Physics* (Singapore: McGraw-Hill International Edition)
- Resnick R, Halliday D and Krane K 2002 *Physics* vol 1 5th edn (New York: Wiley)

IOP Publishing

Thermodynamics and Statistical Mechanics
An introduction for physicists and engineers
Samya Zain

Chapter 12

Formalisms in statistical mechanics

12.1 Main ideas in statistical physics

As we saw in chapter 11, keeping track of each molecule in a sample of just one mole ($\approx 6.022 \times 10^{23}$ molecules) becomes very difficult if not outright impossible. However, we still want to use our knowledge of matter at the microscopic (molecular) scale to understand the macroscopic nature of matter.

As discussed before, the main goal of statistical physics is to take the over-elaborate details of quantum mechanics and the generalities of classical mechanics and explain physical phenomena found in nature. The main way to do this is using what is called the '*fundamental averaging postulate*' detailed below.

12.1.1 Statistical averaging postulates

For any given macrostate there are an enormous number of possible 'microstates' or the number of ways of arranging the members. Microstates give the most detailed specification of the ensemble/assembly, for example, in classical kinetic theory of gases, the microstate will specify the (vector) positions and momentum of each of the N gas particles in the system. This will give $6N$ coordinates that we have to define at all times under consideration for every single particle in the system. This leads us to two main issues:

1. Microstates for most systems contain too much information to keep track of, both manually or using computers.
2. If we were somehow able to keep track of microstates at a certain time, things get worse again because the system changes its microstate very quickly, for example, one mole of a typical gas changes its microstate about 10^{32} times a second! How do you keep track of this ever changing situation?

Clearly, we need some sort of averaging over microstates. This is one of those rare times where quantum mechanical consideration is much easier to evaluate than classical mechanical considerations. In quantum mechanics, a microstate by definition

doi:10.1088/978-0-7503-3083-1ch12

is a quantum state of the whole ensemble which can be described by a single N-particle wave function, that contains in it all the possible information about the state of the whole system.

Please recall from chapter 11, that quantum states are discrete. So even when a macrostate has a large number of possible microstates consistent with it, the number of microstates is still just a finite number. We called this number the multiplicity (Ω) of the system. According to statistical mechanics we cannot say that the probability of one microstate is greater than the probability of another microstate in an ensemble. This is called the averaging postulate and is the basis of the whole of statistical physics. The formal definition of the averaging postulate can be written as:

'All accessible microstates are equally probable.'

Please note that this averaging postulate is treated as an assumption in statistical physics, but it is important also to note that it is a very reasonable assumption! Based on the averaging postulate we can define the following two averages, the time-average and the ensemble average.

1. **Postulate 1: Time-averages:**

   The first postulate is about '*time averages*'. We know that in order to make any physical measurement we have to have a non-zero time measurement, which is in general given by [Δt]. In this time (Δt) the system will have passed through many, many microstates. Suppose the physical measurement we want to know is the pressure (p) on walls of a container due to an enclosed gas. The reason why we need pressure (p) averaged over time is because observable fluctuations of pressure are small over time (Δt).

   Assume that all microstates are equally likely and readily accessible. Then the time-average argument indicates that averaging over all microstates is necessary. This means we can replace a time average of one system by the instantaneous average over a large number of systems. A good way to think about time averaging in statistical mechanics is that we can keep track of the location of a single student during his or her four years on campus during his or her undergraduate time or equivalently we can average the entire student body location over a certain time Δt, say one hour. The time-average postulate says that both of these measurements will be equal.

2. **Postulate 2: Ensemble-averages:**

   The second argument is to treat the postulate as a *'confession of ignorance'*, which is a common strategy used in quantum mechanics. Since we do not know which microstate the system is in at the time of interest, we can average equally over all possibilities, i.e., over all microstates and call this the *'ensemble average'*. In other words, we can say that given a long enough time one particle state goes at least once through each of the allowed microstates. Another way to think about it is to say that if a system is selected at random, the probability it is found in a particular quantum state is equal for all possible available quantum states. This is also called the '**principle of equal *a priori* possibilities**'.

It is interesting to note that we can combine postulate 1 and postulate 2, to state that the single isolated system of actual interest spends equal amounts of time, given a long enough period, in each of the available quantum states; this is called the '**quantum ergodic hypothesis**'.

12.1.2 Statistical distributions

Basic to any consideration of statistical methods is the concept of a '*distribution*', which is systematically arranging objects in a system into meaningful categories or classifications. In our daily life we tend to organize things around us, from clothes in our closets arranged by colors to a child with his savings, arranging his change in stacks of pennies, nickels and dimes for easy counting. Once the coins are stacked or indeed listed in a table, by their value it becomes easier to count the total amount of money in the piggy bank as done in the example below.

Worked example 12.1. Suppose Sami from our story of sugar cubes in chapter 2 wants to pay his friend back for his sweat and sticky treat by buying something for his friend. In order to do that he has to figure out how much money he has to spend. So Sami opens his piggy bank and finds that he has many different coins in his piggy bank that he has to count. To make his life simpler he decides to arrange the coins by order of value and decides to stack all the coins. Once stacked, Sami realizes he has 5 quarters, 10 dimes, 20 nickels, and 40 pennies.

During the stacking process, he realizes that the most natural thing for him to do is to put the values of coins into a bar graph where he places the value of the coins on the x-axis (independent variable) and the number of coins on the y-axis (dependent variable) as shown in figure 12.1.

Another way he expresses the values is by putting them in a table as shown in table 12.1. Please keep in mind, both table 12.1 and figure 12.1 represent the '*distribution*' of coins in the piggy bank.

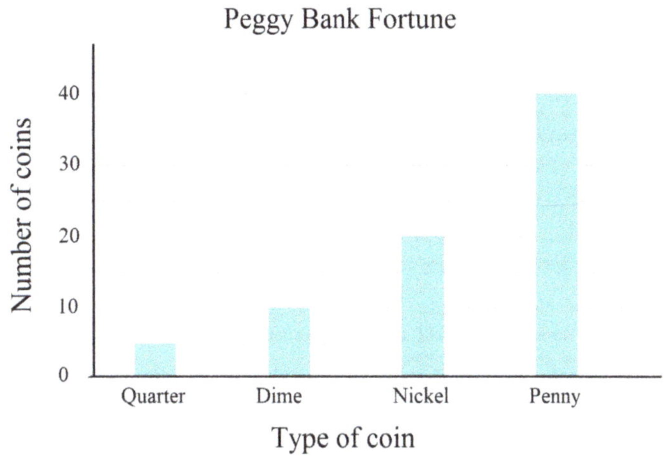

Figure 12.1. Example: Sami and his piggybank.

Table 12.1. Coin counting.

States	Number	Numerical Value	Relative Frequency
Quarter	5	125c	5/75 = 0.066
Dime	10	100c	10/75 = 0.133
Nickel	20	100c	20/75 = 0.266
Penny	40	40c	40/75 = 0.533
	75	365c	1

In this example, once Sami has distributed the coins into stacks he does not necessarily need to keep track of every single penny in the penny-stack or every single nickel in the nickel-stack. He just needs to worry about the stacks as a whole. If he starts worrying about every single penny or every single dime, it will get very tedious very quickly.

Similarly, in statistical mechanics the knowledge at the microstate level (information about each coin) for most situations is too much information to keep track of and so we restrict our curiosity to merely a '*distribution*' specification (information about each coin-stack). The distribution involves assigning individual energies, also called private energies, to each of the N particles of the system, which is the only sensible (and indeed possible) situation for an ensemble of negligibly interacting particles[1].

A distribution in statistical mechanics then relates to the energies of individual particles in the system. For each microstate of the ensemble of N identical negligibly interacting particles, each particle is considered to be in an identifiable one-particle state. A distribution is intermediate in detail between macrostate and microstates in a system. For a distribution, we choose not to investigate '*which*' particles are in the particular state, we instead specify the '*total number*' of particles in the state.

The two alternate definitions of a distribution are as follows:

1. **Definition 1: distribution in states:**

   This is a set of numbers (n_1, n_2, … , n_j) in particular states with particular energies, where n_1 is the number of particles in state 1 with energy E_1, and n_2 is the number of particles in state 2 with energy E_2, and so on. n_j is called the '**distribution number**' and is defined as the number of particles in state j, which has energy E_j. The shorthand for the whole set of distribution numbers is n_j. This is similar to the apartments on each floor in the example from chapter 11, figure 11.3.

2. **Definition 2: distribution in levels:**

   This is a set of numbers (n_1, n_2, … , n_i, …) for which the typical number (n_i) is defined as the number of particles in level (i), which has energy (E_i) and

[1] Negligibly interacting in statistical mechanics means that interacting energies between the particles are much smaller than the individual energies of the particles. Actually, thermodynamic systems must have some interactions between their particles, otherwise the system will never reach equilibrium.

degeneracy (g_i)[2] is the number of states belonging to that level. The shorthand for the whole set of distribution numbers is (n_i).

12.2 Formalisms in statistical mechanics

Before the introduction of quantum mechanics, various scientists like Maxwell, Boltzmann, Gibbs, tried to apply the ideas of classical physics to explain statistical mechanics and these methods are collectively known as classical statistics or Maxwell–Boltzmann (MB) statistics. The classical statistical method, derived on the basis of purely classical physics arguments, was successful in explaining pressure, temperature, etc, of gaseous systems but could not explain experimental results like the energy distribution in black-body radiation, specific heat at low temperature, etc. For explaining such phenomena Bose, Einstein, Fermi, and Dirac made use of newly discovered quantum theories. The new statistics are known as quantum statistics and can be divided into the following categories:

1. Bose–Einstein (BE) statistics.
2. Fermi–Dirac (FD) statistics.

Quantum statistics uses two major results of quantum mechanics; one, the Pauli exclusion principle, which limits the number of particles occupying any state of energy (E) and second, which is a direct result of Schrödinger's equation, that there is a finite number of states in any energy interval, $[E, E + dE]$.

In order to figure out the differences between the various formalisms in statistical mechanics let's start with discussing the various ways of counting microstates.

12.2.1 Counting microstates

When counting the microstates of a system we can divide our analysis broadly into two categories, one for distinguishable particles and the other for indistinguishable particles.

12.2.1.1 Distinguishable particles

Combinatorial analysis for distinguishable particles encompasses three significant cases. I will consider each case individually.

1. **Case 1:** N identical, distinguishable particles placed in M different containers with a limit of one particle per container requires $N \leqslant M$. The first particle may be placed in any of the M available containers, the second particle can now be only placed in $(M - 1)$ available containers and so on. Hence, the number of ways that N identical, distinguishable particles can be placed in M different containers with a limit of one particle per container is given as:

$$\Omega_1 = M(M-1)(M-2)......(M-N+1) = \frac{M!}{(M-N)!} \tag{12.1}$$

[2] Degeneracy: when the energy level corresponds to more than one independent state, the level is called degenerate.

2. **Case 2:** For N identical, distinguishable particles placed in M different containers such that each container holds exactly N_i (a particular number) of particles implies that the total number of permutations of N particles is $N!$. However within the container we are concerned only with the total number of particles rather than their identities and hence the permutations of particles become irrelevant. This implies that the number of permutations ($N!$) over-counts the number of ways and hence the number of ways for N identical, distinguishable particles to be placed in M different containers such that each container holds exactly N_i particles is given as:

$$\Omega_2 = \frac{N!}{\prod_{i=1}^{M} (N_i)!} \tag{12.2}$$

where $\prod_{i=1}^{M}$ is the running product of the factorials.

3. **Case 3:** N identical, distinguishable particles placed in M different containers with no limit on the number per container, implies each particle can be placed in any of the M containers. This means the number of ways N identical, distinguishable particles can be placed in M different containers with no limit on the number per container is given as:

$$\Omega_3 = M^N \tag{12.3}$$

Please note: this situation is identical to the one given in example 11.7, and is the basis of **Boltzmann statistics**.

12.2.1.2 Indistinguishable particles

Combinatorial analysis for indistinguishable particles encompass two significant cases discussed below.

1. **Case 1:** Rearranging N identical, indistinguishable particles among M different containers with a limit of one particle per container, results in the same arrangement regardless of how the actual arrangement is carried out. In this case Ω over-counts the number of ways by a factor of $N!$. Hence, the number of ways N identical, indistinguishable particles can be placed in M different containers with a limit of one particle per container is mathematically given as:

$$\Omega_4 = \frac{M!}{N!(M-N)!} \tag{12.4}$$

Please note: this situation is identical to the one we saw in example 11.2, and is the basis of **Fermi–Dirac statistics** for particles called *Fermions*. Examples of fermions include electrons, neutrons, and protons, which are spin-1/2 particles, and helium atoms ($^4He_2$) and lithium nuclei (^7Li), which have spin-3/2.

2. **Case 2:** For a fully unconstrained case, where N identical, indistinguishable particles can be placed in M different containers with no limits on number of

particles per container, we cannot use the same approach that we used above for identical, indistinguishable particles.

In this case, if distinguishable particles are labeled as 1, 2, 3, ... , N and arranged in a row with the identified M containers separated by partitions, statistically we know that regardless of their actual arrangement, the maximum number of rearrangements among N and $(M - 1)$ particles is given as $= (N + M - 1)!$. Interchanging the partitions will not result in any new arrangements, which implies we have over-counted by a factor $(M - 1)!$. Similarly, because the N particles are actually indistinguishable, we have again over-counted by a factor of $N!$ as in Ω_4 in equation (12.4). Hence the number of ways N identical, indistinguishable particles can be placed in M different containers with no limits on particles per container is given mathematically as,

$$\Omega_5 = \frac{(N + M - 1)!}{N!(M - 1)!} \tag{12.5}$$

This is the basis for the **Bose–Einstein statistics** used for defining particles called *Bosons*. Photons with spin one and helium atoms ($^4He_2$) with spin zero are examples of bosons.

Worked example 12.2. Determine the number of ways of placing two balls in three containers for each of the following; Boltzmann statistics, Fermi–Dirac statistics and Bose–Einstein statistics.

Solution: In this case $N = 2$ balls and $M = 3$ containers. So we can use the formulas we developed in the previous section to calculate the number of ways of placing the two balls into three containers.

(a) **Boltzmann statistics:** In Boltzmann statistics the balls are distinguishable and containers do not have a limit on occupancy. We can use equation (12.3)

$$\Omega_3 = M^N = 3^2 = 9.$$

(b) **Fermi–Dirac statistics:** In Fermi–Dirac statistics the balls are indistinguishable and containers have a limit of one ball per container. We can use equation (12.4)

$$\Omega_4 = \frac{M!}{N!(M - N)!} = \frac{3!}{2!1!} = 3.$$

(c) **Bose–Einstein statistics:** In Bose–Einstein statistics the balls are indistinguishable and containers have no limit on occupancy. We can use equation (12.5).

$$\Omega_5 = \frac{(N + M - 1)!}{N!(M - 1)!} = \frac{4!}{2!2!} = 6.$$

These are shown in figure 12.2.

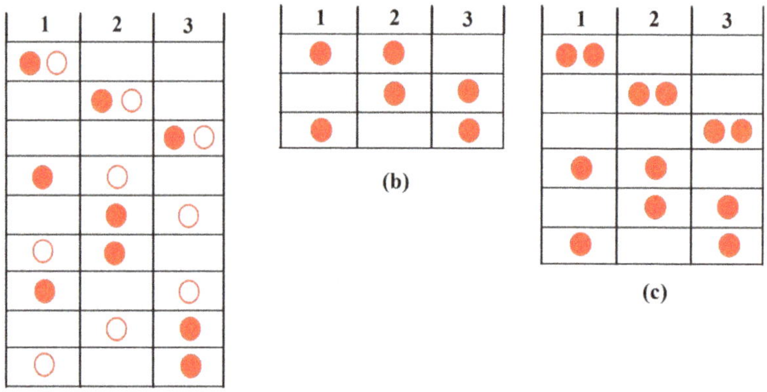

Figure 12.2. Example 12.2. Number of ways of placing two balls in three containers. (a) Boltzmann statistics, (b) Fermi–Dirac statistics, (c) Bose–Einstein statistics.

Worked example 12.3. Suppose a thermodynamical ensemble has seven independent particles that have access to two energy levels. Determine the number of microstates if the particles are bosons or fermions, given one particular particle distribution for the system and the associated degeneracy for each level as,

1. $N_1 = 3$, $M_1 = 4$.
2. $N_2 = 4$, $M_2 = 6$.

Solution:

1. **Fermions:** The number of ways a single energy level can be constructed for a fermion is given by Fermi–Dirac statistics as,

$$\Omega_i = \frac{M_i!}{N_i!(M_i - N_i)!}$$

Therefore, for each energy level we get,

$$\Omega_1 = \frac{4!}{3!(4 - 3)!} = 4$$

$$\Omega_2 = \frac{6!}{4!(6 - 4)!} = 15$$

This implies the number of microstates corresponding to this macrostate when the particles are fermions is,

$$\Omega_i = \Omega_1 \times \Omega_2 = 60$$

2. **Bosons:** The number of ways that a single energy level can be constructed for bosons is given by the Bose–Einstein statistics.

$$\Omega_i = \frac{(N_i + M_i - 1)!}{N_i!(M_i - 1)!}$$

Therefore, for each energy level we get,

$$\Omega_1 = \frac{(3+4-1)!}{3!(4-1)!} = 20$$

$$\Omega_2 = \frac{(4+6-1)!}{4!(6-1)!} = 126$$

This implies the number of microstates corresponding to this macrostate when the particles are bosons is,

$$\Omega_i = \Omega_1 \times \Omega_2 = 2520$$

12.2.2 Quantum mechanics and statistical properties

The main problem in statistical methods is to count the number of microstates corresponding to a particular distribution. A microstate relates to the quantum state of all N identical indistinguishable particles. When the particles in the system are identical, we cannot define '*which*' particle is in a particular state. Microstates are completely specified only by stating how many particles are in each state.

In section 12.2.1, we defined fermions and bosons based on the number of ways of arranging the particles or the multiplicity (Ω). Quantum mechanics provides more ways to distinguish between fermions and bosons. In this section we will redefine fermions and bosons based on quantum mechanical consideration.

12.2.2.1 Symmetry rules

In quantum mechanics, fermions and bosons obey different rules based on how many particles are allowed to be in each state. In a system of two identical indistinguishable particles we can define the quantum state as a two-particle wave function, represented as $\Psi(1, 2)$, where 1 denotes all the coordinates (space and spin) of particle 1 and 2 denotes all the coordinates (space and spin) of particle 2. For $\Psi(1, 2)$ to be a valid two-particle wave function two conditions of quantum mechanics must be satisfied;

1. $\Psi(1, 2)$ must be a valid solution of the Schrödinger equation.
2. $\Psi(1, 2)$ must also satisfy the symmetry requirements for the interchange of the two special labels. Suppose our exchange operator is given as ($\hat{P}$), then this condition leads us to two possibilities,
   (a) **Symmetric:** The wave function $\Psi(1, 2)$ can be symmetric with respect to the interchange of the labels, given as,

   $$\hat{P}\Psi(1, 2) = \Psi(2, 1) = +\Psi(1, 2)$$

   which is another definition of *bosons*.
   (b) **Anti-symmetric:** The wave function $\Psi(1, 2)$ can be anti-symmetric with respect to the interchange of the labels, given as,

   $$\hat{P}\Psi(1, 2) = \Psi(2, 1) = -\Psi(1, 2)$$

   which is another definition of *fermions*.

Please note that both of these states are called the 'eigenstate's of the exchange operator' ($\hat{P}$).

According to these systematic rules, particles that are symmetric with respect to the exchange labels are called *bosons* where those that are antisymmetric with respect to the exchange of labels are called *fermions.*

12.2.2.2 Occupation rules

Another way to define fermions and bosons is to think about these particles in terms of the number of particles per quantum state, or in other words whether particles are subject to Pauli's exclusion principle or not. In fact, Pauli's exclusion principle states that some types of particles do not share the same quantum states while other types of particles can share quantum numbers, hence can be in the same quantum state. A quantum number is a value that is used when describing the energy levels available to a state, and they include principal quantum number (n), azimuthal quantum number (l) magnetic quantum number (m_l) and the spin quantum number (s), (n, l, m_l, s). The quantum numbers for a state are calculated by solving Schrödinger's equation for the wave functions of the state. For more details about quantum numbers please refer to a quantum mechanics textbook.

Particles that are subject to Pauli's exclusion principle are called *fermions* and particles that are not subject to Pauli's exclusion principle are called *bosons.* Pauli's exclusion principle states that no two identical fermions can exist in the same quantum state, whereas, bosons can share a quantum state.

In summary we can say that in the orbital model of non-interacting particles, we have the following occupancy rules:

1. An orbital can be occupied by none or a maximum of one fermion of the same kind, as determined by the Pauli's exclusion principle.
2. An orbital may be occupied by any number of bosons of the same kind.

12.2.2.3 Distribution functions

When the chemical potential (μ) and temperature (T) of a particular system are fixed, the mean occupation number of any state (say the ith state) is only a function of energy (E_i). It is therefore convenient to consider the distribution function ($f(E)$) which is defined to be the mean occupation number of the state with energy E.

1. **Fermions:** The distribution function for fermions was introduced in 1926 by Enrico Fermi and Paul Dirac and is called the Fermi–Dirac function. The Fermi–Dirac function provides the probability of occupancy of energy levels by fermions and is given mathematically as,

$$\langle n \rangle = f(E) = \frac{1}{e^{(E-\mu)/kT} + 1} \tag{12.6}$$

The right-hand side of this equation is called the **Fermi factor**.

2. **Bosons:** the distribution function for bosons was first introduced in 1924 CE by Bose who recognized the existence of a collection of identical and indistinguishable particles and applied this distribution function to photons. Later, in 1925 CE, Einstein generalized this function to be applicable to

atoms. The Bose–Einstein distribution function provides the probability of occupancy of energy levels by bosons and it is mathematically given as,

$$\langle n \rangle = f(E) = \frac{1}{e^{(E-\mu)/kT} - 1} \tag{12.7}$$

The right-hand side of this equation is called the **Bose factor**.

12.2.3 Types of statistics

In this section I summarize the properties of the various statistics.

12.2.3.1 Maxwell–Boltzmann (MB) statistics

The basic postulates of Maxwell–Boltzmann statistics are,

1. The associated particles are distinguishable.
2. Each energy state can contain any number of particles.
3. The total number of particles in the entire system is constant.
4. The total energy of all the particles in the entire system is constant.
5. Particles are spin-less.

Examples includes gas molecules at high temperature and low pressure.

12.2.3.2 Bose–Einstein (BE) statistics

Particles called bosons follow the Bose–Einstein statistics and the basic postulates of Bose–Einstein statistics are,

1. The associated particles are identical and indistinguishable.
2. Each energy state can contain any number of particles.
3. The total energy and total number of particles of the entire system is constant
4. The particles have zero or integral spin, i.e., $0\hbar$, $1\hbar$, $5\hbar$, $50\hbar$, etc, where $\hbar$ is the unit of spin.
5. The wave function of the system is symmetric under the positional exchange of any two particles, denoted as (Ψ_{sym}).

Examples includes photon, phonon, and all mesons (π, η, etc) which are all bosons.

12.2.3.3 Fermi–Dirac (FD) statistics

Fermions follow Fermi–Dirac statistics and the basic postulates of Fermi–Dirac statistics are as follows,

1. Particles are identical and indistinguishable.
2. Total energy and total number of particles of the entire system is constant.
3. Particles have half-odd integral spin, $\frac{1}{2}\hbar$, $\frac{3}{2}\hbar$, $\frac{37}{2}\hbar$, etc, where $\hbar$ is the unit of spin.
4. Particles obey Pauli's exclusion principle, i.e., no two particles in a single system can have the same value for each of the four quantum numbers. In other words, a single energy state can contain at best a single particle with appropriate spin.

5. The wave function of the system is anti-symmetric under the positional exchange of any two particles, denoted as ($\Psi_{antisym}$).

Examples: electrons, protons, neutrons, and all hyperons (Λ, Σ, Ω), etc.

We can draw a flow chart of statistical mechanics, as done in figure 12.3, and can summarize the formalisms of statistical mechanics, as done in table 12.2.

Worked example 12.4. Imagine that a particle can exist in one of two states, which we can label as $|0\rangle$ and $|1\rangle$. Consider two such particles and describe their joint state by a product state, say $|0\rangle|1\rangle$. Then $|0\rangle|1\rangle$ describes the total state in which the first particle is in state $|0\rangle$ and the second article is in state $|1\rangle$. What are the possible states for this system if the two particles are (a) distinguishable, (b) indistinguishable but classical particles, (c) indistinguishable bosons, (d) indistinguishable fermions?

Solution:

1. **Distinguishable particles:** for distinguishable particles, there are four possible states:

$$[|0\rangle|0\rangle],\ [|0\rangle|1\rangle],\ [|1\rangle|0\rangle],\ [|1\rangle|1\rangle]$$

2. **Indistinguishable but classical particles:** for indistinguishable particles there is no way to tell the difference between state $|0\rangle|1\rangle$ or state $|1\rangle|0\rangle$, hence there will be only three possible states,

$$[|0\rangle|0\rangle],\ [|1\rangle|0\rangle],\ [|1\rangle|1\rangle]$$

3. **Indistinguishable bosons:** for indistinguishable zones there will be three states again. Clearly, $|0\rangle|0\rangle$ and $|1\rangle|1\rangle$ are eigenstates of the exchange operator, however, states $|0\rangle|1\rangle$ and $|1\rangle|0\rangle$ are not. So in order to find a third state we have to make a linear combination of the form $\left(\frac{1}{2}(|0\rangle|1\rangle + |1\rangle|0\rangle)\right)$[3]. This means for indistinguishable bosons the three possible states will be;

$$[|0\rangle|0\rangle],\ \left[\frac{1}{2}(|0\rangle|1\rangle + |1\rangle|0\rangle)\right],\ [|1\rangle|1\rangle]$$

4. **Indistinguishable fermions:** Pauli's exclusion principle states that no two fermions can be in the same quantum state, which implies that states $|0\rangle|0\rangle$ and $|1\rangle|1\rangle$ are not allowed, therefore only one state is possible:

$$\left[\frac{1}{2}(|0\rangle|1\rangle - |1\rangle|0\rangle)\right]$$

[3] this is an example of quantum mechanical entanglement, two particles are entangled because if particle one is in state $|1\rangle$ then particle two has to be in state $|0\rangle$ and vice versa.

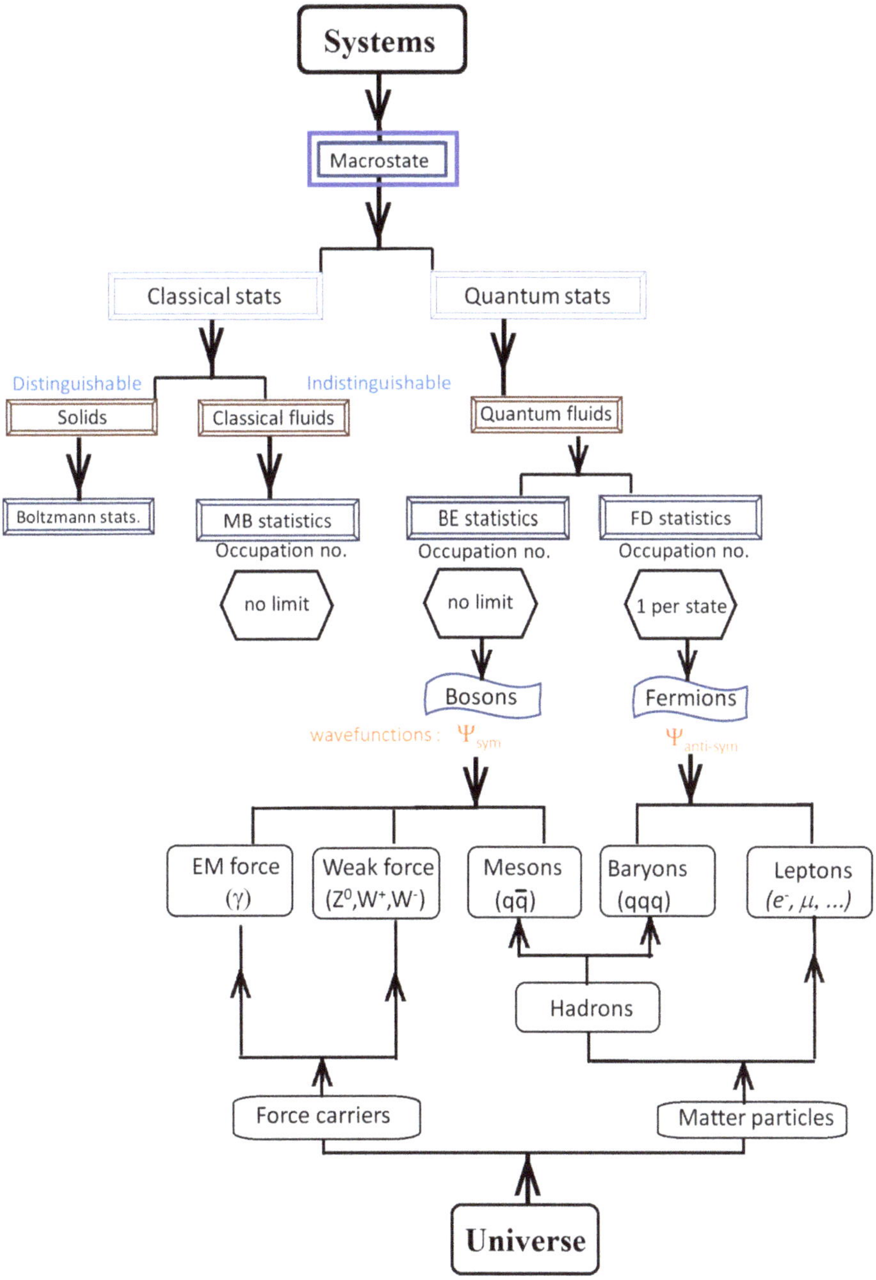

Figure 12.3. Flow chart for statistical mechanics. 'q' represents quark and 'q̄' represents anti-quark; mesons have quark-anti-quark pairs whereas baryons have three quarks.

Table 12.2. Summarizing formalisms in statistical mechanics.

	MB-Statistics	BE-statistics	FD-statistics
	Classical	Quantum	Quantum
1.Type of particles	Distinguishable	Indistinguishable	Indistinguishable
2.Number of particles per state	Unlimited	Unlimited	One per state
3.Number of particles in system	Constant	Constant	Constant
4.Total energy of all particles	Constant	Constant	Constant
5.Spins of particles	Spin-zero	Zero or integral spin	Half-integral spin
6.Quantum wave function		symmetric, Ψ_{sym}	anti-symmetric, $\Psi_{antisym}$
7.Examples:	Gas molecules (at high T and low p)	Bosons	Fermions

Concept questions 12.1

1. What is the difference between a fermion and a boson? Explain with examples.
2. How many types of fermions are there?
3. How is quantum statistical mechanics different from classical statistical mechanics?

Exercises 12.1

1. Determine the number of ways of placing three balls in three numbered containers for Boltzmann statistics, Fermi–Dirac statistics, Bose–Einstein statistics.
2. Suppose you have a system made of two solids, labeled solid A and solid B. Each solid is assumed to contain 10 oscillators and share a total of 20 units of energy. Calculate,
   (a) The number of different macrostates available to this system.
   (b) The number of microstates available to this system.
   (c) Assuming that the system is in thermal equilibrium, the probability of finding all the energy in solid B.
   (d) Assuming that the system is in thermal equilibrium, the probability of finding half the energy in solid B.

12.3 Steps toward statistical descriptions of the macrostate

In statistical mechanics, an ensemble is used to represent the possible states of a mechanical system. Consider an isolated system made of a fixed number of particles (N) that are identical and interacting weakly, contained in a fixed volume (V) with fixed internal energy (E). There are essentially four steps toward statistical descriptions of this macrostate.

1. Solve problems having only one particle.
2. Count/list all possible distributions.
3. Count the microstates corresponding to each distribution.
4. Find the average distribution.

The next step is to apply the statistical method outlined above to real thermodynamical systems. In order to do this we need to talk about properties of assemblies of negligibly interacting identical particles. Materials are broadly divided into two types, namely, solid ensembles and fluid ensembles (including liquids and gases).

1. **Solid ensemble:** A solid ensemble is made of N identical atoms in a simple solid. Particles in a solid ensemble are distinguishable from each other, not by the identity of the atom (since all atoms are identical in a solid) but by the distinct location of each lattice site. In reality a solid ensemble can be thought of as a set of N lattice sites.
2. **Fluid ensemble:** Unlike solids, in which the particles are fixed at their respective lattice sites, in a fluid ensemble the identical particles are usually gas molecules that move in random paths and constantly collide with each other. It is not possible to tell one gas particle from another, for example, it is not possible to put red paint on one and green paint on another molecule to tell them apart and then follow their individual trajectories and properties over time. Or indeed color all 6.022×10^{23} molecules in a single mole of a sample and keep track of every individual particle! Additionally, in quantum mechanics, molecules are considered to be identical and indistinguishable and hence cannot in principle be painted anyhow.

12.3.1 General thermodynamical (or equilibrium) ensembles

There are three basic types of equilibrium ensembles that can be defined for any system bound by a finite volume (V) with a number of particles (N) and energy (E)[4], with a simple form. They include micro-canonical, canonical and grand-canonical ensembles and are shown in figure 12.4.

The most common ensemble encountered while studying statistical mechanics is the canonical ensemble. The grand canonical ensemble is used when dealing with quantum systems, whereas the micro-canonical ensemble is not used much because of the difficulty in identifying and evaluating the accessible microstates. I will briefly discuss these here, however please refer to a thicker textbook for more mathematical details and a more robust study.

1. **Micro-canonical ensemble (for isolated systems)**: A micro-canonical ensemble describes a system with exact number of particles with precisely defined energies. Hence a system described by a micro-canonical ensemble has N, E, V fixed. Suppose a particular isolated system is described in the

[4] It is the total energy not energies of individual particles.

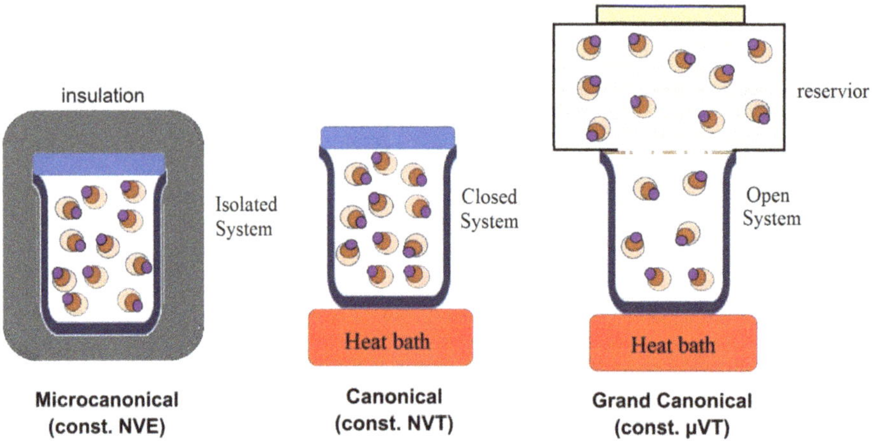

Figure 12.4. Statistical ensembles.

energy range between $[E, E + \delta E]$. From the postulates of statistical mechanics, we know that all microstates are equally probable and that the probability of the system to be in any microstate is a constant, as long as the total energy of the system is in the range between $[E, E + \delta E]$. The difficulty in this ensemble lies in identifying the correct set of possible microstates. For example, every particle in an ideal gas in the micro-canonical ensemble is isolated and does not interact with other particles. If each gas particle is considered a 'system', the micro-canonical ensemble would consist of only those particles with kinetic energies between $[E, E + \delta E]$ and therefore will be made up of only the particles with a particular velocity. Picking out these particular particles from all the particles in any given sample is an extremely difficult chore if not completely impossible.

2. **Canonical ensemble (for closed systems)**: The canonical ensemble was first described by Boltzmann in 1884 CE but was developed by Gibbs in 1902 CE. The canonical ensemble describes a system that is in thermal equilibrium with a heat reservoir at some specific temperature (T). Fixed variables for a canonical ensemble are N, T, V.

   The system can exchange energy with a heat reservoir, which means a canonical ensemble contains states of varying energy but the system has a precise number of particles that stays constant. This means the system and heat reservoir can exchange energy through heat, but no work will be done by the system on the reservoir or vice versa. The different states in the ensemble have different probabilities depending on their total energy.
3. **Grand canonical ensemble (open systems)**: The grand canonical ensemble is the realistic representation of the canonical ensemble with no restrictions to the number of particles in the system. Hence, the grand canonical ensemble describes a system with uncertain numbers of particles, that is in thermal and chemical equilibrium with thermodynamic reservoirs. These reservoirs have a precise temperature and precise chemical potential for various types of particle, as seen in figure 12.4.

A grand canonical ensemble can exchange both energy and particles with a heat bath of specified temperature (T) and chemical potential (μ). The different states in the ensemble are accorded different probabilities depending on their total energy and total number of particles. In a grand canonical ensemble the fixed variables are μ, T, V.

The basic example for grand canonical ensemble is found in a chemical reaction, where the number of each molecular species is not conserved. In a chemical reaction, the number of each type of atom is conserved, however, the number of each type of molecule is not necessarily conserved. Other examples in nature where number of particles is not conserved include blackbody radiation, particle annihilation and decays of particles.

Important cases where thermodynamic ensembles do not give identical results include microscopic systems, large systems at a phase transition and large systems with long-range interactions. In these cases, the correct thermodynamic ensemble must be chosen carefully. However, for systems that are made up of a large number of particles, all three ensembles tend to give identical behaviors. It is then simply a matter of mathematical convenience which ensemble is used.

12.3.2 Partition function

In physics, a partition function (denoted as '$\mathcal{Z}$') describes the statistical properties of the system in thermal equilibrium. Partition functions are defined as needed and may have many physical meanings depending on the situation. It expresses the number of thermally accessible states that a system provides to carriers, for example electrons. Partition functions are functions of state variables such as temperature T, volume V, chemical potential μ, etc. Partition functions are dimensionless; they are just pure numbers. Each partition function is constructed to represent a particular statistical ensemble, described in section 12.3.1.

Fundamentally, the partition function tells us the number of microstates accessible to our system in a given ensemble. For example, a canonical partition function applies to the canonical ensemble when the system exchanges heat energy at fixed T, V, N with the environment, whereas a grand canonical partition function applies to the grand canonical ensemble when the system can exchange both heat and particles at fixed T, V, μ with the environment.

12.3.2.1 Derivation of partition function

In cases with more than two identical particles, the easiest way to find the partition function is to find the grand partition function $\mathcal{Z}$ for a system of either fermions or bosons, while not fixing the total number of particles when we do this.

Assume we have a system of particles and there are certain possible quantum states in which to place our particles and the energy cost of putting a particle into the ith state is given by E_i. Then the partition function for our system is given as,

$$\mathcal{Z} = \sum_i e^{\frac{\mu N_i - E_i}{kT}} \tag{12.8}$$

where i denotes the state of the system. If we put n_i particles into the ith state, then n_i is called the *occupation number* of the ith state. A particular configuration of our system is given by the product:

$$\left[e^{\frac{\mu-E_1}{kT}}\right]^{n_1} \times \left[e^{\frac{\mu-E_2}{kT}}\right]^{n_2} \times \cdots = \sum_{n_\alpha} \prod_{\{\alpha\}} e^{\frac{n_\alpha}{kT}(\mu-E_\alpha)} \tag{12.9}$$

where $\{n_\alpha\}$ denotes a set of occupation numbers allowed by the symmetry of the particle. This means for fermions $\{n_\alpha\} = \{0, 1\}$ and for bosons $\{n_\alpha\} = \{0, 1, 2, 3, \ldots\}$.

12.3.2.2 Number of particles in each energy level

The number of particles in each energy level is given as: $\langle n_\alpha \rangle = -kT\left(\frac{\partial}{\partial E_\alpha} ln \mathcal{Z}\right)$, using the value of $\mathcal{Z}$ we can write the number of particles in each energy level as,

$$\langle n_\alpha \rangle = \frac{1}{e^{(E_i-\mu)/kT} \pm 1} \tag{12.10}$$

When μ and T are fixed for a particular system, equation (12.10) shows that the mean occupation of the αth state $\langle n_\alpha \rangle$ is only a function of the energy E_α. It is therefore convenient to consider the distribution function $f(E)$, defined to be the mean occupation of the state with energy (E). Distribution functions $f(E)$ in this case are given as:

$$\begin{aligned} \text{For fermions: } \langle n \rangle = f(E) &= \frac{1}{e^{\frac{E-\mu}{kT}} + 1} \\ \text{For bosons: } \langle n \rangle = f(E) &= \frac{1}{e^{\frac{E-\mu}{kT}} - 1} \end{aligned} \tag{12.11}$$

which are the same expressions we saw in section 12.2.2.3.

12.3.2.3 Probability of a macrostate

The probability of the system being in a particular macrostate can be found by just adding up the probabilities for all the microstates it contains. For simplicity, let's assume the only macroscopic variable of interest is energy (E). Then the probability of a macrostate is given as:

$$P(E) = \frac{1}{\mathcal{Z}} \Sigma e^{(-\frac{E}{kT})} \tag{12.12}$$

The sum is taken over every microstate contained in the macrostate. If there are other macroscopic variables, just use the appropriate exponential factor.

We note that every term in the sum in equation (12.12) has exactly the same value, so instead of using the sum we can just replace it by the number of microstates, in other words, the multiplicity (Ω) and the probability can be rewritten as:

$$P(E) = \Omega \frac{e^{(-\frac{E}{kT})}}{\mathcal{Z}} \tag{12.13}$$

Please note, that is just another way of measuring the number of microstates that make up the system. We know **entropy** of the macrostate (S) is given as,

$$S \equiv k\ \ln(\Omega) \tag{12.14}$$

in which case we can rewrite Ω as,

$$\Omega = e^{\frac{S}{k}}$$

Put this value in equation (12.13) and simplify to get the probability of a macrostate as:

$$P(E) = \frac{1}{\mathcal{Z}} e^{(-\frac{(E-TS)}{kT})} \tag{12.15}$$

Worked example 12.5. Suppose a particle is in equilibrium with a heat reservoir at 300 K. This particle can exist in three states with energies (−0.1 eV), (0 eV) and (+0.5 eV). Calculate,

1. The partition function for this particle.
2. The probability that this particle is in each of the three energy states given above.

Solution: At 300 K we can show $kT = 0.026$ eV. Using this we can find,

1. The partition function for this particle, using equation (12.8), using $\mu = 0$:

$$\begin{aligned}\mathcal{Z} &= e^{-0.1/0.026} + e^{0/0.026} + e^{0.1/0.026} \\ &= 0.021\,4 + 1 + 46.805 \\ &= 47.82\end{aligned}$$

2. The probability for this particle to be in each of the three states:
   Probability for this particle to be in state 1:

$$P_1 = \frac{0.021\,4}{47.82} = 0.045\%$$

   Probability for this particle to be in state 2:

$$P_2 = \frac{1}{47.82} = 2.09\%$$

   Probability for this particle to be in state 3:

$$P_3 = \frac{46.805}{47.82} = 97.87\%$$

Worked example 12.6. Suppose the entropy (S) of an energy level is mathematically given as $k\ln(n)$, where n is the number of degenerate states for the level. Then find the probability of finding an atom in any particular energy level in terms of Helmholtz free energy (F).

Solution: If a particular energy level has n degenerate states, then the probability of the system $P(E)$ being in that level is the product of n with the probability of being in any one of the states. Mathematically, this is given as,

$$P(E) = nP(S) \tag{12.16}$$

From the relationship of entropy, $S = k\ln n$, we can find $n = e^{S/k}$. We also know that

$$P(S) = \frac{1}{\mathcal{Z}}e^{-E/kT}$$

And we can clearly see from equation (12.16),

$$P(E) = e^{S/k}.\ \frac{1}{\mathcal{Z}}e^{-E/kT}$$

and since $F = E - TS$,

$$P(E) = \frac{1}{\mathcal{Z}}e^{-F/kT}$$

which means the probability of finding an atom in any particular energy level is found in terms of Helmholtz free energy (F).

Worked example 12.7. Suppose we have a system in which each particle may occupy any of the 10 available single-particle states within a box. For the sake of simplicity lets assume each state has zero energy. Then find,

1. The partition function of the system if the box contains one particle.

   Solution: For a single particle, the partition function is,

   $$\mathcal{Z} = \mathcal{Z}_1 = \Omega_1 = 10$$

   The partition function equals the number of states, each with energy = 0, available to the particle.
2. The partition function of the system if the box contains two distinguishable particles.

   Solution: For two distinguishable particles, the partition function,

   $$\mathcal{Z} = \mathcal{Z}_1^2 = \Omega_1^2 = 10^2 = 100.$$

   This means that each particle independently has 10 available states.
3. The partition function of the system if the box contains two identical bosons.

   Solution: For two identical bosons, there are 10 ways to put both bosons in the same particle state plus there are (10)(9)/2 = 45 ways to put them into different single particle states. This means that they have a total of 55 different ways to put them into the available states. Hence, the partition function,

   $$\mathcal{Z} = 55.$$

4. The partition function of the system if the box contains two identical fermions.

   Solution: For two identical fermions there are only 45 allowed systems states since the 10 systems states from the above case where both particles in the same state are not allowed for fermions.

Concept questions 12.2

1. What are the types of ensemble in statistical mechanics?
2. What are the main differences between micro-canonical, canonical and grand canonical ensembles?
3. What is the physical meaning of partition function in statistical mechanics?

Exercises 12.2

1. Suppose a particle is in equilibrium with a heat reservoir at 300 K. This particle can exist in only two states with energies (−0.05 eV) and (+0.05 eV). Calculate,
   (a) The partition function for this particle.
   (b) The probability for this particle to be in each of the two states.
2. Compare the probabilities of the ground state being occupied for a system made up of bosons at 300 K, versus a system made up of fermions for the following situations:
   (a) If its energy is 1 eV less than μ.
   (b) If its energy is 100 eV less than μ.
   (c) If its energy is 1000 eV less than μ.

Further reading

- Hill T 2013 *An Introduction to Statistical Thermodynamics* (North Chelmsford, MA: Courier Corporation)
- Kittle C and Kroemer H 1980 *Thermal Physics* 2nd edn (San Fransisco, CA: W.H. Freeman and Company)
- Mandl F 1988 *Statistical Physics* 2nd edn (the Manchester Physics Series) (New York: Wiley)
- Reif F 1985 *Fundamentals of Statistical and Thermal Physics* (Singapore: McGraw-Hill International Edition)
- Resnick R, Halliday D and Krane K 2002 *Physics* vol 1 5th edn (New York: Wiley)
- Schroeder D 2000 *Thermal Physics* (Reading, MA: Addison-Wesley Longman)
- Sears F and Salinger G 1975 *Thermodynamics, Kinetic Theory and Statistical Thermodynamics* 3rd edn (Reading, MA: Addison-Wesley Publishing Company)

Chapter 13

Some applications of statistical mechanics

Quantum statistics is the study of dense systems at the quantum levels. In this chapter I will describe some phenomena that can be described by quantum statistics. I will divide this study into two major categories; quantum solids and quantum fluids (including liquids and gases).

13.1 Quantum statistical solids

In this section I will discuss three main examples where quantum statistics is helpful in understanding natural phenomena.

13.1.1 Einstein model of a solid

The Einstein model of a solid was originally proposed by Einstein in 1907 CE. To explain solids in Einstein's model, Einstein proposed that a solid should be considered to be made up of a large number of identical quantum oscillators with quantized energy units, as seen in figure 13.1.

The Einstein solid is a model of a solid based on two assumptions:

1. Each atom in the lattice of a solid is considered to be an independent quantum harmonic oscillator. In a three-dimensional solid, each atom can oscillate along three independent axes x, y, and z. This means when there are N oscillators in the sample there are actually only $N/3$ atoms[1].
2. All atoms oscillate with the same frequency (f).

The Einstein solid is composed of single-frequency quantum harmonic oscillators; in general, all the oscillations will have one common frequency (f). The quantum harmonic oscillator gives us a series of equally spaced quantized states for each oscillator in an Einstein solid, where the separation between the levels is given by hf,

[1] Examples of quantum oscillators also include the vibrational motions of diatomic and polyatomic gas molecules.

doi:10.1088/978-0-7503-3083-1ch13

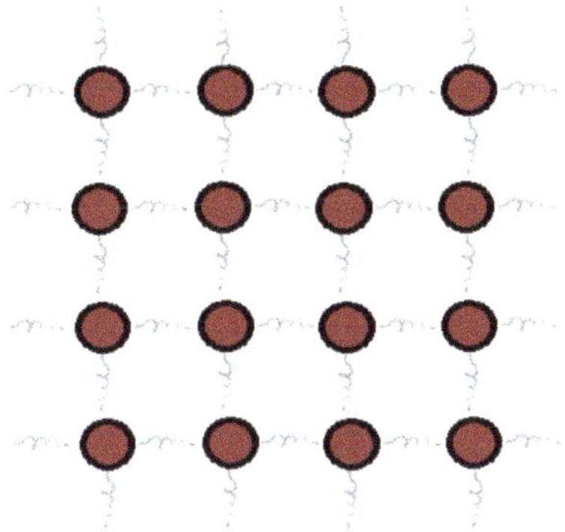

Figure 13.1. A very common way to visualize oscillation of atoms in a solid.

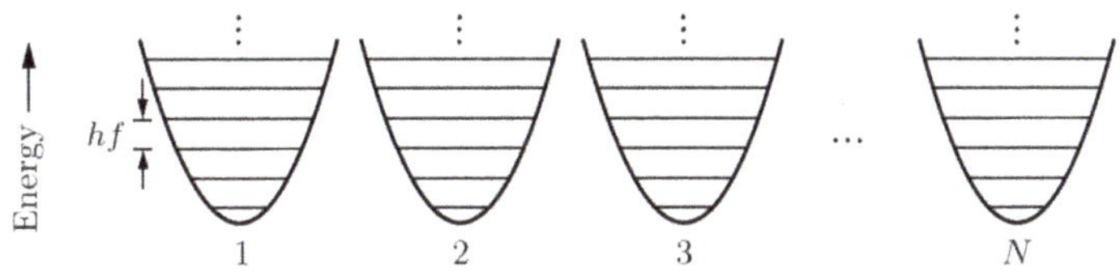

Figure 13.2. An Einstein solid as a collection of N oscillators with the same frequency.

where h is Planck's constant (6.63×10^{-34} J s) and f is the frequency of the oscillator, as shown in figure 13.2. The frequency of the oscillator is just the same as the frequency of a mass-spring system given as:

$$f = \frac{1}{2\pi}\sqrt{\frac{k}{m}}$$

where k is the spring constant and m is the mass attached. Figure 13.2, visualizes a group of uniformly spaced oscillators in a solid; please note in this case we ignore any interaction between the oscillators.

The concept of an Einstein solid is very helpful in determining the multiplicity in the distribution of energy among the available energy states of the system. All energy levels are considered to be equally probable within the constraint of having 'q' units of energy and 'q' oscillators. The general formula for multiplicity of an Einstein solid with 'N' oscillators and 'q' energy units is:

$$\Omega(N, q) = \frac{(q + N - 1)!}{q!(N - 1)!} = \binom{q + N - 1}{q} \tag{13.1}$$

Worked example 13.1. Consider a small Einstein solid with only three oscillators ($N = 3$) and four energy units ($q = 4$). Find the various microstates of this system.

Solution: We can use equation (13.1),

$$\Omega(N, q) = \frac{(q + N - 1)!}{q!(N - 1)!} = \frac{(4 + 3 - 1)!}{4!(2)!} = 15.$$

This means there are 15 microstates of this system, detailed in table 13.1. In the figure, the number 4 corresponds to energy quanta ($E = 4hf$), 3 corresponds to quanta ($E = 3hf$) and so on. We can also visualize this as shown in figure 13.3.

13.1.1.1 The entropy of the Einstein solid

The entropy of the Einstein solid is expressed in terms of the multiplicity (Ω) as,

$$S = k\,\ln(\Omega(N,\,q)) = k\,\ln\frac{(q + N - 1)!}{q!(N - 1)!}$$

In reality, an Einstein solid contains a very large number of oscillators (N) and energy units (q). In this case the above equation can be written as,

$$S = k\,\ln(\Omega(N,\,q)) \approx k\,\ln\frac{(q + N)!}{q!(N - 1)!}$$

Using the Stirling approximation ($\ln N! = N\ \ln N - N$) and the properties of logs as referenced in appendix B, we can simplify and write the entropy of the Einstein solid as,

$$S = k\,\ln\left(\frac{qe}{N}\right)^N = Nk\,\ln\left(\frac{qe}{N}\right)$$

Table 13.1. Quantum microstates for Einstein solid with $N = 3$ and $q = 4$.

Oscillator	1	2	3	1	2	3	1	2	3
Energy:	4	0	0	3	1	0	2	2	0
	0	4	0	3	0	1	2	0	2
	0	0	4	1	3	0	0	2	2
				0	3	1	2	1	1
				1	0	3	1	2	1
				0	1	3	1	1	2

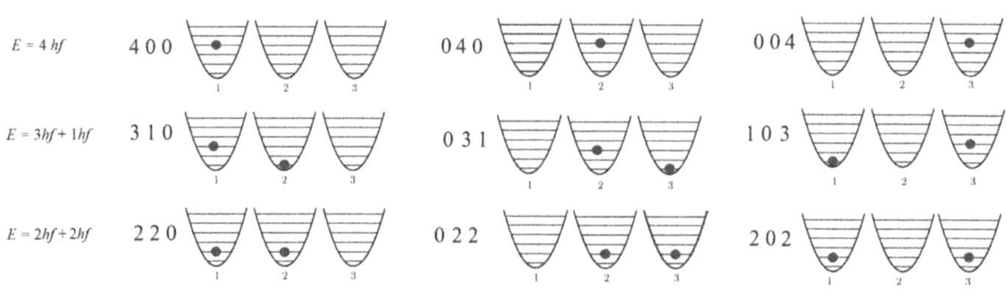

Figure 13.3. Visualizing some microstates in an Einstein solid with $N = 3$ and $q = 4$.

which can also be further simplified knowing that $\ln(e) = 1$, to give us the entropy of the Einstein solid as,

$$S = Nk \ln\left(\frac{q}{N}\right) + Nk \tag{13.2}$$

13.1.1.2 Heat capacity of Einstein solid

Classical mechanics says that the specific heat of a solid is independent of the temperature. However, in the Einstein model of solids, the heat capacity of an Einstein solid is mathematically found to be,

$$C_V = 3Nk\frac{(E/kT)^2 e^{E/kT}}{(e^{E/kT} - 1)^2} \tag{13.3}$$

where N is the number of atoms and $E = hf$ is the units of energy for identical oscillators.

1. When $kT \gg E$, the heat capacity is approximately equal to $3Nk$, which is a numerical constant.

   $$C_V = 3Nk$$

   This is called the '*Dulong–Petit law*'.
2. Below $kT \approx E$, the heat capacity becomes smaller as the temperature goes down and it approaches zero as temperature goes to absolute zero. This agrees with experimental data to the first approximation.
3. Equation (13.3) predicts that the heat capacity goes to zero exponentially in the limit $T \to 0$ K. However, experimentation shows that in reality C_V is directly proportional to T ($C_V \propto T^3$), given by,

   $$C_V \approx \frac{12Nk\pi^4}{5}\left(\frac{T}{T_D}\right)^3$$

   where T_D is called the '**Debye temperature**'. This was explained by Peter Debye in 1912 CE and is shown in section 13.1.2.3.

13.1.2 Phonons: Debye theory of solids

As we saw in the section above, the heat capacity of an Einstein solid was predicted to go to zero exponentially as the temperature is reduced. However experimentation showed that in reality C_V is directly proportional to T^3 ($C_V \propto T^3$). This is because atoms in a crystal do not vibrate independently of each other. Movement in one atom inevitably causes vibrations in its neighboring atoms. The amount of vibration depends on the frequency of oscillation and this behavior is quite complicated. In the low-frequency modes of oscillation, large groups of atoms move in-phase with their neighbors, whereas in high-frequency modes, large groups of atoms move out-of-phase with their neighbors.

We know from section 13.1.1 that energy is proportional to the frequencies of the modes of vibration ($E = hf$). This means units of energy (quanta) come in different sizes depending on the frequency of vibration (f). Which means as we lower the temperature, the high-frequency modes disappear; this is commonly called the *freeze-out of higher modes*. However, we do not see this at low temperatures. What we observe experimentally is that at even very low temperatures some low-frequency modes remain active. This is why the heat capacity (C_V) goes to zero less dramatically than what was predicted in the Einstein model of solids.

13.1.2.1 Phonons

The modes of oscillations of the solid crystal are very similar to the modes of oscillations of the electromagnetic field in vacuum. Debye realized that low-energy excitations of a solid material were not oscillations of a single atom, as in the Einstein model, but rather are collective modes propagating through the material. He considered such vibrations as sound waves, and used Quantum theory to introduce the concept of a particle called a '*phonon*'. The concept of phonons is analogous with the concept of photons in the electromagnetic wave. Phonons in quantum theory are considered to be *quasi-particles* with definite energies and directions of motion. The energy of a phonon, like that of a photon, is given as,

$$E = hf = h\frac{v_s}{\lambda}$$

where f is the frequency of oscillation, v_s is the speed of sound wave and λ is the wavelength. It is important to keep in mind that, unlike electromagnetic radiation in a box, a phonon will not have unbounded frequency. The frequency of a phonon is limited by the medium through which it propagates.

Phonons are bosons and as such they obey Bose–Einstein statistics and the occupation number of bosons in a state with energy E is given as:

$$n(E) = \frac{1}{e^{E/kT} - 1}$$

13.1.2.2 Difference between sound and light

There are some differences between sound and light waves:

1. The speed of sound waves is much lower than light waves.
2. Sound waves are limited by the stiffness and density of the medium of propagation. Light waves do not need a medium for propagation.
3. Sound waves cannot have wavelengths shorter than twice the atomic spacing. Light waves can have arbitrarily short wavelengths.
4. Sound waves can be transversely or longitudinally polarized. Light waves are always transversely polarized.

13.1.2.3 The Debye temperature

In the Debye model, the heat capacity depends on temperature and is defined using only one parameter called the '*Debye temperature*' (T_D). In terms of N, the total volume V, the volume for each allowed wave vector in the three-dimensional space, the Debye temperature T_D is defined as:

$$T_D = \frac{hv_s}{2k}\left(\frac{6N}{\pi V}\right)^{\frac{1}{3}} \tag{13.4}$$

where v_s is the speed of sound through the material, h is the Planck's constant and k is the Boltzmann constant. It is important to note that since the Debye temperature T_D is defined in terms of constants, it will not change once the values are specified. Below the Debye temperature, the heat capacity of a crystal changes with the cube of the temperature. Above the Debye temperature, the heat capacity of the same crystal is constant (temperature-independent).

In general, heavier atoms have lower Debye temperatures than lighter atoms. This is because the velocity of sound decreases as the density increases. The Debye temperatures of several substances are listed in table 13.2.

Table 13.2. The Debye temperatures of various substances.

Aluminum = 428 K	Iron = 470 K	Silicon = 645 K	Tungsten = 400 K
Lead = 105 K	Silver = 225 K	Carbon = 2230 K	Copper = 343.5 K
Nickel = 450 K	Gold = 165 K	Platinum = 240 K	Titanium = 420 K

Worked example 13.2. The speed of sound in gold (Au) is approximately 3240 m s^{-1}. Use this value to find the Debye temperature of gold atoms. Given, atomic number of gold is 79, and atomic mass is 196.96 a.m.u. or g mol^{-1}, density of gold is 19 300 kg m^{-3}.

Solution: The volume occupied by one mol of gold atoms is,

$$V = \frac{\text{mass}}{\text{density}} = \frac{0.196\,96 \text{ kg mol}^{-1}}{19\,300 \text{ kg m}^{-3}} = 1.021 \times 10^{-5} \text{ m}^3 \text{ mol}^{-1}$$

The theoretical Debye temperature is then calculated using equation (13.4) and knowing that $N = N_A = 6.022 \times 10^{23}$, Avogadro's number, $h = 6.63 \times 10^{-34}$ Js, $k = 1.38 \times 10^{-23}$ J K^{-1}, Debye temperature is found to be approximately 160.8 K.

13.1.3 Magnets and magnetization

Magnetism is the force exerted by magnets when they attract or repel each other and is caused by the motion of electric charges. The magnetic field is the area around a magnet that has magnetic force. All materials are made of atoms containing positive nuclei and negative electrons. The electrons revolve around the nucleus while spinning like a top. This movement of electrons generates a tiny electric current which causes each electron to act like a tiny microscopic magnet.

In most atoms, electrons are found in pairs and since electrons are fermions, they are subject to Pauli exclusion principle. Which means that in the electron pair the individual spins of electrons must be in the opposite directions. The opposite spins of electrons cause the pair's magnetic field to disappear and become ineffective.

Materials such as cloth or paper are very weakly magnetic because they have an equal numbers of electrons in their atoms spinning in opposite directions which cancels out the overall magnetism of the material.

However, materials that have some unpaired electrons will have a net magnetic field and will react to an external applied field. These include substances such as iron and nickel. It is interesting to note that these materials may not be magnetic by nature but can be magnetized. For example, iron can be magnetized by rubbing it with a magnet. When iron is rubbed with a magnet the south-seeking poles of the atoms in the iron line up in the same direction, whereas the north-seeking poles align in the opposite direction. This shift creates a force and causes the iron to become magnetic, which in turn creates a magnetic field around the iron.

Other materials can be magnetized by an electric current. When electricity runs through a coil of wire, it produces a magnetic field around it. However, this magnetization is temporary and the magnetic field disappears instantaneously once the electric current is turned off. These materials are called '*electro-magnets*'.

13.1.3.1 Dipoles

Suppose a system composed of N spin-$\frac{1}{2}$ particles is immersed in a magnetic field pointing in the positive z-direction. Each particle in the system acts like a compass needle and experiences a torque[2] ($\vec{\tau} = \vec{r} \times \vec{F}$) that tends to align the magnetic dipole moment with the external magnetic field. This behavior of the particles is why we call this material a '**magnet**'.

According to quantum mechanics, components of a particle's dipole moments are quantized. A spin-$\frac{1}{2}$ particle can only be quantized in two values, which we will call *up* and *down*. *Up*-state will be the state that aligns naturally with the external magnetic field, whereas *down*-state will be the state that aligns naturally opposite to the external magnetic field. To flip a dipole from the *up*- to *down*-state we would have to add energy to the system that is numerically equal to $2\mu_B\vec{B}$, where $\vec{B}$ is the strength of the external magnetic field and μ_B is a constant which is called the 'effective strength' of the compass needle. Please note that other textbooks may use μ for the strength, but in this book I have used the symbol μ for chemical potential, so I will use the symbol μ_B for the 'effective strength' of the compass needle.

13.1.3.2 Multiplicity of dipoles

If we define N_{up} to be the number of elementary dipoles that point up and N_{dn} to be the number of elementary dipoles that point down and consider that at a certain time 't' these numbers are fixed then the total number of dipoles (N) is the sum of dipoles as,

$$N = N_{\text{up}} + N_{\text{dn}}.$$

[2] Torque is defined as ($\vec{r} \times \vec{F}$) and is a force applied to an object that tends to cause rotation. Torque is a vector quantity and is what causes an object to acquire angular acceleration.

This system has one microstate for each possible value of N_{up} from 0 to N. The multiplicity will hence be the same as the one we got in equation (11.12) which I will repeat here for this particular case:

$$\Omega(N_{up}) = \binom{N}{N_{up}} = \frac{N!}{N_{up}!(N - N_{up})!} = \frac{N!}{N_{up}!N_{dn}!} \tag{13.5}$$

13.1.3.3 Total energy of the system

The total energy of the system is determined by calculating the total number of *up*-dipoles and the total number of *down*-dipoles. Therefore, finding the macrostate of the system is the same as specifying its total energy.

In order to twist the little dipoles in a magnetic material, the external magnetic field has to exert a torque. For an external field pointing up, the *up*-dipole will have less-energy than the *down*-dipole. This is because less energy will be required to align it with the external field. Similarly, for an external field pointing down, the *down*-dipole will have less-energy than the *up*-dipole. For the sake of symmetry, we say that the energy of the dipole that points up is $(-\mu_B B)$ and the energy for the dipole that points down is $(\mu_B B)$. Then the total energy of the system will be given as:

$$E = E_{up} + E_{dn} \tag{13.6}$$

Worked example 13.3. Starting from the basic definition of energy in equation (13.6), show that the total energy of the system can be written as,

$$E = \mu_B B(N - 2N_{up})$$

Solution: Assuming that the energy of the dipole that points up is $(-\mu_B B)$ and that for the dipole that points down it is $(\mu_B B)$, we can write for a system, energy (E) is,

$$\begin{aligned} E &= E_{up} + E_{dn} \\ &= -\mu_B B N_{up} + \mu_B B N_{dn} \\ &= -\mu_B B(N_{up} - N_{dn}) \\ &= \mu_B B(N_{up} - (N - N_{up})) \\ &= -\mu_B B(2N_{up} - N) \end{aligned}$$

This means that the total energy of the system is,

$$E = \mu_B B(N - 2N_{up}) \tag{13.7}$$

13.1.3.4 Magnetization

Magnetization (M) is defined as the total magnetic moment of the system. Each *up*-dipole has a magnetic moment $(+\mu_B N_{up})$ and each *down*-dipole has a dipole moment $(-\mu_B N_{dn})$, so the total magnetization can be written as:

$$\begin{aligned} M &= M_{\mathrm{up}} + M_{\mathrm{dn}} \\ &= \mu_B(N_{\mathrm{up}} - N_{\mathrm{dn}}) \\ \Longrightarrow M &= -\frac{E}{B} \end{aligned} \tag{13.8}$$

13.1.3.5 Types of magnetic materials

In this section I will briefly define the various types of magnetic materials.

1. **Paramagnet (temporary magnets):** Paramagnetic materials are weakly attracted by an externally applied magnetic field (say by a magnet) and form internal, induced magnetic fields in the direction of the applied magnetic field, as seen in figure 13.4. Most elements and some compounds are paramagnetic. In paramagnets, the magnetic moment induced by the applied field is linear and is rather weak.

   Materials that have at least one unpaired electron can be made magnetic. However, paramagnets loose the induced magnetization as soon as the externally applied magnetic field is removed. This happens because thermal motion randomizes the spin orientations very quickly and the total magnetization drops to zero pretty much as soon as the applied magnetic field is removed.

   Curie law:

   For an electronic two-state parameter, knowing e is the electron's charge and m_e is the mass of the electron, the numerical value of the constant μ_B is given as:

$$\begin{aligned} \mu_B &\equiv \frac{eh}{4\pi m_e} \\ &= 9.27 \times 10^{-24}\ \mathrm{J\ T^{-1}} = 5.78 \times 10^{-5}\ \mathrm{eV\ T^{-1}} \end{aligned} \tag{13.9}$$

   If we take $B = 1$ T, then $\mu_B B = 5.8 \times 10^{-5}$ eV, and at room temperature, $kT \equiv 1/40$ eV, so we can say that $\mu_B B \ll kT$. Under these conditions, the magnetization becomes:

$$M \equiv \frac{N\mu_B^2 B}{kT} \tag{13.10}$$

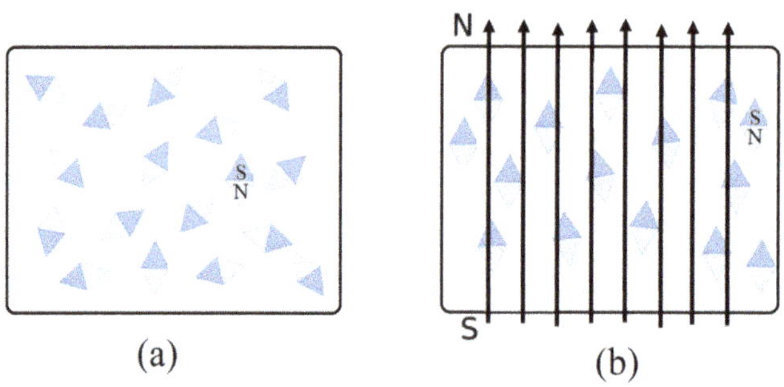

Figure 13.4. Orientation in paramagnetic material, (a) when magnetic field is not applied, (b) when magnetic electric field is applied.

We see in equation (13.10), in paramagnets, the strength of the external magnetic field is directly proportional to the magnetic moment induced. So as the external magnetic field increases, the magnetic moment induced also increases linearly. It is interesting to note that in equation (13.10), $\frac{N\mu_B^2 B}{k}$ are all constants, so it can be rewritten as:

$$M \equiv C\frac{B}{T} \tag{13.11}$$

where C is a material-specific constant called Curie constant. $M \propto 1/T$ was discovered experimentally by Pierre Curie and is known as **Curie's Law**. Curie's law holds for at the high-temperature and low magnetization limit for all paramagnets. However, it does not hold in the low-temperature and high-field region, at these regions magnetization becomes saturated and magnetic dipoles all align with the external applied magnetic field. Increasing the magnetic field will not increase the total magnetization because we will not be able to further align dipoles.

It is important to keep in mind that strong paramagnetism decreases as the temperature increases because thermal energy induces a greater random motion of the atomic magnets, which causes de-alignment of the atomic magnets. On the other hand, weak paramagnetism, found in many metallic elements in the solid state, is independent of temperature. This is because an applied magnetic field affects the spin of some of the loosely bound conduction electrons in these materials.

2. **Ideal paramagnet:** When there are no interactions between dipoles, each dipole only experiences torque form the external field. In this case we say that the system is an ideal paramagnet.
3. **Diamagnet:** Materials such as copper, silver, and gold, can be magnetized by the application of an external magnetic field. Such materials are called diamagnets. Magnetic properties arise in diamagnetic materials because of the realignment of the electron paths due to the influence of an external magnetic field. In diamagnetic materials all electrons are paired so there is no permanent net magnetic moment per atom. This is why diamagnetism is a weak form of magnetism and diamagnetic materials only remain magnetic until the external field is applied and disappears as soon as the external field is removed.

   When a diamagnetic material is brought close to a material that produces a magnetic field, the diamagnetic material repels it. This repulsive force is very weak but we can enhance it by placing the diamagnetic material in a strong magnetic (or electromagnetic) field. At this point the diamagnet overcomes Earth's gravitational pull and causes the diamagnetic material to float in the air. This is called '*diamagnetic levitation*'.
4. **Ferromagnet (permanent magnets):** Ferromagnetism is the basic mechanism by which certain materials form permanent magnets. Ferromagnetism, derived

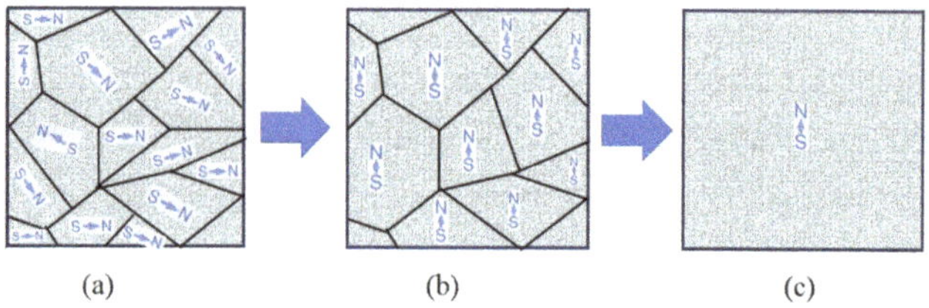

Figure 13.5. (a) An unmagnetized piece of iron (or other ferromagnetic material) has randomly oriented domains. (b) When magnetized by an external field, the domains begin to align. (c) Final overall magnetization.

from the Latin word for iron, *ferrum*, is generally associated with elements such as iron, cobalt, nickel, and some alloys containing these elements. Lodestone, discovered in nature about 2000 years ago, has the ability to acquire magnetic properties and hence is often called a 'natural ferromagnet'. Figure 13.5 shows how an unmagnetized piece of iron can be magnetized in the presence of an external field.

The magnetic properties of elements mentioned above are due to the presence of unpaired electrons in their outermost valence shells. These substances are much more paramagnetic than the rest of the elements and because the attraction experienced by ferromagnetic materials is much stronger, it is easily observable. Ferromagnetic materials are widely used in electric motors, generators, transformers, telephones, and loudspeakers, etc.

13.1.3.6 Differences between magnetic materials

It is important to note some vital differences between the various magnetic materials:

1. Diamagnetic materials are repelled by a magnetic field, whereas both paramagnetic and ferromagnetic materials are attracted by a magnetic field.
2. Paramagnetism is a stronger magnetic behavior but is present in few particular materials, whereas diamagnetism is a weak magnetic behavior that can be induced in nearly all materials.
3. Ferromagnetism is a permanent phenomena in certain materials, whereas diamagnetism can easily be suppressed in the presence of stronger magnetic properties.
4. The magnetic fields created by paramagnetic materials are in the opposite direction to the applied external magnetic field, whereas the magnetic field produced by diamagnetic materials is in the same direction as the applied magnetic field.
5. Above the Curie temperature, the thermal energy overcomes the internal aligning forces of the material and the spontaneous magnetization of the ferromagnetic material vanishes and the material becomes a paramagnetic material. The Curie temperature varies widely for different materials; for iron it is about 770 °C (1043 K), whereas it is only about 20 °C (293 K) for gadolinium.

Worked example 13.4. The magnetic properties of a substance can be determined by examining its electronic configuration. If the substance has unpaired electrons, the substance is paramagnetic and if all electrons in the substance are paired, the substance is then diamagnetic.

1. Boron (B) is a chemical element with atomic number 5 which is produced entirely by cosmic ray spallation and supernovae. Are boron atoms paramagnetic or diamagnetic?

   Solution: The boron (B) atom has the electron configuration: $1s^2 2s^2 2p^1$. It has one unpaired electron so boron is paramagnetic.
2. Zinc is the second-most-abundant trace mineral in your body and is a vital mineral that your body uses. It helps the immune system fight off invading bacteria and viruses. Are zinc atoms paramagnetic or diamagnetic?

   Solution: The zinc (Zn) atom has the electron configuration: $1s^2 2s^2 2p^6 3s^2 3p^6 3d^{10} 4s^2$. Zinc has no unpaired electrons so zinc is diamagnetic.

Concept questions 13.1

1. Is it okay to say 'each phonon is a *collective mode* caused by the motion of every atom in the lattice.'?
2. Are phonons quantized? Explain in your own words.
3. What are the different types of magnetization?

Exercises 13.1

1. Find the number of microstates and then list them for a small Einstein solid when the number of oscillators is $N = 4$, and energy units is ($q = 2$).
2. Calculate the multiplicity of an Einstein solid with 40 oscillators and 40 units of energy. How long do you think it will take you to list all of them manually? Please do not try to list them by hand!
3. Are the following materials paramagnetic or diamagnetic?
   (a) Oxygen atoms.
   (b) Chlorine atoms.
   (c) Bromine atoms.

13.2 Quantum statistical fluids

In this section we will discuss some of the examples of quantum statistical fluids.

13.2.1 Photons: black-body radiation

The term black body was first introduced by Gustav Kirchhoff, a German physicist, in 1860 CE. A black body is an object that absorbs all the electromagnetic radiation incident upon it. Then in order to stay in thermal equilibrium, a black body emits equivalent radiation at the same rate as it absorbs it, so a black body is also a perfect radiator of radiation. The radiated energy from the black body does not depend on the type of incident radiation. A black body at room temperature appears black, as most of the energy it radiates is in the infrared region of the electromagnetic spectrum and hence cannot be seen by the human eye.

Radiation emitted by a black body is called '*Black-body radiation*' or 'cavity radiation'. The radiated energy is considered to be produced by a standing wave or by resonant modes of the radiating cavity. Black-body radiation can be modeled in terms of modes in absorbing cavity, which gives us the Planck radiation formula and hence the Planck radiation curve, which for low temperatures gives us the Rayleigh–Jeans law.

Black-body radiation depends on the wavelength and peaks at shorter wavelength and higher temperatures according to the Wien displacement law, which provides a tool for measuring the surface temperature of the stars and other objects.

Classical Physics:
In classical physics electromagnetic radiation is treated as a 'continuous field' that permeates all space. For a black body we have to consider what happens to the electromagnetic radiation when it is confined inside a box of length (L) held at a certain temperature (T), for example in an oven. Inside the box only those modes will exist that correspond to standing wave wavelengths (λ). Each mode is a harmonic oscillator with a defined classical frequency 'f' given by, $f = \frac{c}{\lambda}$, where c is the speed of light.

Like any other mechanical oscillator, each electromagnetic standing wave has two degrees of freedom with average energy:

$$E_{\text{av}} = 2 \times \frac{1}{2}kT = kT \tag{13.12}$$

This equation says that energy is directly proportional to the temperature ($E \propto T$); this means as T increases E increases linearly. Since the total number of possible oscillators in the classical electromagnetic field is infinite, which implies that classically the total thermal energy is infinite as well.

Also, according to classical physics, the amount of radiation emitted at a given frequency is proportional to the number of modes and all modes have an equal chance of being produced. Classical physics also says that the number of modes is proportional to the square of the frequency, as shown in figure 13.6. However, this predicted increase in the radiated energy with frequency does not happen experimentally. For example, you are not subjected to a blast of electromagnetic radiation every time you open the oven door to check on the biscuits in the oven. Rather, we find that after a certain frequency, the energy goes down. This disagreement between classical theory and experiment is called the '**ultraviolet catastrophe**', a term coined in 1911 CE by Paul Ehrenfest.

Quantum physics:
The solution to the ultraviolet catastrophe comes from quantum mechanics. In quantum mechanics, harmonic oscillators have discrete energies, given by the

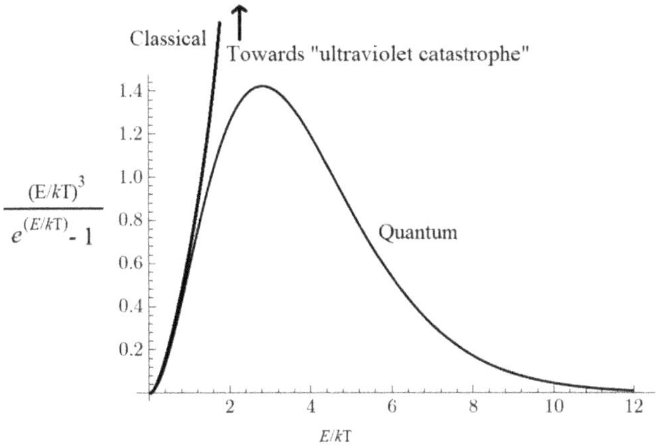

Figure 13.6. Classical versus quantum spectrum for black-body radiation.

formula $E = hf$ with values 0, hf, $2hf$, $3hf$, ... and the partition function for a single quantum harmonic oscillator is given as,

$$\mathcal{Z} = \frac{1}{1 - e^{-hf/kT}} \tag{13.13}$$

and the average energy is given as,

$$E_{\text{av}} = \frac{hf}{e^{hf/kT} - 1} \tag{13.14}$$

Assuming that the energy is in units of ($E = hf$), the average number of units of energy in the oscillator will be,

$$\langle n \rangle = \frac{1}{e^{hf/kT} - 1} \tag{13.15}$$

This formula is called the '*Planck's distribution*' and it gives us the number of photons present in a single mode of the electromagnetic field, called the '*single-particle state*'. According to this formula the short wavelengths with energies much greater than kT ($E \gg kT$) are exponentially suppressed and can be assumed not to exist. This is called the '*freeze out*' of short wavelengths and this contributes to the solution of the ultraviolet catastrophe.

13.2.1.1 Distribution function of photons

Units of energy (kT) in electromagnetic fields can be thought of as particles, called 'photons'. Photons are bosons, so they follow the Bose–Einstein statistics and any number of photons can be placed in any mode or wave pattern of the field. All photons travel at the same speed, which is the speed of light (c).

Number of photons is given by the Bose–Einstein distribution and knowing that the chemical potential is zero:

$$N_{\mathrm{BE}} = \frac{1}{e^{\frac{E}{kT}} - 1} = \frac{1}{e^{hf/kT} - 1} \tag{13.16}$$

13.2.1.2 Wien displacement law

As we saw above, the Planck's distribution tells us how many photons are in a single mode of the electromagnetic field, called the *single-particle state*'. The energies of the photons are given by, $E = pc = hnc/2L$. The average energy in a particular mode is mathematically given by multiplying the energy with the occupancy of that mode, where the occupancy is given by the Planck distribution. The total energy per unit volume is given by,

$$\frac{U}{V} = \int_0^\infty \frac{8\pi E^3}{(hc)^3}\left(\frac{1}{e^{E/kT} - 1}\right)dE \tag{13.17}$$

The integrand in equation (13.17) is the energy density per unit photon energy and is called the *spectrum of the photons* and is represented by $U(E)$ as,

$$U(E) = \frac{8\pi E^3}{(hc)^3}\left(\frac{1}{e^{E/kT} - 1}\right) \tag{13.18}$$

This function was first derived by Planck and is called '*Wein's Law*'. It gives the relative intensity of the radiation as a function of photon energy.

Experimentally, we measure the temperature inside an oven by letting out a bit of radiation and noting its color. For example, a temperature of 1500 K gives a spectrum that peaks at 0.36 eV in the near infrared region. Photons in the visible light region have energies in the range of (2–3) eV.

Total energy

In order to find the total energy we can integrate equation (13.17) from zero to ∞. I will leave the actual derivation for the reader to try and provide the result here.

$$\frac{U}{V} = \frac{8\pi^5(kT)^4}{15(hc)^3} \tag{13.19}$$

This expression looks complicated but when you really look at it really isn't. All terms on the right-hand side are constant except T so you can simply write,

$$\frac{U}{V} = (\text{const})T^4 \tag{13.20}$$

The most important feature in this expression is that energy per unit volume is proportional to the fourth power of temperature. This means when you double the temperature of your oven, the amount of electromagnetic energy inside the oven increases by a factor of $2^4 = 16$. Numerically, the total electromagnetic energy inside a typical oven is quite small; at baking temperature of about 375 °F the energy per unit volume is approximately 3.5×10^{-5} J m^{-3}, which is very, very small when compared to the thermal energy of air inside the oven.

Entropy of a photon gas

The easiest way to find the entropy of a photon gas is to start with the heat capacity:

$$C_V = \frac{\partial U}{\partial T} = \text{(const) } 4\ T^3$$

where const $= \frac{8\pi^5 k}{15}\frac{k^4 V}{(hc)^3}$ and the entropy is defined as,

$$S(T) \equiv \int_0^T \frac{C_V}{T} dT$$

After a bit of integration, and putting in the value of the heat capacity, the total entropy for a photon gas is found to be,

$$S(T) = \frac{32\pi^5 k}{45}\left(\frac{kT}{hc}\right)^3 V \tag{13.21}$$

13.2.1.3 Stefan–Boltzmann law

Black-body radiation summed over all wavelengths gives the Stefan–Boltzmann Law, which is the basis for radiation heat transfer. To calculate the number of photons emitted from a blackbody, we start with the photon gas contained in a box with a small hole, with area A, that leaks some photons. The amount of photons escaping the whole of area (A) must be coming from a volume given in spherical coordinates by $(Rd\theta)(R \sin\theta d\phi)(cdt) = R^2 \sin\theta d\theta d\phi(cdt)$. The amount of energy that escapes through the hole in time interval (dt) is,

$$= \int_0^{2\pi} \int_0^{\pi/2} \frac{A \cos\theta}{4\pi}\left(\frac{U}{V}\right) cdt \sin\theta \tag{13.22}$$

Which can be integrated and simplified to give,

$$= \frac{A}{4}\left(\frac{U}{V}\right) cdt$$

This means that the amount of energy that escapes is proportional to the area of the hole (A) and the duration of the time interval (dt). We can get the power per unit area per unit time interval ($dt = 1$) by putting the value of $\left(\frac{U}{V}\right)$ from equation (13.19) and simplifying,

$$\frac{P}{A} = \frac{c}{4}\left(\frac{U}{V}\right) = \frac{c}{4}\left(\frac{8\pi^5 (kT)^4}{15(hc)^3}\right) = \sigma T^4$$

where σ is called the *Stephan–Boltzmann constant*, which is numerically equal to $\frac{2\pi^5 (k)^4}{15(hc)^3} = \sigma = 5.67 \times 10^{-8}$ W m^{-2} K^{-4}. Hence, power per unit area can be simply written as,

$$\text{Power per unit area} = \sigma T^4 \tag{13.23}$$

This is called *Stephen's Law* and it was first discovered in 1879 CE. Five years later, in 1884 CE, Boltzmann derived the T^4 behavior from theory by applying classical thermodynamic reasoning to a box filled with electromagnetic radiation, and using Maxwell's equations to relate pressure to energy density. So these days this law is called the '*Stefan–Boltzmann Law*'.

13.2.2 Bosons: Bose–Einstein condensation

Bose–Einstein condensation (BEC) is a quantum phenomenon that was first observed in dilute atomic gases in 1995. However, the story of BEC does not start there. The story begins when Satyendra Nath Bose wrote a letter in 1924 CE to Albert Einstein in which he hypothesized about the existence of a physical law that describes how light and matter interact. It was understood at this time that light is carried by particles called photons, which can be described as particles or waves. However, Bose described a new approach by which photons and other particles like photons could be analyzed. Bose and Einstein gave us a new tool called the Bose–Einstein statistics and the prediction of new materials, called Bose–Einstein condensates, at very low temperatures.

In 1937 CE, physicists discovered superfluidity in an isotope of helium when it was cooled to 2.17 K. However, scientists of the time could not agree if this superfluid consisted of BECs or not. The first unequivocal demonstration of a BEC, using a gas of rubidium-87 atoms, came in 1995 by Carl Wieman and Eric Cornell. This discovery was reconfirmed within a few months by Wolfgang Ketterle, by achieving BEC in sodium atoms[3]. Since then more condensates have been created using atoms of hydrogen, lithium, potassium, cesium, calcium, strontium, chromium, and ytterbium.

A Bose–Einstein condensate forms in atoms that act like bosons at temperatures very close to absolute zero (0 K). When boson atoms are cooled to just above 0 K, they abruptly condense into the ground state and form a condensate. Since bosons follow the Bose–Einstein statistics, they are not limited by number per state, so all bosons end up in the ground state in the condensate. At this time boson atoms will lose their individuality and behave as one large collective super-atom[4]. This exotic behavior of bosons provides us with an opportunity to study quantum effects on a large scale, instead of dealing with individual atoms. A Bose–Einstein condensate looks like a dense little lump in the bottom of a magnetic trap; think of it as a drop of water condensing out of damp air onto a cold bowl.

For atoms confined to a cubical box of length (L) and volume $V = L^3$, the energy of the ground state is given as:

$$E_0 = \frac{3h^2}{8mL^2}. \tag{13.24}$$

[3] In 2001, Wieman, Cornell, and Ketterle shared the Nobel prize in physics for their groundbreaking work on BECs.

[4] By the way, this is similar to how photons become indistinguishable in a laser beam.

At any temperature, the average number of atoms (N_0) is given by the Bose–Einstein distribution, mathematically expressed as:

$$N_0 = \frac{1}{e^{\frac{E_0-\mu}{kT}} - 1}. \tag{13.25}$$

When T is sufficiently low, N_0 is quite large. When N_0 is large the argument of the exponential has to become zero, which is possible when the chemical potential is $\mu = E_0$. This means at $T = 0$, the chemical potential $\mu = E_0$, and when T is greater than zero, μ is a little bit less than E_0.

Condensation temperature:
Bose–Einstein condensation occurs abruptly when the temperature is lowered to a certain critical transition temperature. This transition temperature is called the *condensation temperature* (T_{cond}), while the ground-state atoms themselves are called the *condensate*. Numerically, the condensation temperature is not as low as we may have guessed. A single particle in a box of volume (V) is most likely to be found in its ground state when kT is of the order of E_0. However, if we place a large number of identical bosons in the same box, we can get most of them in the ground state at temperatures just less than T_{cond}, which is much higher. The condensation temperature is numerically expressed as,

$$T_{\text{cond}} = (0.527)\frac{h^2}{2\pi km}\left(\frac{N}{V}\right)^{2/3} \tag{13.26}$$

where N is the total number of atoms. The number of atoms (N_0) in terms of the condensation temperature in the ground state is given as,

$$N_0 = N\left[1 - \left(\frac{T}{T_{\text{cond}}}\right)^{3/2}\right] \tag{13.27}$$

Worked example 13.5. A sample of 10^4 atoms of rubidium-87 are confined in a cubical box with sides 10^{-5} m.

1. Calculate the ground state energy.
2. Calculate the condensation temperature.
3. Calculate the number of atoms in the ground state when $T = 0.9T_{\text{cond}}$.

Solution:

1. The ground state energy can be calculated using equation (13.24):

$$E_0 = \frac{3h^2}{8mL^2} = \frac{3}{8}\frac{(6.63 \times 10^{-34}\ \text{Js})^2}{8(87 \times (1.66 \times 10^{-27}\ \text{kg}))(10^{-5}\ \text{m})^2} = 1.14 \times 10^{-32}\ \text{J}.$$

2. From equation (13.26), the condensation temperature is,

$$T_{\text{cond}} = (0.527)\frac{h^2}{2\pi km}\left(\frac{N}{V}\right)^{2/3}$$

$$= (0.527)\frac{(6.63 \times 10^{-34}\ \text{Js})^2}{2\pi(1.380\,649 \times 10^{-23}\ \text{J K}^{-1})(87)(1.66 \times 10^{-27}\ \text{kg})}\left(\frac{10^4}{(10^{-5}\text{m})^3}\right)^{2/3}$$

$$\Longrightarrow T_{\text{cond}} = 8.6 \times 10^{-8}\ \text{K}$$

3. The number of atoms in the ground state when $T = 0.9T_C$, calculated by equation (13.27).

$$N_0 = N\left[1 - \left(\frac{T}{T_C}\right)^{3/2}\right]$$

$$= (10^4)\left[1 - \left(\frac{0.9T_C}{T_C}\right)^{3/2}\right]$$

$$\Longrightarrow N_0 = 1460$$

13.2.3 Fermions: Fermi gas

Fermions are particles with half-integer spin that follow Fermi–Dirac statistics. Fermions are subject to the Pauli exclusion principle, which states that no quantum state can be occupied by more than one fermion with an identical set of quantum numbers. Therefore, at low temperatures a Fermi gas cannot condense into a Bose–Einstein condensate like bosons do. However, weakly-interacting Fermi gases might form Cooper pairs that are discussed in section 13.2.5.

The state of a fermion depends on the 'orientation' of spin with respect to an applied magnetic field and on its location in phase space. For spin s, there are $(2s + 1)$ spin states or 'orientations' of s. In the absence of a magnetic field, these $(2s + 1)$ states all have the same energy and they increase the number of states available to the fermion.

The statistical mechanics of the Fermi gas follows directly from the grand canonical partition function which gives the expected number of fermions in energy state (E) as,

$$n_{\text{FD}}(E) = \frac{1}{e^{(E-\mu)/kT} + 1} \tag{13.28}$$

and the internal energy is,

$$U = \int n(E)g(E)dE \tag{13.29}$$

where $g(E)$ is numerically given as $\left(= 2\pi\left(\frac{2m}{h^2}\right)^{3/2} E^{1/2}\right)$.

It is important to keep in mind that: one, the Fermi energy is actually an energy difference, usually a difference in the kinetic energy, since for non-interacting particles the potential energy is zero, whereas the Fermi level is a total energy level including kinetic energy and potential energies; and two, the Fermi energy is defined only at absolute zero temperature, while the Fermi level is defined for any temperature (T).

A Fermi gas is called '**degenerate**' when its temperature is low when compared to the Fermi energy (E_F). When T is much less than E_F, the orbitals of energy lower than the Fermi energy will be almost entirely occupied and higher energy orbitals will be almost entirely vacant. Remember, orbitals are considered to be occupied when they contain one fermion. Hence at absolute zero (T = 0 K), the statistical properties of the Fermi gas are given simply by,

$$n_{\mathrm{FD}}(E) = \frac{1}{e^{(E-E_\mathrm{F})/kT}+1} = \begin{cases} 0, & \text{when } E > E_\mathrm{F}, \\ 1, & \text{when } E < E_\mathrm{F} \end{cases} \tag{13.30}$$

where E_F is called the '*Fermi energy*' and is numerically given as,

$$E_F = \frac{\hbar^2}{2m}(3\pi^2 n)^{2/3} \tag{13.31}$$

The expected number of fermions in energy state (E) is shown in figure 13.7. A Fermi gas is said to be '**non-degenerate**' when the temperature is high relative to E_F (figure 13.8).

13.2.3.1 Fermions: Fermi level and Fermi function

At $T = 0K$, the Fermi gas is in its ground state, which is formed by filling up the lowest energy states until all N fermions in the system are accommodated. Since there are two spin states, one occupied by a spin-up fermion (s_z = 1/2) and the other by a spin-down fermion (s_z = −1/2), allowed per space state, this requires $N/2$ space states in phase space. This 'sea' of occupied states is often called the '*Fermi sea*'.

The term '*Fermi level*' describes the top of the occupied electron energy levels at absolute zero temperature (T = 0 K). Electrons at absolute zero pack themselves into the lowest available energy states and build up a '*Fermi sea*' of electron energy states. At higher temperatures, characterized by the Fermi function, $f(E)$, that arises from Fermi–Dirac statistics, a certain fraction of electrons may exist above the Fermi level. The Fermi function $f(E)$, gives the probability that a given available electron energy state will be occupied at a given temperature as,

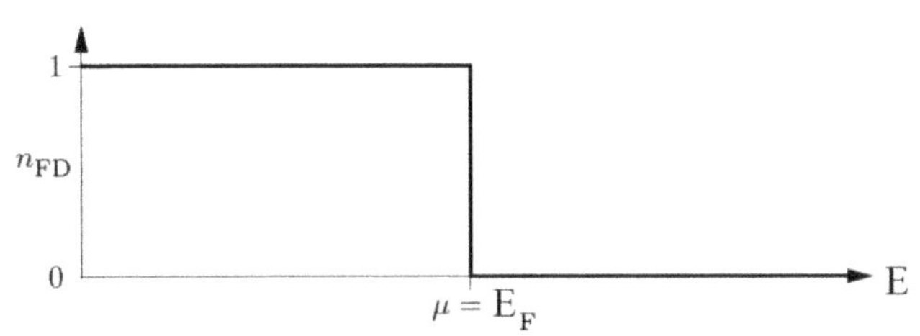

Figure 13.7. The Fermi–Dirac distribution at absolute zero temperature. At T = 0, the Fermi–Dirac distribution equals 1 for all states with $E_F < \mu$ and equals 0 for all states with $E_F > \mu$.

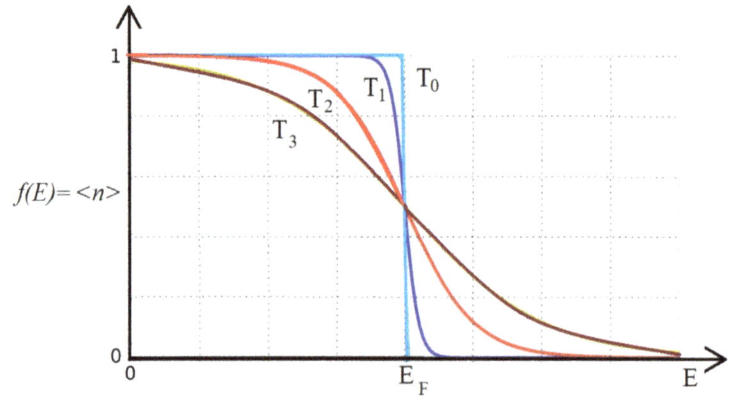

Figure 13.8. The Fermi–Dirac distribution at various temperatures, where $T_3 > T_2 > T_1 > T_0$.

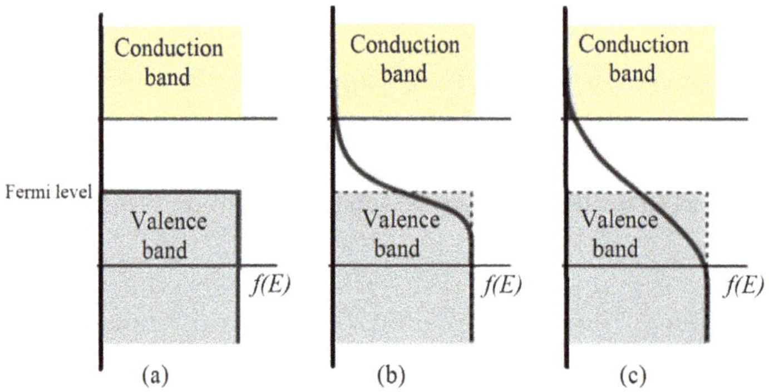

Figure 13.9. Fermi level for semiconductors: (a) At 0 K no electrons are above the Fermi level. (b) As the temperature increases from 0 K, some electrons jump to the conduction band. (c) At higher temperatures, more electrons jump to the conduction band.

$$f(E) = \langle n \rangle = \frac{1}{e^{(E-E_{\mathrm{F}})/kT} + 1} \tag{13.32}$$

From this equation we see that around room temperature, most levels up to the Fermi level (E_{F}) are filled and few electrons have energies above the Fermi level. For example, for copper at room temperature (298 K) the Fermi level is $\approx$7 eV and thermal energy kT is $\approx$0.03 eV. From these values we see Fermi function at room temperature is approximately 1 up to the Fermi level and rapidly goes to zero above the Fermi level.

In figure 13.9(a), we see that there are no electrons above the Fermi level for a semiconductor at 0 K. This is because no electron in the system has energy above the Fermi level. As the temperature increases, figure 13.9(b) the electrons gain thermal energy and some reach the conduction band and contribute to the electric current. These electrons are called '*free electrons*' and in metals, the velocity of free-electrons is given as,

$$v = \sqrt{\frac{2E_F}{m}}$$

Figure 13.9(c) shows the effect of raising the temperature of the semiconductor. As the temperature increases, the number of electrons in the conduction band also increases. This makes physical sense; raising the temperature of a substance takes the electrons to higher levels. This increases the occupation numbers for energy states above the Fermi energy and lowers the occupation numbers for states just below the Fermi energy.

13.2.3.2 Fermions: ground state of Fermi gas in 3D

The Fermi energy is the energy of the highest filled orbital at absolute zero. It is determined by requiring that the system in the ground state holds N electrons and each orbital is filled with one electron up to the Fermi energy given by the formula,

$$E_F = \frac{\hbar^2}{2m}(3\pi^2 n)^{2/3} \tag{13.33}$$

This relates the Fermi energy to the electron concentration $\frac{N}{V} \equiv n$.

13.2.3.3 Fermions: Fermi temperature

From the Fermi energy (E_F) we can find the corresponding temperature, which is called the '*Fermi temperature*' (T_F) and is defined as

$$kT_F \equiv E_F.$$

1. **Case 1: for low temperatures** ($T \ll T_F$): The Fermi distribution deviates from the values at $T = 0$ mainly in the neighborhood of E_F in a layer of thickness kT. Particles at energies of order kT below the Fermi energy are excited to energies of order kT above the Fermi energy.
2. **Case 2: for higher temperatures** ($T \gg T_F$): At temperatures higher than Fermi temperature, the Fermi distribution approaches the Maxwell–Boltzmann distribution and the quantum nature of the constituent particles becomes irrelevant.

13.2.3.4 Fermions: Fermi gas in metals

The alkali metals copper, silver, and gold have one valence electron per atom, which becomes the conductive or the free electron in a metal. The concentration of conductive electrons in a metal equals the concentration of atoms in that metal. This conductive electron quantity can be evaluated by the density and the crystal lattice structure of the metal.

If the conductive electrons in the metal act as a free fermion gas and knowing $n(\equiv N/V)$, which is the number density of conduction electrons in metals and is approximately 10^{28} to 10^{29} electrons per m^3, the value of Fermi energy in any metal is given as,

$$E_F = \frac{\hbar^2}{2m}(3\pi^2 n)^{\frac{2}{3}} \approx 1.602\,2 \times 10^{-18}\ \text{J}. \tag{13.34}$$

We can calculate T_F, from this value of Fermi energy. We find that the Fermi temperature is of the order of 10^6 K, which is much higher than the temperature at the surface of the Sun (5778 K). Metals generally boil before reaching this temperature under normal atmospheric pressure and thus for any practical purpose, a metal can be considered a Fermi gas at absolute zero temperature as a first approximation!

Worked example 13.6.

a. Calculate the Fermi energy for copper. Given that the density (ρ) of copper is 8960 kg m^{-3} and its atomic weight is 0.0635 kg mol^{-1}.
   Solution: To find the Fermi energy we use the formula:

$$E_F = \frac{\hbar^2}{2m}(3\pi^2 n)^{2/3}$$

knowing, $\hbar = 1.055 \times 10^{-34}$ J s, $m = 9.11 \times 10^{-31}$ kg, and 1 eV $= 1.60 \times 10^{-19}$ J and the electron concentration (n) can be calculated as,

$$n = \frac{N}{V} = \frac{\text{atoms}}{\text{Volume}} = \frac{N_A}{m/\rho} = \frac{N_A\rho}{m}$$

$$= \frac{(6.022 \times 10^{23}\ \text{mol})(8960\ \text{kg m}^{-3})}{0.0635\ \text{kg mol}^{-1}} = 8.49 \times 10^{28}\ \text{m}^3$$

We find Fermi energy as,

$$E_F = 11.2 \times 10^{-19}\ \text{J} \approx 1\ \text{eV}.$$

b. What is the corresponding electron velocity?
   Solution:

$$E_F = \frac{1}{2}mv^2$$

will give us,

$$v = \sqrt{\frac{2E_F}{m}} = 1.6 \times 10^6\ \text{ms}^{-1}$$

c. Calculate the Fermi temperature?
   Solution: The Fermi temperature can be found as,

$$T_F = \frac{E_F}{k} = \frac{11.2 \times 10^{-19}\ \text{J}}{1.38 \times 10^{-23}\ \text{J K}^{-1}} = 8.1 \times 10^4\ \text{K}.$$

13.2.4 Fermions: electrons in white dwarf stars

The first white dwarf, Sirius B, was observed in 1915 CE by Walter Adams. He found that the spectrum of the stellar object orbiting around Sirius was similar to the star Sirius and he named it Sirius B. The mass of the newly discovered star was found by applying third Kepler's third law to be between $[0.75 - 0.95]M_{Sun}$, with an effective temperature of 8000 K and a radius, R_{SirusB} 5850 km.

The white dwarf star is essentially a degenerate electron gas mixed in with nuclei to balance the charge and to provide the gravitational attraction that holds the star together. Usually, white dwarfs are composed of helium, carbon and oxygen. When the nuclear material in the core of a star has been burnt, the core of the star contracts to a point where the distance between nuclei becomes comparable with the dimensions of the nuclei, it is called a neutron star. In a white dwarf, the distance between nuclei is about 100 times the diameter of a nucleus. At this time, external orbits of the electrons are squeezed off, and the density of the star increases. As a consequence of this process a dense core of nucleons forms, which is immersed in a degenerate gas of free electrons (figure 13.10).

White dwarf stars tend to be extremely hot by our standards. It is estimated that about 10% or more of all the stars out there consist of white dwarf stars. White dwarf stars have masses comparable to that of the Sun (M_{Sun}), but have radii of the order of 10^4 km. The mass and radius of the Sun are,

$$\begin{aligned} M_{Sun} &= 1.9 \times 10^{24}\ \text{kg} \\ R_{Sun} &= 6.9 \times 10^{5}\ \text{km} \end{aligned} \tag{13.35}$$

A typical white dwarf is half as massive as the Sun, yet only slightly bigger than the size of Earth. Earth has an average density of 5.4×10^3 kg m^{-3}, whereas an Earth-sized white dwarf has a density of1×10^9 kg m^{-3}. This means a white dwarf is 200 000 times as dense as Earth. Atoms under these densities are entirely ionized into nuclei and free electrons, and the electron gas becomes a degenerate gas. The number density

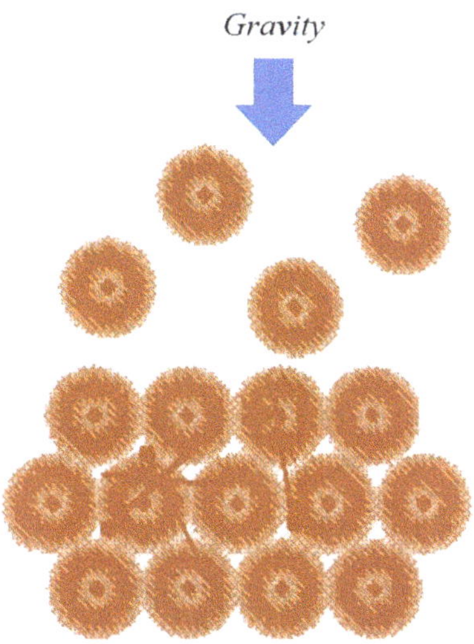

Figure 13.10. White dwarf: protons and neutrons free to move around, however, the electrons run out of room to move around and they prevent further collapse of the star.

of electrons in a white dwarf is of the order of 10^{36} electrons m^{-3} and their Fermi energy is:

$$E_F = \frac{\hbar^2}{2m_e}\left(\frac{3\pi^2(10^{36})}{1\ m^3}\right)^{2/3} \approx 4.81 \times 10^{-14}\ J$$

Electron degeneracy pressure[5], which is a direct consequence of the Pauli exclusion principle, stops the gravitational collapse of a star if its mass is below the Chandrasekhar limit[6] of 1.44 solar masses. This is the pressure that prevents a white dwarf star from collapsing. However, above the Chandrasekhar limit a star may collapse to form a black hole or a neutron star if there is no significant thermally generated pressure to prevent the collapse.

13.2.5 Superfluids and Cooper pairs

13.2.5.1 Cooper pairs

American physicist Leon Cooper in 1956 CE described a quantum effect by which a very small, almost negligible attraction between the electrons in a metal causes a paired state of electrons to have a lower energy than the Fermi energy. This causes the electrons to pair up and these pairs of electrons are called '*Cooper pairs*'. According to quantum mechanics, Cooper pairing effect is due to electron–phonon interactions, where the phonon causes the collective motion of the positively-charged lattice of the metal.

So even though Cooper pairing is a quantum mechanical effect, it can also be seen classically. An electron in a metal has the characteristics of a free particle; it is repelled by other electrons and is attracted to the positive ions that make up the lattice of the metal. At sufficiently low temperatures, this attraction moves the ions slightly toward the electron, increasing the positive charge density of the lattice in the vicinity, which in turn attracts other electrons toward the lattice. At far way distances, this attraction between electrons caused by the displaced lattice overcomes the repulsion between same-charged electron and causes the electrons to pair up (figure 13.11).

Please note that in this case we are talking about long-range interactions, so the electrons in a Cooper pair are not necessarily close together and be many hundreds of nanometers apart. The energy of the pairing interaction is quite weak and even a small amount of thermal energy can break them up; this means a significant number of the electrons in Cooper pairs only exit at very low temperatures.

We also know that electrons are fermions and have spin-1/2, whereas the total spin of a Cooper pair is integer (0 or 1) so it is a '**composite boson**'. Therefore, many Cooper pairs are allowed to be in the same quantum state. This tendency for all

---

[5] Electron degeneracy pressure results from the same mechanism that provides the electron orbital structure of matter, with the nucleus in the middle and electrons in orbit around it. For bulk matter with no net electric charge, the attraction between electrons and nuclei exceeds the mutual repulsion of electrons plus the mutual repulsion of nuclei, so without electron degeneracy pressure, the electrons would collapse into a single nucleus.

[6] The concept of a limiting mass for white dwarfs was first introduced by Chandrasekhar in a paper published in 1931 in *The Astrophysical Journal*, 74 n.1, 81, 'The maximum mass of ideal white dwarfs'.

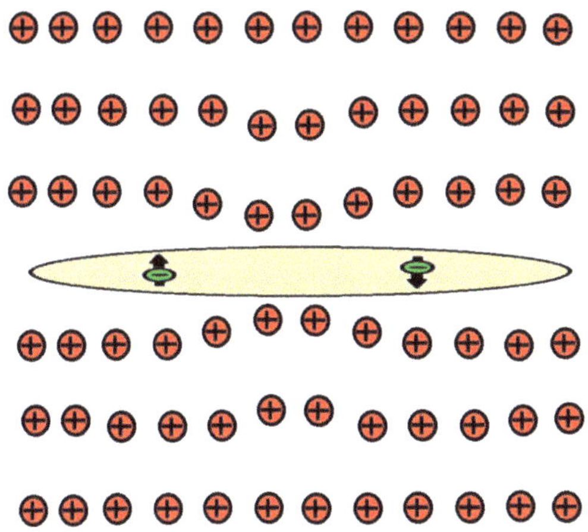

Figure 13.11. Cooper pairs: In classical materials, spin-opposite pairs form due to the interaction with lattice vibrations (phonons), which can give rise to an attractive force. Figure not drawn to scale.

Cooper pairs in a body to 'condense' into the same ground quantum state is responsible for superconductivity.

13.2.5.2 Superfluids

A substance is called a '*superfluid*' when it has zero viscosity and hence zero internal resistance. This means superfluids flow without any loss of kinetic energy. If a superfluid is stirred, it forms a vortex that continues to rotate forever because it has zero internal resistance.

Many metals become super conducting at low temperatures. The transition temperature (T_{sc}) is typically a few kelvin[7]. In recent years, super conductivity has been of interest again because we discovered 'high-temperature superconductivity' (T_{sc} = 100 K) in certain oxide materials. It has been shown that Cooper pairs are again involved in superconductivity. However, the precise mechanism for the attractive interaction is a matter of much controversy and uncertainty.

Superfluidity occurs in two isotopes of helium, namely helium-3 (^{3}He) with atomic mass of 3.02 a.m.u. and helium-4 (^{4}He) with atomic mass of 4.00 a.m.u. Helium-4 has two protons, two neutrons, and two electrons, with an overall spin of zero, which makes it a boson, and follows the Bose–Einstein statistics, whereas, with one fewer neutrons, helium-3 has an overall spin of one half, which makes it a fermion, which follows the Fermi–Dirac statistics.

Superfluidity in Bose–Einstein systems

An ideal Bose–Einstein gas (made up of say helium-4) at low temperatures becomes ordered by having all of its particles enter the same ground state. The assembly ground state is simply the systematic combination which has all N individual particles in the same, lowest, one-particle state. This type of coherent occupation of

[7] Note: non-valent metals do not become superconductive.

a single quantum state by all the particles gives a valuable picture of superfluidity observed in liquid helium-4 below 2.17 K.

Superfluidity in Fermi–Dirac systems

It is even more exotic to find superfluidity in Fermi–Dirac systems. A Fermi gas (made up of say helium-3) becomes ordered in a very different way, with the assembly of the ground state being dominated by the Pauli exclusion principle. Since all one-particle states can now be only singly occupied, the assembly ground state has full occupation for the one-particle states with energies up to the Fermi-energy $\mu(0)$, and zero occupation for those with higher energies. This provides a well-ordered (zero-entropy) state, but with a considerable energy ($0.6N\mu(0)$) for an ideal gas. This is called '*zero-point energy*'. Nature does not like this and would like to lower the zero-point energy. One way to do this is for a magnetic transition to occur. Another way is for a transition to a superfluid state to occur which causes superfluidity in Fermi–Dirac systems.

Concept questions 13.2

1. What is a degenerate gas?
2. What is the difference between Fermi level and Fermi energy?
3. Explain the behavior of bosons versus fermions at low temperatures.

Exercises 13.2

1. Calculate the Fermi temperature for the electron gas at the center of the Sun, given the mass of electron as 9.1×10^{-31} kg, concentration of electrons is 10^{32} per m^3. Show that this is close to the actual temperature of the order 10^7 K.
2. Our Sun has a radius of 6.96×10^8 m.
   (a) How much power is radiated by the Sun given the temperature of the Sun's surface is 5700 K. Hint: $\sigma = 5.67 \times 10^{-8}$ W m^{-2} K^{-4}.
   (b) What is the wavelength of maximum intensity of solar radiation?
3. Calculate the condensation temperature for 2×10^{10} atoms of atomic hydrogen that is cooled till its density becomes 1.8×10^{14} atoms cm^{-3}.

Further reading

- Hill T 2013 *An Introduction to Statistical Thermodynamics* (North Chelmsford, MA: Courier Corporation)
- Kittle C and Kroemer H 1980 *Thermal Physics* 2nd edn (San Fransisco, CA: W.H. Freeman and Company)
- Mandl F 1988 *Statistical Physics* 2nd edn (The Manchester Physics Series) (New York: Wiley)

- Reif F 1985 *Fundamentals of Statistical and Thermal Physics* (Singapore: McGraw-Hill International Edition)
- Schroeder D 2000 *Thermal Physics* (Reading, MA: Addison-Wesley Longman)
- Sears F and Salinger G 1975 *Thermodynamics, Kinetic Theory and Statistical Thermodynamics* 3rd edn (Reading, MA: Addison-Wesley Publishing Company)

IOP Publishing

Thermodynamics and Statistical Mechanics

An introduction for physicists and engineers

Samya Zain

Appendix A

Conversion factors:

Mass

1 slug = 14.59 kg

Length (L)

1 mm = 0.1 cm = 0.001 m

1 ft = 12 in = 2.54 cm

1 yd = 0.9144 m

1 mi = 1 609.3 m = 5280 ft

1 nautical mile = 6076 ft = 1.852 km

Time

1 in^2 = 645.2 mm^2

1 day = 24 h = 86 400 s

Area

1 ft^2 = 144 in^2 = 0.092 90 m^2

Pressure

1 Pa = 1 N m^{-2} = 1 kg m^{-1} s^{-2}

1 lb in^{-2} (psi) = 144 lb ft^{-2} = 6895 Pa

1 atm = 101.325 kPa = 1.013 25 bar = 14.70 psi

= 760 mm Hg [0 °C] = 10.332 56 m H_2O [4 °C]

Volume

1 ft^3 = 0.028 32 m^3

1 gal = 3.7854 liter = 0.1337 ft^3

1 in^3 = 16 390 mm^3 =1.639×10^{-5} m^3

Velocity

1 mi h^{-1} = 1.467 ft s^{-1} = 1.609 km h^{-1} = 0.4470 m s^{-1}

1 knot = 1 nautical mile h^{-1} = 1.689 ft s^{-1}
= 0.5144 m s^{-1}

Acceleration

1 ft s^{-2} = 0.304 8 m s^{-2}

g = 9.81 m s^{-2} = 32.17 ft s^{-2}

Energy

1 J = 1 N m = 1 kg m^2 s^{-2} = 0.737 562 lbf ft

1 cal (Int.) = 4.186 81 J

1 Btu (Int.) = 1.055 056 kJ = 778.169 3 lbf ft

1 erg = 1.0×10^{-7} J

1 eV = $1.602\,2 \times 10^{-19}$ J

Work-Energy: 1 ft lb = 1.356 J = 1.356 N m

Temperature (T)

$T_C = T_K - 273.15 = \frac{9}{5}(T_F - 32)$

$T_F = T_R - 459.67 = 1.8T_C + 32$

$T_R = 1.8T_K$

1 K = 1 °C = 1.8 R = 1.8 °F

1 R = $\frac{5}{9}$ K

(*Continued*)

Power	**Density**
1 ft lb s^{-1} = 1.356 W	1 slug ft^{-3} = 515.4 kg m^{-3}
1 hp = 550 ft lb s^{-1} = 746 W	
universal gas constant	**Angle**
R = 8.314 51 kJ $kmol^{-1}$ K^{-1} = 1.985 kcal $kmol^{-1}$ K^{-1}	1 rad = $180/\pi$ deg = 57.30 deg
= 82.06 atm L $kmol^{-1}$ K^{-1}	1 rev = 2π rad = 360°
	1 rev min^{-1} (rpm) = 0.1047 rad s^{-1}

Thermodynamic symbols (table A.1):

Table A.1. Summary of various thermodynamical symbols.

Symbol	Property	Units
H	Enthalpy	kJ
h	Specific enthalpy	kJ kg^{-1}
P	Pressure	1 bar ≈ 1 atm = 0.1 MPa
s	Specific entropy	kJ kg^{-1} K^{-1}
t_s	Temperature at saturation	
u	Specific internal energy	kJ kg^{-1}
v	Specific volume	m^3 kg^{-1}
ϵ	Static dielectric constant	dimensionless
μ	Viscosity	10 ?6 kg s^{-1} m^{-1} = MPa s
λ	Thermal conductivity	mW K^{-1} m^{-1}
ρ	Density	kg m^{-3}
σ	Surface tension	kg s^{-2} = N m^{-1}
	Specific entropy of vaporization	kJ kg^{-1} K^{-1}

Gas laws:

Boyles law	$pV = \text{const}$
Charles's law	$\frac{V}{T} = \text{const}$
Gay-Lussac's law	$\frac{p}{T} = \text{const}$
Avogadro's principle	$\frac{n}{V} = \text{const}$

Universal gas constant:
8.314 J K^{-1} mol^{-1} = 0.082 1 L atm K^{-1} mol^{-1} = 6.24×10^4 L mm Hg K^{-1} mol^{-1} = 1.99 cal K^{-1} mol^{-1} = 8.314 Volt Coulomb K^{-1} mol^{-1}.

IOP Publishing

Appendix B

Logarithmic scale: Logarithmic scale is also called a *compressed scale*, because it allows us to plot numbers with a big variation in values on a smaller, more manageable scale. In other words, logarithmic scales reduce wide-ranging quantities onto manageable scales. The frequency response of the ear is generally expressed on a logarithmic scale, because our sense of hearing perceives equal ratios of frequencies as equal differences in pitch.

We define one type of logarithm called 'log base 2' and denoted $\log_2$ and defined for any numbers A and C as,

$\log_2 C = A$ is the same as $2^A = C$.
$\log_2 8 = 3$ is the same as $2^3 = 8$.
$\log_2 16 = 4$ is the same as $2^4 = 16$.

Another type that is commonly used is called 'log base 10' and denoted $\log_{10}$ and defined for any numbers A and C as,

$\log_{10} C = A$ is the same as $10^A = C$.
$\log_{10} 1000 = 3$ is the same as $10^3 = 1000$.
$\log_{10} 10\ 000 = 4$ is the same as $10^4 = 10\ 000$.

Natural logarithm: Logarithms typically use a base of 10, while natural logarithms use a base of e, where e represents a mathematical constant equal to 2.718 28 known as the natural exponent. The natural logarithm denoted as ln, is the inverse function of e^x, For example, $\log_e(10) = \ln(10) = 2.30$, whereas, $\log_{10}(10) = \ln(10) = 1$.

doi:10.1088/978-0-7503-3083-1ch15

Logarithmic properties:

Product rule	$\log_a(xy) = \log_a(x) + \log_a(y)$
Quotient rule	$\log_a\left(\frac{x}{y}\right) = \log_a(x) - \log_a(y)$
Power rule	$\log_a(x)^y = y \log_a(x)$
Equality rule	If $\log_a(x) = \log_a(y)$, then $x = y$
Reciprocal rule	$\log_a\left(\frac{1}{x}\right) = -\log_a(x)$
Log of *e*	$\ln(e) = 1$
Log of 1	$\log(1) = 0$
ln of *e* raised to the *x* power	$\ln(e^x) = x$
***e* raised to the ln power**	$e^{\ln(x)} = x$

IOP Publishing

Appendix C

C.1 Rates of processes of heat transfer

The rate at which heat flows from a hot to a cold object depends on many factors, but the main factor is the *mechanism* of heat transfer under consideration. Some processes for heat transfer are discussed here.

1. **Conduction:** Heat transfer by direct contact at the molecular level is called *conduction*. Conduction occurs across solids, liquids and gases. In solids, heat is conducted by lattice vibrations, in metals conduction is via conduction free-electrons, discussed in chapter 13. Conduction electrons carry both energy and electric current. In a liquid or gas, energy is transferred through molecular collisions, where fast-moving molecules impart energy to slow-moving molecules via collisions.

   Law of heat conduction:
   Assume a system with two segments, a hot segment at T_2 separated from a cold section T_1, where heat flows from the hot segment to the cold segment. The amount of heat (Q) transferred is directly proportional to the area (A) of the pane that separates the two segments, the time interval of the transfer (Δt) and the actual temperature difference between the segments, $T_2 - T_1 = \Delta T$. The amount of heat (Q) transferred is inversely proportional to the thickness of the pane (Δx). Summarizing these, we get:

   $$Q \propto \frac{A\Delta t\Delta T}{\Delta x} = k_t\frac{A\Delta t\Delta T}{\Delta x} \tag{C.1}$$

   where k_t is called *thermal conductivity* of the material. The SI units of thermal conductivity are watts per meter per kelvin (W m^{-1} K^{-1}), and the thermal conductivity of some common materials are air 0.026 W m^{-1} K^{-1}, wood 0.08 W m^{-1} K^{-1}, and water 0.6 W m^{-1} K^{-1}.

doi:10.1088/978-0-7503-3083-1ch16

Hence the *law of heat conduction* can be written as the rate of heat flow as,

$$\frac{Q}{\Delta t} = -k_t A \frac{dT}{dx} \tag{C.2}$$

The negative sign is added to remind us that the temperature change is opposite to the flow of heat.

Worked example C.1. Suppose a given window glass pane has an area of 0.5 square meter and a thickness of 3 mm. If the temperature on the right side of the pane is 20 °C and temperature on the left side of the pane is 0 °C, what is the rate of flow heat through the pane? Assume the conductivity of the pane is 0.8 W m^{-1} K^{-1}.

Solution: The rate of flow heat through the pane will be:

$$\frac{Q}{\Delta t} = \frac{(0.8\ \mathrm{W\ m^{-1}\ K} - 1)(0.5\ \mathrm{m^2})(293 - 273)\ \mathrm{K}}{0.003\ \mathrm{m}} = 2666\ \text{watts}. \tag{C.3}$$

2. **Convection:** In thermodynamics, '*convection*' is heat transfer by convection and includes fluid movement both during bulk motion, called *advection*, and by the motion of individual particles, called *diffusion*.
   (a) **Advection:**Advection is the transport of a substance, usually liquids and gases, for example, the transport of silt in a river by bulk water flow. During advection, a fluid transports quantities, such as energy or mass, via the bulk motion of the fluid. Advection does not occur in rigid solids.
   (b) **Diffusion:** The word diffusion comes from the Latin word, '*diffundere*', which means 'to spread out'. Diffusion is the movement of molecules or atoms from a region of high concentration to a region of low concentration. Diffusion results in mixing or mass transport, without requiring bulk flow of material.
3. **Radiation:** When objects are separated by empty space, say in the case of Earth–Sun system, the only possible heat transfer mechanism is '*radiation*'. The amount of energy a hot object radiates depends on temperature of the object. An object with temperature T and surface area A radiates energy in a given time (t) given by the Stefan–Boltzmann law of radiation, defined as,

$$Q = \sigma A t e T^4 \tag{C.4}$$

where σ is Stefan–Boltzmann constant = 5.67×10^{-8} J s^{-1} $m^{-2}K^{-4}$. e is called '*emissivity*', which is a measure of the fraction of incident radiation energy absorbed and radiated by the object. Emissivity is a property of the material under consideration. Objects that reflect a lot of energy have emissivities close to 0, whereas black bodies have emissivities close to 1.

Appendix D

Intensive and extensive properties

The number of interesting thermodynamic variables has been growing, and now we have to find a way to organize them. One simple way of doing this is to organize them by the ones that double by simply doubling the amount of the material. The thermodynamic quantities that double by doubling the material are called '*extensive properties*' and these include things like V, m, N, S, U, H, F, and G. The thermodynamic quantities that do not double by doubling the material are called '*intensive properties*' and these include things like T, p, μ, and density.

Some properties:

1. Extensive (×) Intensive = Extensive quantity (volume × density = mass)
2. Extensive (÷) Extensive = Intensive quantity ($m \div V$ = density)
3. Extensive (+) Extensive = Extensive property ($H = U + pV$) where the product pV is extensive as well.
4. Intensive (+) Intensive = Intensive property.
5. Extensive (×) Extensive = unknown; if you ever get this there is a good chance that you have some error in your calculation.
6. Extensive (+) Intensive = not allowed. Hence you will never see ($G + \mu$) even though they have the same units.

doi:10.1088/978-0-7503-3083-1ch16

Appendix E

E.1 Information theory

Information theory can be viewed as simply a branch of applied probability theory. Information theory says that the more we know about a particular subject, the less new information we will be able to know about it. If an event is likely to occur we are not surprised when it happens, for example the Sun rising in the east in the morning is not a surprise to anybody. On the other hand, if an event is improbable, it will be much more informative if this event happens. This means that the content of information is the reciprocal of the probability of the event.

E.1.1 Entropy in information theory

Entropy (denoted S) is a concept used in thermodynamics, statistical mechanics and information theory. The change in entropy in thermodynamics is the ratio of the heat transfer (Q) to the absolute temperature (T) at which the process takes place:

$$\Delta S = \frac{Q}{T}$$

Entropy in statistical mechanics is defined as the log of multiplicity (Ω), where Ω is the number of ways of arranging a system.

$$S = k\ln(\Omega)$$

Entropy in information theory is a measure of unpredictability of the state, or more specifically it is a measure of unpredictability of the average information content of a system. Consider the example of a coin toss for a fair coin. The probability of heads is the same as the probability of tails; this means the entropy of the coin toss is at maximum at 50%. There is no way to predict the outcome of the coin toss ahead of time. In such a coin toss there are two possible outcomes that occur with equal probability and learning the actual outcome contains only one bit of information. This means this coin toss has one bit of entropy.

doi:10.1088/978-0-7503-3083-1ch18

On the other hand, if we toss a coin that has two heads and no tails, the coin toss will result in the same result, head only. This situation has zero entropy as the outcome can be predicted perfectly every time.

E.1.2 Shannon's entropy

Shannon's entropy is named after Claude E Shannon, who is known as the 'father of information theory'. Information theory was born in Shannon's classic papers during the 1940s, that contained the basic results for simple memoryless sources. Shannon's original work focused on memoryless sources whose probability distribution did not change with time and whose outputs were drawn from a finite alphabet or the real number line.

The connection between thermodynamics and information theory was first made by Ludwig Boltzmann in the 1870s and expressed by his famous equation called **Boltzmann's formula** as,

$$S = k\ln(W)$$

where S is the thermodynamic entropy of a particular macrostate, W is the number of microstates that give the same macrostate, and k is Boltzmann's constant. W is counted using the formula for permutations,

$$W = \frac{N!}{\prod_i N_i!}$$

where i covers all possible microstates.

In Boltzmann's formula, it is assumed that each microstate is equally likely, so that the probability (P_i) of a given microstate is given as $W = 1/P_i$. When these probabilities are substituted into the above expression we get the Shannon entropy as,

$$S = -k\sum_{i=1}^{n} \ln(P_i)$$

where P_i is the probability of state i.

In information theoretic terms, the information entropy of a system is the amount of 'missing' information needed to determine a microstate, given the macrostate. The entropy of a message per bit multiplied by the length of that message is a measure of how much total information the message contains. Shannon's theorem also says that when using a data compression system, some messages may come out shorter but at least one must come out longer due to what is called the pigeonhole principle. The pigeonhole principle states that if n items are put into m containers, with $n > m$, then at least one container must contain more than one item.

www.ingramcontent.com/pod-product-compliance
Ingram Content Group UK Ltd.
Pitfield, Milton Keynes, MK11 3LW, UK
UKHW051906100926
8556IPUK00005B/780